101 Solved Civil Engineering Problems

Michael R. Lindeburg, P.E.

Professional Publications, Inc.
Belmont, CA 94002

In the ENGINEERING REFERENCE MANUAL SERIES
Engineer-In-Training Reference Manual
Engineering Fundamentals Quick Reference Cards
Engineer-In-Training Sample Examinations
Mini-Exams for the E-I-T Exam
1001 Solved Engineering Fundamentals Problems
E-I-T Review: A Study Guide
Diagnostic F.E. Exam for the Macintosh
Fundamentals of Engineering Video Series: Thermodynamics
Civil Engineering Reference Manual
Civil Engineering Quick Reference Cards
Civil Engineering Sample Examination
Civil Engineering Review Course on Cassettes
101 Solved Civil Engineering Problems
Seismic Design of Building Structures
Seismic Design Fast
Timber Design for the Civil P.E. Examination
246 Solved Structural Engineering Problems
Mechanical Engineering Reference Manual
Mechanical Engineering Quick Reference Cards
Mechanical Engineering Sample Examination
101 Solved Mechanical Engineering Problems
Mechanical Engineering Review Course on Cassettes
Consolidated Gas Dynamics Tables
Fire and Explosion Protection Systems
Electrical Engineering Reference Manual
Electrical Engineering Quick Reference Cards
Electrical Engineering Sample Examination
Chemical Engineering Reference Manual
Chemical Engineering Quick Reference Cards
Chemical Engineering Practice Exam Set
Land Surveyor Reference Manual
1001 Solved Surveying Fundamentals Problems
Land Surveyor-in-Training Sample Examination
Engineering Economic Analysis
Engineering Law, Design Liability, and Professional Ethics
Engineering Unit Conversions

In the ENGINEERING CAREER ADVANCEMENT SERIES

How to Become a Professional Engineer
The Expert Witness Handbook—A Guide for Engineers
Getting Started as a Consulting Engineer
Intellectual Property Protection—A Guide for Engineers
E-I-T/P.E. Course Coordinator's Handbook
Becoming a Professional Engineer
Engineering Your Start-Up
High-Technology Degree Alternatives
Metric in Minutes

101 SOLVED CIVIL ENGINEERING PROBLEMS

Printed in the United States of America

Professional Publications, Inc.
1250 Fifth Avenue, Belmont, CA 94002
(415) 593-9119

Current printing of this edition: 2

Library of Congress Cataloging-in-Publication Data
Lindeburg, Michael R.
101 solved civil engineering problems / Michael R. Lindeburg.
p. cm. -- (Engineering reference manual series)
ISBN 0-912045-64-7
1. Civil engineering--Problems, exercises, etc. I. Title.
II. Title: One hundred one solved civil engineering problems.
III. Series.
TA159.L52 1994
624'.076--dc20 94-9618
CIP

TABLE OF CONTENTS

PREFACE

The problems in *101 Solved Civil Engineering Problems* are the most representative and realistic one-hour civil P.E. exam problems that I could write. The wording, illustrations, and data in these problems are original. However, the problems are based on the same subjects and concepts that have been important in previous civil engineering P.E. exam problems.

Examination problems are not made public after the exams are administered. NCEES keeps the exam problems secret so that they can be used time and time again. Please do not expect to find actual exam problems in this book, because there are none.

This book should be used throughout your review. Each time you have finished reviewing a particular subject, you can try your hand at the problems in the corresponding chapter of *101 Solved Civil Engineering Problems*. If you are reasonably familiar with the nomenclature, terminology, and solution methodology for most of the problems in that chapter, and if your own solution to each problem does not take significantly more than one hour, you are probably ready to move on to the next review subject.

Although the problems in this book are intended to illustrate important civil engineering subjects, there is no guarantee that any subject covered in this book will ever appear on an engineering licensing examination. I hope you will not base your review solely on what you find in *101 Solved Civil Engineering Problems*. The scope of a book limited to 100 or so problems is too narrow to use as the pilot of your review program.

Since NCEES never releases the official solutions for the engineering community to review and comment on, it is difficult to predict how a particular type of problem should be solved. The solutions in this book are generally the shortest and most straight-forward possible. However, there are often several ways to approach a given problem, and different levels of sophistication are possible.

From past experience, I know the examination sometimes requires a theoretical (as opposed to a heuristic) solution, and sometimes a code-based (as opposed to a theoretical) solution is expected. Knowing which approach to take requires you to have some test-taking skills as well as a good background in engineering. You will have to read each problem statement carefully to pick up the clues.

Although the civil P.E. exam is now 50% multiple-choice problems, there are no problems in this format in *101 Solved Civil Engineering Problems*. However, I have included many problems that have numerous (up to 10) parts. These multipart problems illustrate the level of difficulty and "interconnectedness" of the multiple choices in the civil P.E. exam problems.

I had a lot of help with this book, but the responsibility for its content, quality, and correctness is mine. I would appreciate your taking the time to bring any errors to my attention so I can take care of them prior to the next printing.

Michael R. Lindeburg, P.E.
Belmont, CA
April, 1994

ACKNOWLEDGMENTS

I have wanted to bring out a book of representative civil licensing exam problems ever since I wrote *101 Solved Mechanical Engineering Problems* (a similar book for mechanical engineers) in 1988. The writing, editing, and production work on this book have now spanned five years. However, this book would have been even longer in coming were it not for the assistance of many people.

Many of the solutions in this book are the work of a team of professors from Northeastern University. I checked the team's work and calculations, filled in missing steps in solutions, added explanatory text, and edited the solutions for consistency. I acknowledge the significant contribution made by this team, which consisted of Dennis Bernal, Ph.D. (Concrete Design, Masonry Design, Steel Design, Timber Design); Mark D. Evans, Ph.D., P.E. (Soils/Soil Mechanics, Foundations); Peter G. Furth, Ph.D. (Traffic); Ali Touran, Ph.D., P.E. (Engineering Economic Analysis, Surveying); and Irvine W. Wei, Ph.D. (Wastewater, Water Supply). I would also like to thank Daniel E. Medina, Ph.D., from Greenhorne & O'Mara, Inc., for his work on the solutions for Fluid Mechanics, Hydraulic Machines, Hydrology, and Open Channel Flow.

In Professional Publication's production department, I would like to acknowledge David Bergeron (illustrations), Mary Christensson (typesetting), Kurt Stephan (copyediting), and Jessica R. Whitney (proofreading), for helping to give this book tangible form.

Prior to publication, the manuscript of this book was used by approximately 100 engineers who were preparing for the civil P.E. exam. Their comments were invaluable, and their suggestions have been incorporated into the book. Their names are too numerous to list here, but their contributions have made the book much easier for you to use.

Thanks to you all!

Michael R. Lindeburg, P.E.
Belmont, CA
April, 1994

Notice to Examinees

Do not copy, memorize, or distribute problems from the Principles and Practice of Engineering (P&P) Examination. These acts are considered to be exam subversion.

The P&P examination is copyrighted by the National Council of Examiners for Engineering and Surveying. Copying and reproducing P&P exam problems for commercial purposes is a violation of federal copyright law. Reporting examination problems to other examinees invalidates the examination process and threatens the health and welfare of the public.

1 ENGINEERING ECONOMIC ANALYSIS

ECONOMICS–1

The grass in a public golf course turns brown each summer because the irrigation system has deteriorated to the point where it must be replaced. Three alternatives are available to replace the existing irrigation system. The chosen alternative will be funded by tax-free bonds that pay 8% to the bearer.

Each alternative requires the installation of a pump and piping. The pump is expected to last 10 years and have no salvage value after that time. Replacement pumps of the same design will be used to extend the operation to 30 years. Each pump vendor has guaranteed to provide the replacement pumps at the same cost as the originals. The piping can be expected to last 30 years and will have no salvage value after that period of time.

alternative	A	B	C
initial pump cost	\$7500	\$15,000	\$30,000
initial piping cost	\$15,000	\$15,000	\$22,500
annual maintenance	\$6000	\$4500	\$3000
pumping cost per 100 gal	\$4.50	\$3.75	\$3.00

(a) If the pumping cost is disregarded, which alternative should be chosen?

(b) Is the selection sensitive to the volume of water pumped?

ECONOMICS–2

A small partnership spent three years developing a product. The company spent \$55,000 in the first year of development, \$75,000 in the second year, and \$85,000 in the third year. At the end of the three years, the product was immediately placed into production.

The partners have an average personal tax rate of 46% and want an 18% return (after taxes) on their investment. All partnership profits and losses are passed through to the partners. The development costs are to be depreciated over a three-year period after the start of production using the following percentages: first year, 25%; second year, 38%; and third year 37%. There is no (investment) tax credit. Assume all development costs are depreciable.

The company expects to produce and sell 4000 units each year for the 10 years following the beginning of production. The unit manufacturing cost, including all labor, material, and overhead (but excluding taxes and development costs) is \$60. What amount should be added to the cost of each unit to recover all development costs over the 10-year period?

ECONOMICS–3

Route 420 is currently the only way to get between two cities; it carries 1200 commercial vehicles per day at an average speed of 50 mph. Route 422 is being proposed as a replacement for Route 420.

	Route 420	Route 422
length	9.2 mi	6.5 mi
initial cost	0	\$1,200,000
pavement life	15 yr	15 yr
resurfacing cost	\$300,000	\$260,000

The vehicle operating cost is \$0.22 per mile for both routes. The time savings is \$0.30 per vehicle-minute. Consider costs and savings only for the commercial traffic, and neglect maintenance (vehicle and pavement) and all other factors. The average speed is unchanged. Assume an infinite life and zero salvage value. Use the incremental benefit/cost ratio method with an interest rate of 10%. Which route is superior?

ECONOMICS–4

A disposal site serves a population of 100,000, and this number is not expected to change during the next 10 years. Municipal solid waste (MSW) is collected daily at the rate of 5 lbm per person. This quantity is expected to increase 5% annually. The composition of the MSW and the fraction of each component recovered for recycling are

	fraction in MSW	fraction recoverable
combustible materials (includes plastics)	50%	60%
ferrous materials	8%	90%
glass	15%	80%
aluminum	5%	70%
other (mineral matter, yard waste, other metals, etc.)	22%	0%

The disposal site sells its recoverables at the following prices.

- combustibles: 50% of the price of coal. Coal is selling at \$45/ton. This price is expected to increase at the rate of 4% per year.
- ferrous: 50% of the price of steel. Steel is selling at \$80/ton. This price is expected to increase at the rate of 8% per year.
- glass: \$20/ton. This price is expected to increase at the rate of 2% per year.
- aluminum: \$200/ton. This price is expected to increase at the rate of 12% per year.

(a) What is the annual revenue from recoverables in the first year of operation?

(b) What is the annual revenue from recoverables in the fifth year of operation?

(c) What is the annual revenue from recoverables in the tenth year of operation?

ECONOMICS–5

A municipality intends to purchase new pickup trucks for its public works department and drive them 17,000 miles each year. A bid has been received from a fleet dealer of \$10,000 per truck, with the following guaranteed salvage values.

end of year	salvage value
1	\$8000
2	\$6500
3	\$5000
4	\$3000
5	\$2500
6	\$2000
7	\$1500
8	\$1000

Annual operating costs are expected to be \$1250 for fuel and insurance. Annual maintenance costs are expected to start at \$500 and increase \$100 each year. An interest rate of 10% is used for comparison of alternatives.

(a) What is the total equivalent uniform annual cost of ownership if a vehicle is kept for 8 years?

(b) What is the total present worth if a vehicle is kept for 8 years?

(c) What is the total equivalent uniform annual cost of ownership if a vehicle is kept for 4 years?

(d) What is the total present worth if a vehicle is kept for 2 years?

(e) What is the most economical length of time to keep a vehicle?

Use the following information for the remaining parts of this problem: A local garage has agreed to perform all maintenance during the next 8 years if the municipality pays it \$3500 per vehicle now. This will eliminate the annual maintenance performed by the municipality.

(f) What is the total equivalent uniform annual cost of ownership if a vehicle is kept for 8 years?

(g) What is the total present worth if a vehicle is kept for 8 years?

(h) What is the most economical length of time to keep a vehicle?

ECONOMICS–6

A contractor submitted a bid to an owner-developer to construct a new office building for \$2,400,000, to be paid in a single payment at the completion of construction. As an alternative, the contractor agreed to give the developer three years (starting from the time of completion) to make payments of 50%, 25%, and 25% of the bid amount, with adjustments for interest (figured at 10% per year before taxes) and inflation (figured at 5% per year). The contractor's actual cost of construction is \$2,000,000, which would be payable to subcontractors, employees, and suppliers at the completion of the project. The contractor's income tax rate is 45%. All other accounting conventions are to be neglected.

(a) If the construction bid price is paid off at the time of completion, what is the contractor's after-tax rate of return (ROR)?

(b) If the construction bid price is paid off over three years following completion, what is the contractor's before-tax rate of return?

(c) If the owner-developer depreciates the building over 25 years using straight-line depreciation and zero salvage value, what is the before-tax present worth of the revenue that must be generated to recover the building cost? The owner-developer is accustomed to receiving an after-tax 15% return on his investments. Assume an income tax rate of 45% for the owner-developer, and that the owner-developer pays the contractor in a single payment.

2 FLUID MECHANICS

FLUID MECHANICS–1

As a condition of obtaining a business license, a factory was required by the planning commission to install a private water supply to supplement what it draws from the municipal supply. It did so, using new PVC pipe exclusively. Its reservoir (filled from a local well) is open to the atmosphere. No water is added to or taken from the system at any point between the pipe entrance at point A and the factory at point C. The water temperature is 50°F.

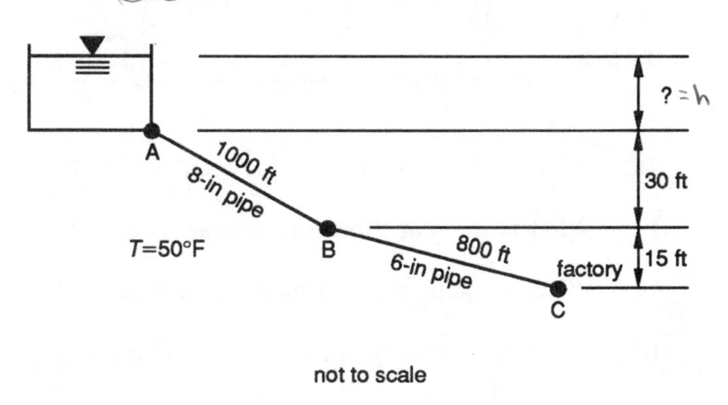

(a) How high must the water level in the reservoir be to maintain a static pressure of 40 psig at the reservoir outlet (point A)?

For the remaining questions, assume the reservoir is 70 ft deep at point A.

(b) What is the maximum flow rate if the static pressure at point C must not drop below 40 psig?

(c) If the factory wants to become self-sufficient, what size pipe should be installed directly from point A to point C (parallel to the existing line) to provide 1900 gal/min at a static pressure of 40 psig?

FLUID MECHANICS–2

An elevated tank feeds a simple pipe system as shown. All pipes are 5- to 10-year-old cast iron. There is a fire hydrant at point C. The minimum allowable pressure at point C is 22 psig for firefighting requirements. All minor losses are insignificant. Neglect friction losses from the tank to point A.

(a) What is the maximum static head at point C?

(b) What is the maximum flow out of the hydrant?

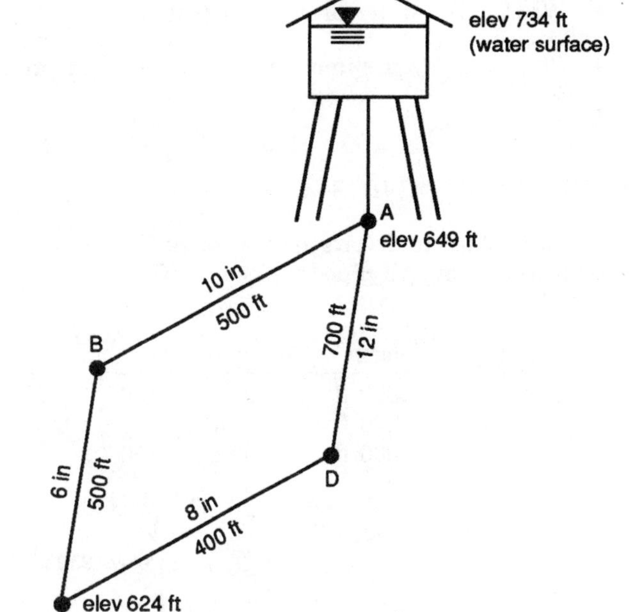

FLUID MECHANICS–3

The water pressure at point D in a new PVC pipe (Hazen-Williams coefficient, $C = 150$) water distribution system is too low.

section	length	diameter	average flow removed
AB	16,000 ft	6 in	140 gal/min at B
BC	14,000 ft	6 in	130 gal/min at C
CD	8000 ft	4 in	70 gal/min at D

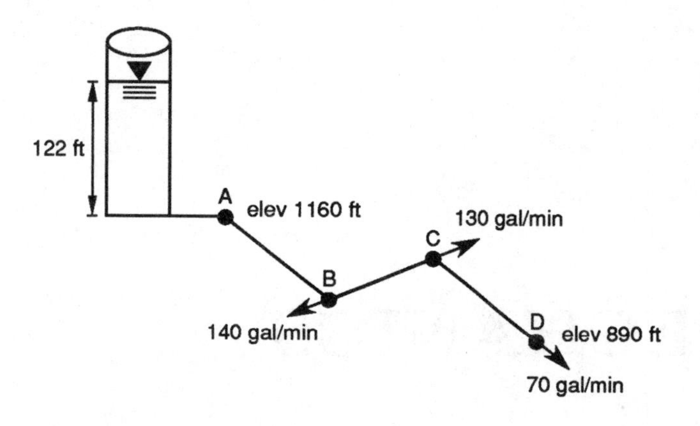

(a) Based on the average flow removed, what is the water pressure at point D?

(b) If the unacceptably low pressure is experienced at the average flows given, what might be the causes?

(c) Suppose the excessively low pressure is experienced only during the hours of daily peak demand. What is the water pressure at point D?

(d) How would you eliminate the low pressure problem?

FLUID MECHANICS–4

Pipes from three reservoirs meet at point D at elevation 1200 ft as shown. All pipes are new PVC.

leg	length	diameter	C
A	5000 ft	12 in	150
B	4000 ft	8 in	150
C	10,000 ft	10 in	150

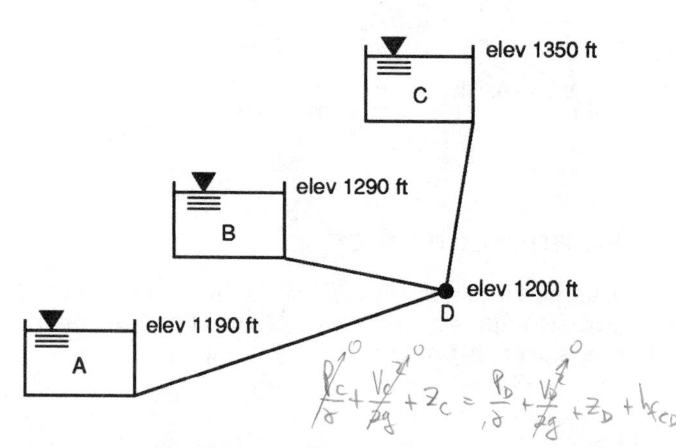

(a) What is the flow direction (away from or into point D) in each leg?

(b) What is the head loss in each pipe between the reservoirs and point D?

FLUID MECHANICS–5

The main water pipeline makes a complete loop around a small community. The pipe is cast iron. There is no elevation change around the loop. Minor losses can be neglected.

section	length	diameter	C
AB	800 ft	8 in	140
BC	1000 ft	8 in	140
AE	1400 ft	6 in	120
ED	2000 ft	10 in	160
DC	1200 ft	6 in	120

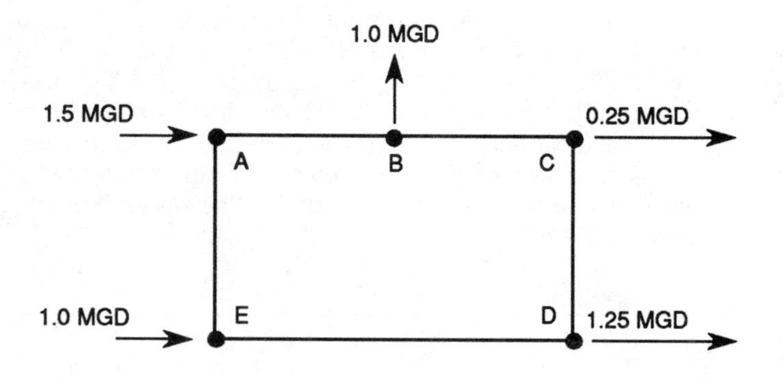

(a) What is the flow quantity between A and B?

(b) What is the flow direction between A and B?

(c) What is the pressure drop between A and B?

(d) What is the flow quantity between B and C?

(e) What is the flow quantity between A and E?

(f) What is the flow quantity between E and D?

(g) What is the flow quantity between C and D?

(h) What is the pressure drop between C and D?

(i) What is the flow direction between C and D?

(j) What is the minimum time to close a valve at point D without experiencing water hammer at point E? The pipe is cast iron between points E and D.

FLUID MECHANICS–6

A pump supplies 50°F water to a residential community through a parallel level (equal-elevation) network as shown. All pipe is 5-year-old cast iron. Minor losses can be neglected. The static pressure drop between the pump and the outlet must not exceed 10 psig.

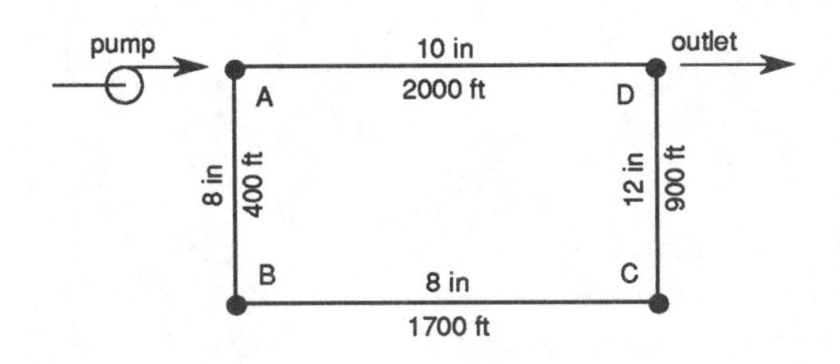

$T = 50°F$

not to scale

(a) What is the maximum flow the pump can operate at without the network exceeding the pressure drop limitation?

(b) What is the flow velocity in leg AD?

(c) What is the pressure drop in leg AD?

(d) What is the pressure drop in leg ABCD?

(e) What is the pressure head (in feet) at the pump discharge?

FLUID MECHANICS–7

A 240-foot-long ductile iron pipe runs between the bottom drains of two sludge processing basins. There are two 90° elbows in the line, and all other minor losses should be disregarded. The surface elevations of the two basins differ by 12 ft. There is no pump, and flow is by gravity from the higher basin to the lower basin. The sludge temperature varies from 45° (in the winter) to 80° (in the summer). Depending on the time of day, the required transfer flow rate varies from a low of 3 MGD to a high of 6 MGD.

(a) What standard-size diameter should be specified for the pipe to connect the sludge basins?

(b) Explain the factors to consider when choosing a diameter.

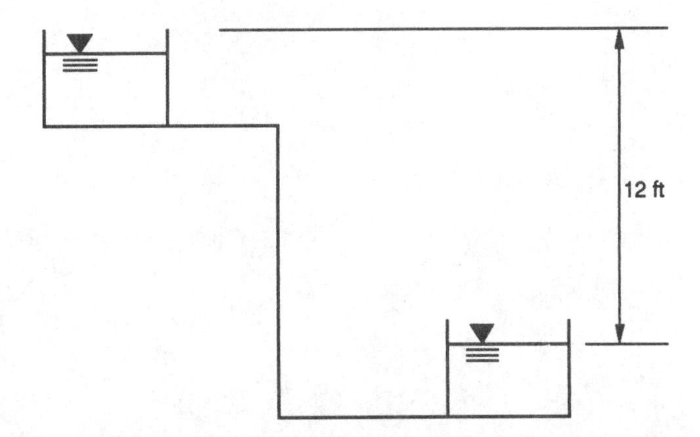

3 HYDRAULIC MACHINES

HYDRAULIC MACHINES–1

A tank supplies water to a small town during the day. At night, when the demand is low, a pump is used to refill the tank. New PVC pipe connects the pump and tank. All minor losses are insignificant. The water temperature is 50°F.

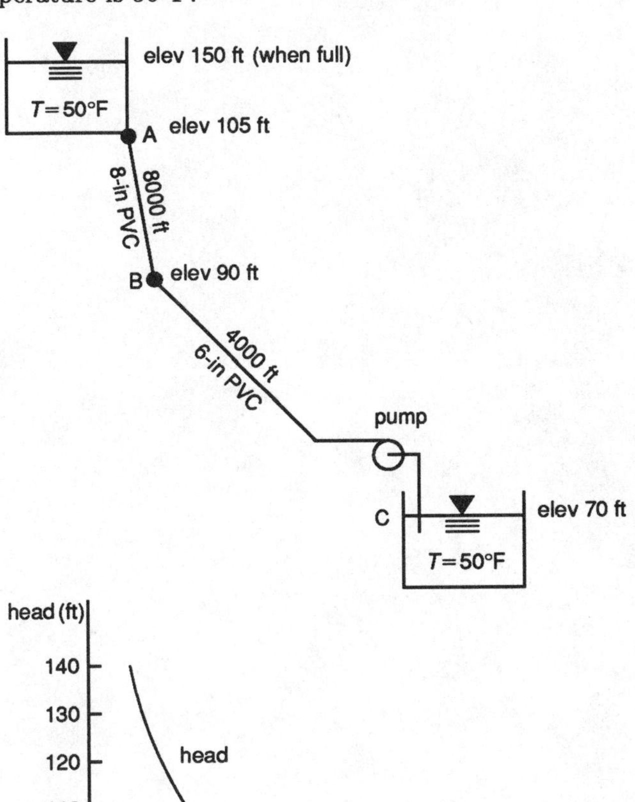

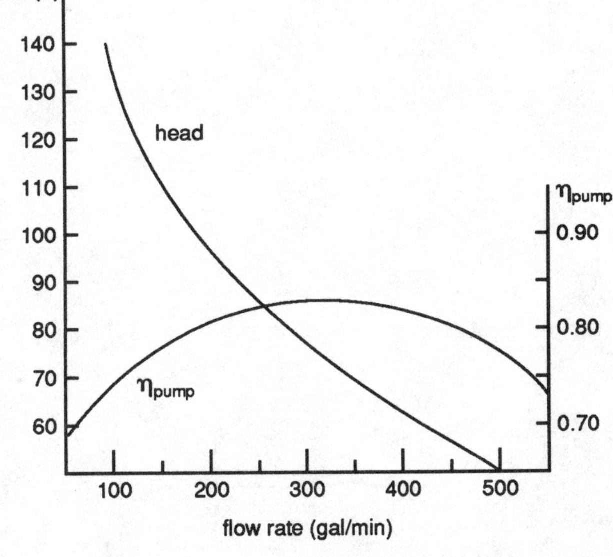

(a) What is the minimum rate at which the tank will be refilled?

(b) If a minimum pressure of 20 psig is needed at a hydrant located at point B, what is the maximum flow rate that can be drawn from the hydrant? Assume that the tank is full and the pump is not in use during firefighting.

HYDRAULIC MACHINES–2

The water surface elevations of two reservoirs differ by 350 ft. The reservoirs are connected by 2200 ft of 12-in steel schedule-40 pipe (Darcy friction factor, $f = 0.02$). The transfer rate (from the lower to the higher reservoir) is 9.5 ft^3/sec. An 83% efficient centrifugal pump is located at the elevation of the lower reservoir drain. Two fully open gate valves provide for pump maintenance and replacement. The water temperature is 50°F. Disregard all other minor losses.

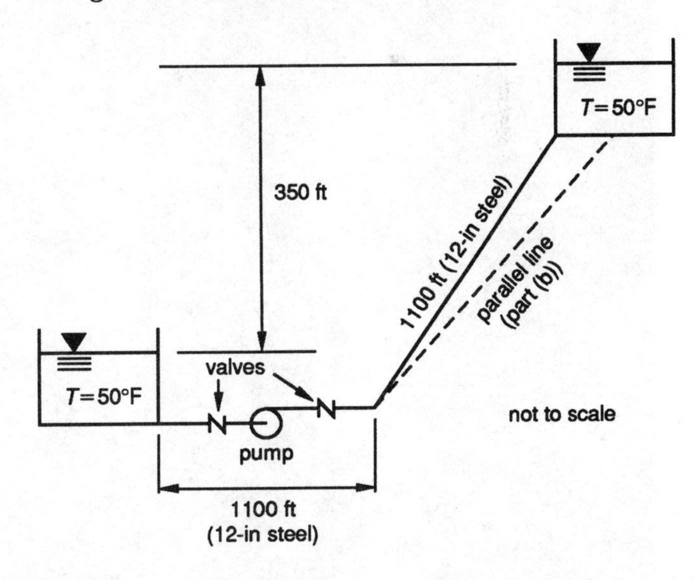

(a) What total motor horsepower is required to pump the water?

(b) If the pump(s) and motor(s) are unchanged, what will be the flow rate if a second 12-in line is placed in parallel with the last half (1100 ft) of the original pipe?

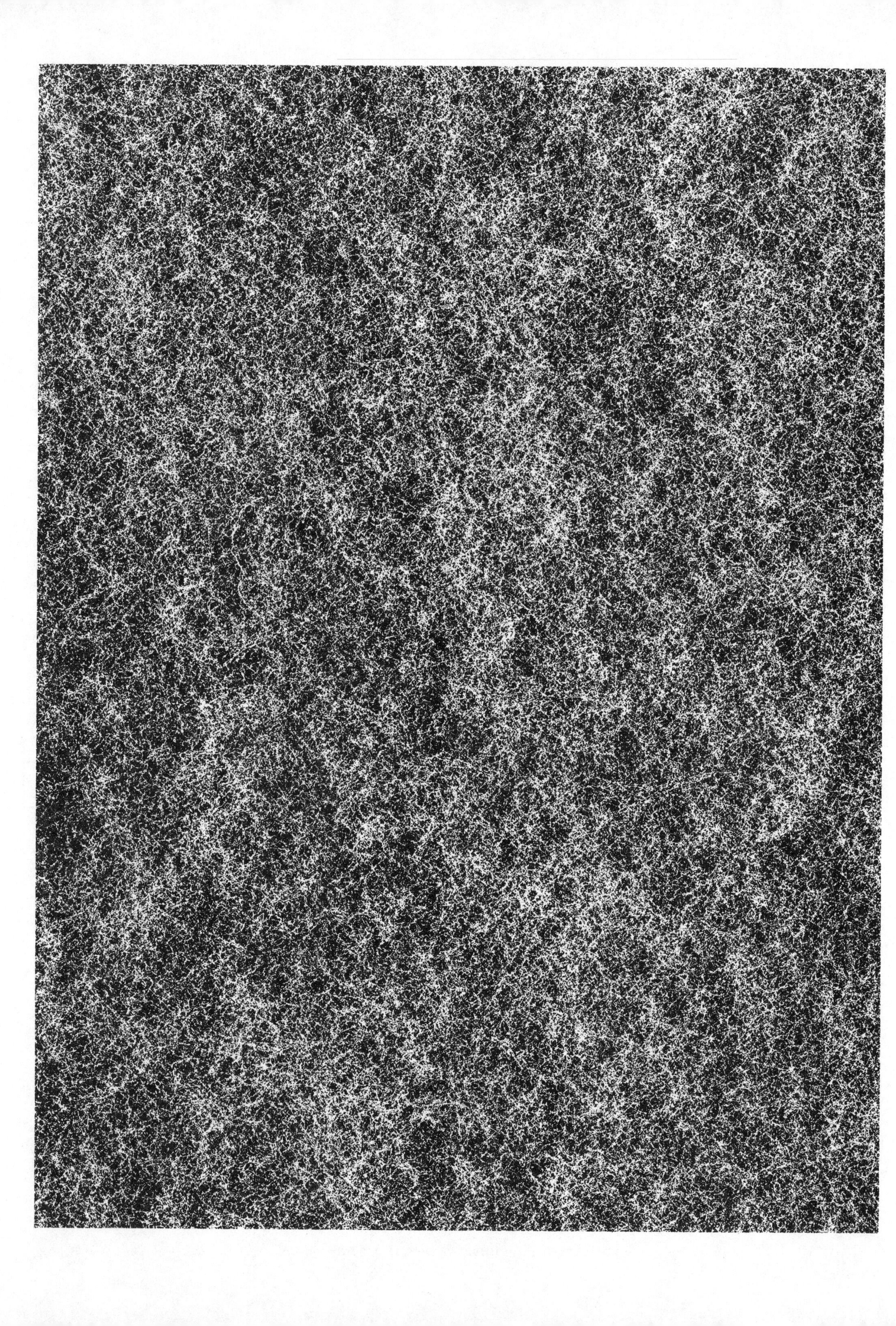

4 OPEN CHANNEL FLOW

OPEN CHANNEL FLOW–1

The cross section of a very long channel with an average geometric slope of 0.0004 ft/ft is shown. At a particular location, the elevations of various points across the channel and flood plain are A, 910.0 ft; B, 905.0 ft; C, 900.0 ft; and D, 909 ft.

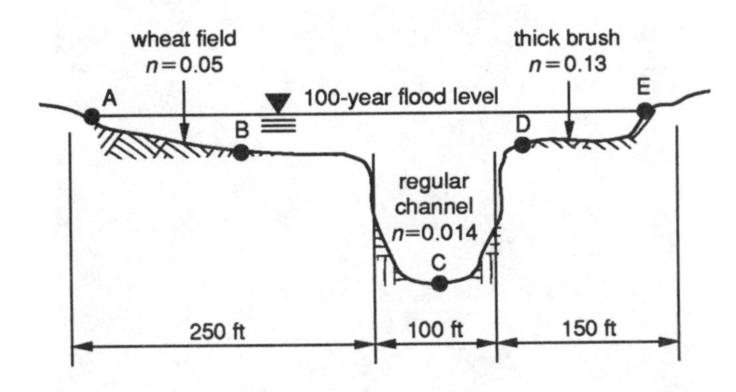

(a) Find the discharge (in cubic feet per second) from a 100-year flood.

(b) Find the average velocity from a 100-year flood.

OPEN CHANNEL FLOW–2

A trapezoidal channel (slope = 0.006 ft/ft; Manning's n = 0.014) has a cross section with a 12-ft-wide base and 1:2 (vertical to horizontal) sides. The flow rate is 2300 ft^3/sec. At a particular point, the channel meets (i.e., joins) a 10-ft-deep stream. The invert elevations of the channel and stream are the same.

(a) What is the normal depth of the channel flow?

(b) What is the critical depth of the channel flow?

(c) Is the flow subcritical or supercritical?

(d) What type of control does the stream represent to the channel?

OPEN CHANNEL FLOW–3

The following questions are not necessarily related.

(a) What is the most efficient cross section?

(b) What is the geometry of a most-efficient trapezoidal channel? Draw a figure and label all sides and angles.

(c) What is the meaning of the term *upstream control*?

(d) What is the meaning of the term *downstream control*?

OPEN CHANNEL FLOW–4

An existing 22-in-diameter PVC pipe (Manning's n = 0.01) is installed on a 2% slope. The flow rate that needs to be carried between the two points separated by 1000 ft is 35 ft^3/sec. However, the pipe does not appear to have such a capacity.

(a) What is the capacity of the 22-in pipe?

(b) What diameter pipe should be installed parallel to the first pipe to carry the excess?

(c) What will be the diameter if the second pipe is installed over a different route with an average slope of 0.025?

(d) If the second pipe starts at elevation 4257 ft, what will be the elevation of the second pipe's invert at the end of the run? Assume the two pipes have the same length.

OPEN CHANNEL FLOW–5

A section of an existing natural channel (Manning's n = 0.030) drops 470 ft as it meanders 8.4 mi through an area scheduled to be developed commercially. The existing channel is basically rectangular in shape, is 25 ft wide, and flows (at maximum) with a depth of 12 ft. The developers want to install an artificial channel 3.7 mi shorter over a more direct route, starting and ending at the same points as the natural channel. The artificial channel will be trapezoidal in shape with a base of 20 ft and sides inclined at 1:1, and it will be earth-lined.

(a) Will the new channel have the capacity to replace the existing channel?

(b) Will erosion be more or less of a problem with the new channel than with the old channel? Why?

OPEN CHANNEL FLOW–6

A contracted measurement weir is to be installed in a 22-ft wide channel. The depth of the weir opening is 16 in, and the bottom of the weir opening is at elevation 479 ft. The bottom of the channel is at elevation 470 ft. For proper measurement and metering, a minimum of 6 in of freeboard is required. The average flow is 6 MGD but varies from a minimum of 3 MGD to a maximum of 12 MGD.

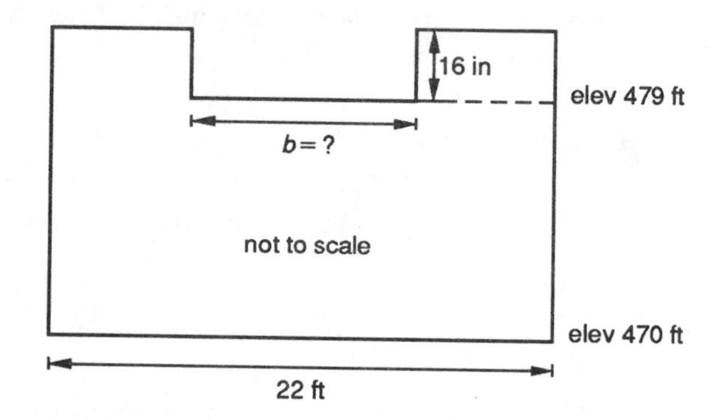

(a) What should be the length of the weir opening?

(b) What is the freeboard at the maximum flow?

(c) How far back from the weir would you place a gauge station in order to measure the true elevation of the water surface in the channel? Explain the reason for your answer.

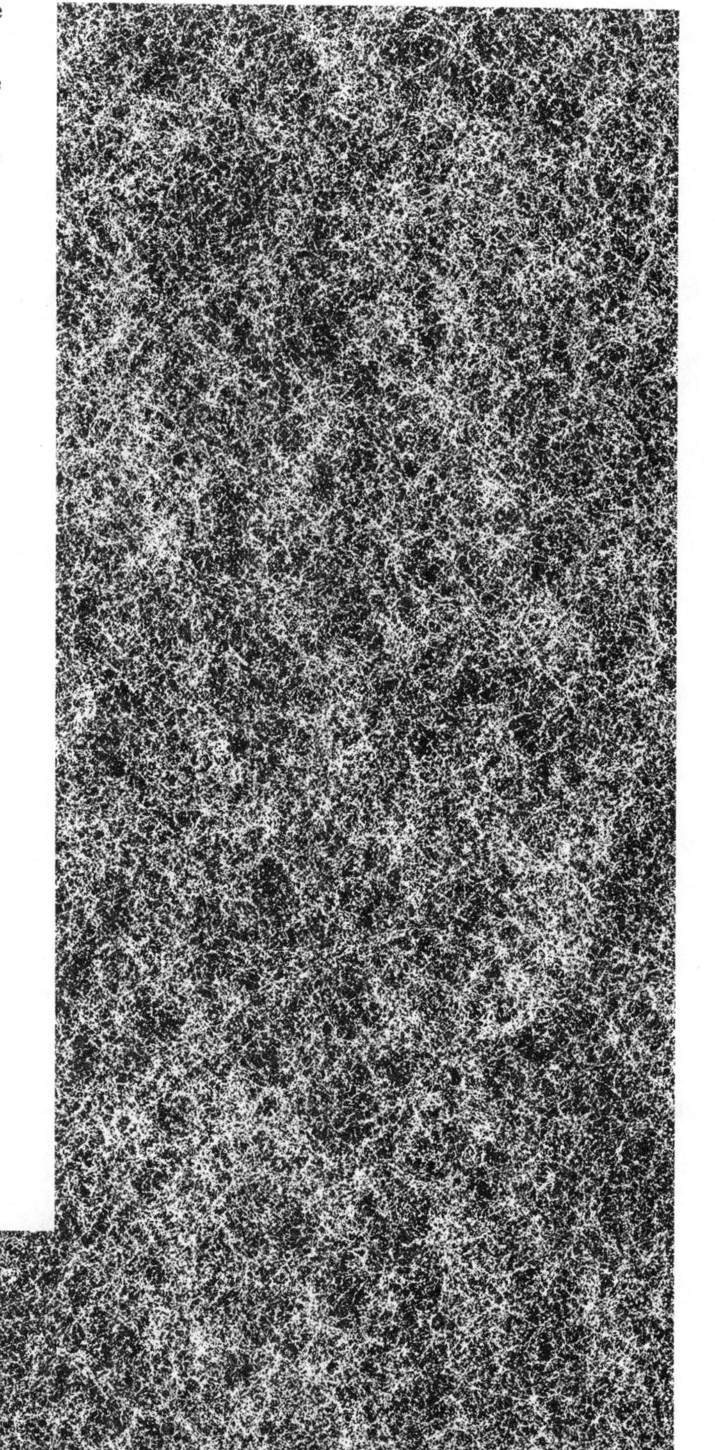

5 HYDROLOGY

HYDROLOGY–1

An unconfined coarse-sand aquifer is 65 ft thick. Its porosity is 30%, and its transmissivity is 120,000 gal/ft-day. A pump draws water for 24 hr from a fully-penetrating production well in the aquifer, after which the pump is shut off. Eight hours after the pump is shut off, an observation well 30 ft from the production well shows a 2.0-ft drawdown.

(a) Estimate the pumping volume immediately before the production well is shut down.

(b) Assume the pumping volume is 800 gal/min immediately before the production well is shut down. What is the steady-state drawdown at the production well?

HYDROLOGY–2

The following equations define the parameters of a hydrograph for a 30-min storm over a 200-ac collection basin.

$$Q = (0.0132)(A^{0.9})(L^{0.3})(C^{1.1})(E^{0.08})$$
$$T_d = (0.44)(A^{1.1})(L^{0.04})(C^{1.2})(E^{0.17})$$
$$T_p = (0.0024)(A^{1.7})(L^{0.04})(C^{0.7})(E^{0.14})$$
$$T_{50} = (0.00367)(A^{1.3})(L^{0.01})(C^{0.7})(E^{0.12})$$
$$T_{75} = (0.0671)(A^{1.2})(L^{0.05})(C^{0.7})(E^{0.11})$$

Q = peak discharge (ft^3/sec)
A = basin area = 200 ac
L = length of channel = 3000 ft
C = average runoff coefficient = 0.35
S = average channel slope = 0.04
E = average imperviousness = 25%
T_d = time from beginning of runoff to end of channel discharge (hours)
T_p = time from beginning of runoff to peak discharge (hours)
T_{50} = time from beginning of runoff to 50% cumulative runoff (hours)
T_{75} = time from beginning of runoff to 75% cumulative runoff (hours)

(a) Draw the unit hydrograph. Indicate Q, T_d, T_p, T_{50}, and T_{75}.

(b) During a 90-min storm in the same basin, 0.3 in fell in the first half hour, 1.5 in fell during the next half hour, and 0.4 in fell during the last half hour. What are Q, T_d, T_p, T_{50}, and T_{75}? (Make any reasonable assumptions.)

HYDROLOGY–3

The entrance to a theme park consists of two adjacent areas, A and B, that are both drained by drop inlets located in the geometric centers of the areas. The drop inlets feed a common collector designed for flow at 2 ft/sec. Area A is an asphalt parking area with an average slope of 1% from all directions to the drop inlet. Area B is primarily a well-maintained lawn with an average slope of 2% from all directions to the drop inlet.

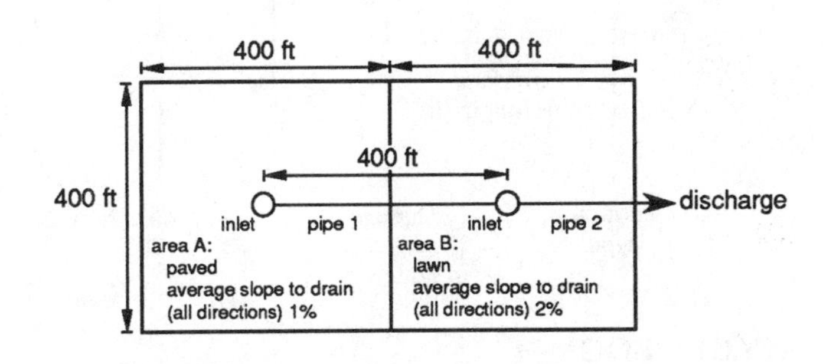

(a) What is the design flow (in ft^3/sec) in pipe 1 for a 10-year storm?

(b) What is the design flow (in ft^3/sec) in pipe 2 for a 10-year storm?

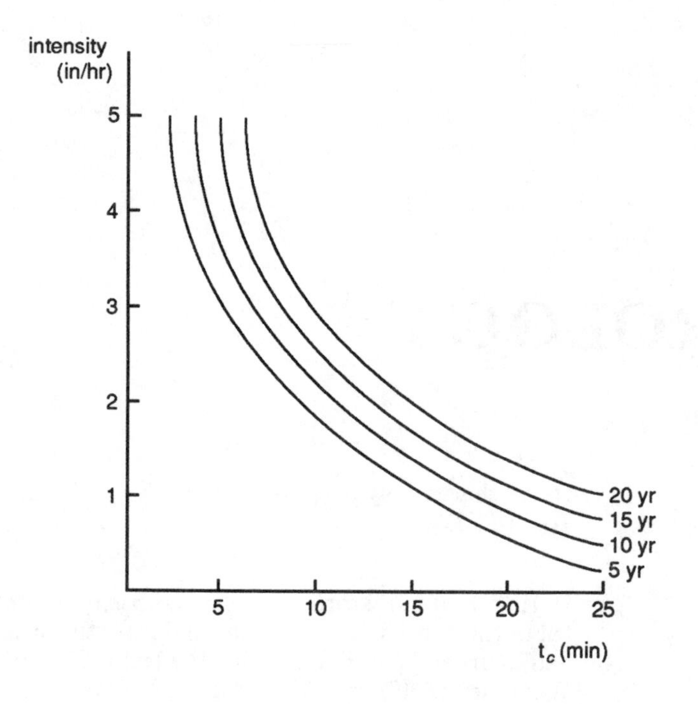

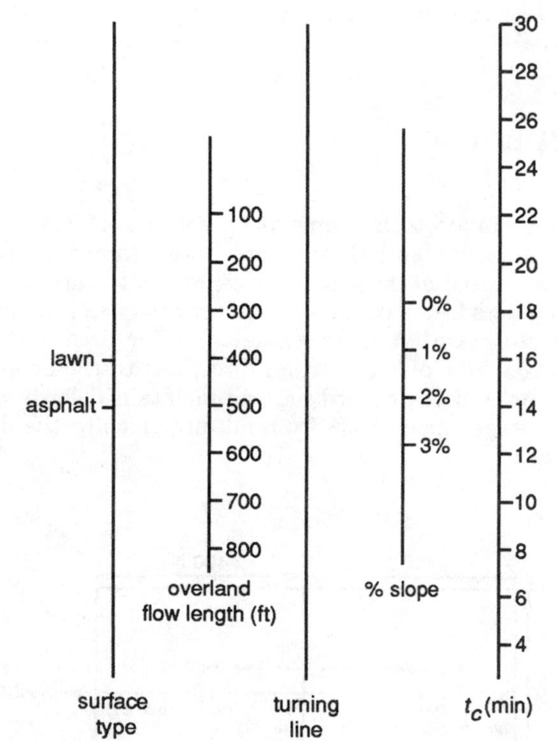

HYDROLOGY–4

Water from a 175-ac light industrial watershed is collected and drained by a trapezoidal open channel. The channel (Manning's roughness coefficient, $n = 0.02$) has a 4.5-ft-wide bottom and 1:1 sides. The channel direction is perpendicular to a road where twin, side-by-side 54-in-diameter corrugated metal pipe (CMP) culverts take the water under the roadway. The average slope of the channel and culverts is 0.75% (i.e., 0.0075 ft/ft). The time for runoff from the farthest part of the watershed to begin contributing to the flow is 35 min.

(a) If curves for 5-, 10-, 25-, 50-, and 100-year floods are available, which would you recommend using? Why?

(b) What rational method runoff coefficient would you recommend? Why?

(c) Using the rational method and assuming the intensity after 35 min is 2 in/hr, what is the runoff?

(d) Are the culverts under inlet or outlet control?

HYDROLOGY–5

Data on intensity and duration have been recorded on storms over the past 45 years. Draw the 10-year intensity-duration-frequency curve.

year	duration (min)	intensity (in/hr)	year	duration (min)	intensity (in/hr)
1	10	3.0	7	20	2.15
2	10	4.1	8	20	3.05
3	10	4.9	9	20	3.55
4	10	5.1	10	20	3.85
5	10	5.5	11	20	4.07
6	10	6.0	12	20	4.4

year	duration (min)	intensity (in/hr)	year	duration (min)	intensity (in/hr)
13	30	1.9	19	40	1.5
14	30	2.4	20	40	2.0
15	30	2.9	21	40	2.3
16	30	3.1	22	40	2.6
17	30	3.3	23	40	2.9
18	30	3.6	24	40	3.1

year	duration (min)	intensity (in/hr)	year	duration (min)	intensity (in/hr)
25	60	1.1	31	10	2.07
26	60	1.5	32	10	3.5
27	60	1.9	33	10	6.8
28	60	2.0	34	20	1.6
29	60	2.15	35	20	2.6
30	60	2.30	36	20	4.9

year	duration (min)	intensity (in/hr)	year	duration (min)	intensity (in/hr)
37	30	1.2	42	40	3.4
38	30	2.1	43	60	0.8
39	30	4.0	44	60	1.25
40	40	1.05	45	60	2.6
41	40	1.75			

HYDROLOGY-6

Two square areas, A and B, drain through pipes into a collection manhole as shown. All pipes have a Manning roughness constant of $n = 0.012$. All slopes are 0.035 ft/ft. Area A is 4 ac and has a rational coefficient of $C = 0.9$. The time to concentration, t_c, for area A is 5 min. Area B is 10 ac and has a rational coefficient of $C = 0.25$. The time to concentration for area B is 18 min. Assume the pipes flow full. Determine the diameters of standard pipe sections 1, 2, and 3 for a 10-year storm.

	intensity (in/hr)			
t_c	5 yr	10 yr	20 yr	50 yr
5 min	6.1	7.0	7.9	8.8
10 min	4.6	5.5	6.4	7.3
15 min	3.5	4.4	5.3	6.2
18 min	2.9	3.8	4.7	5.6
20 min	2.7	3.6	4.5	5.4

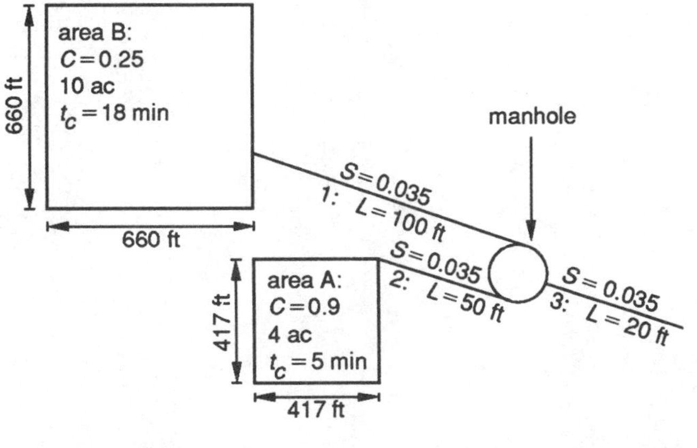

HYDROLOGY-7

A 20-ac pasture (rational coefficient $C = 0.3$, time to concentration $t_c = 15$ min) was originally used for cattle grazing. The pasture drains into a creek that runs along one edge and has barely enough capacity to carry away the water from a 10-year storm. The creek passes through a nearby residential area. When the creek overflows, significant property damage occurs.

The intensity-duration curve for a 10-year storm in this area is given by the following table.

duration (min)	intensity (in/hr)
5	7.7
10	6.9
15	5.8
20	4.7
30	3.5

The pasture has been purchased by a foreign car manufacturer that wants to pave over all 20 ac (rational coefficient $C = 0.90$, $t_c = 10$ min) to provide storage for newly imported cars. The car manufacturer has proposed construction of a detention basin to capture and hold runoff such that the creek's pre-existing capacity is not exceeded during any storm with a frequency of 10 years of less. What should be the capacity in ac-ft of the detention basin?

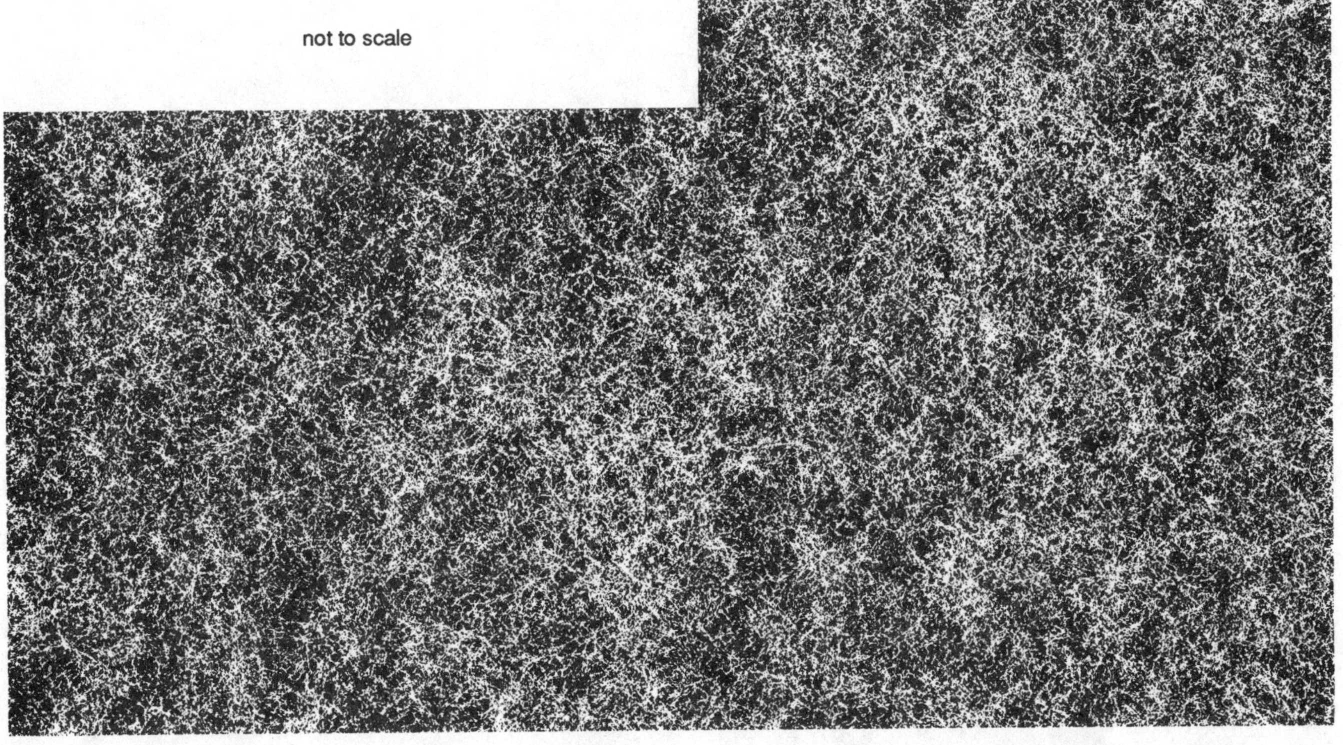

6 WATER SUPPLY ENGINEERING

WATER SUPPLY ENGINEERING–1

A 28-ft × 20-ft (plan dimensions) sand filter (not shown) feeds a clearwell that has plan dimensions of 200 ft × 250 ft. The elevation of the surface of the clearwell is 150 ft and is constant. A pump at elevation 165 ft is driven by a 95% efficient electric motor and removes 3.5 MGD. The pressure gauge at the exit (located at elevation 175 ft) reads 80 psig (static pressure). Minor losses should be ignored.

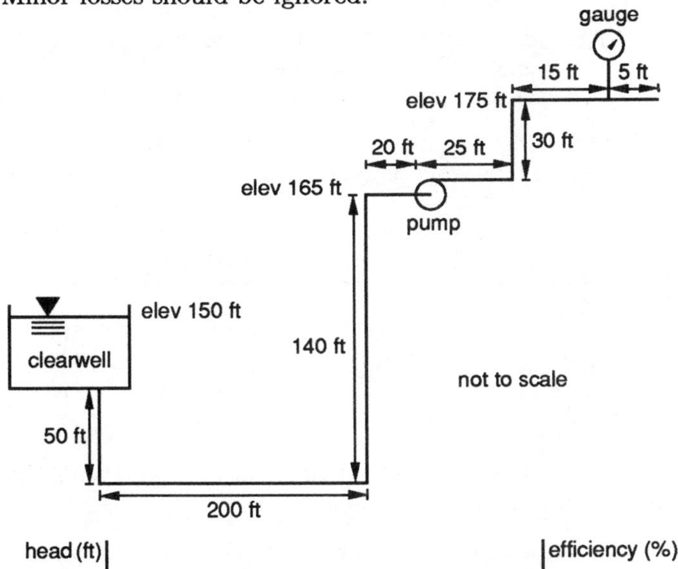

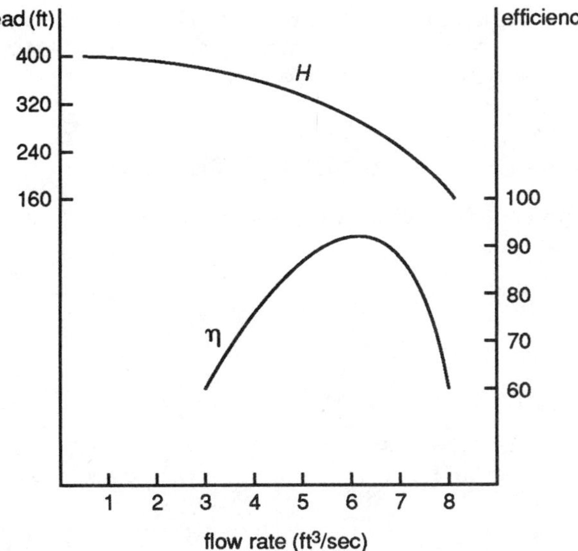

(a) What is the total dynamic head supplied by the pump?

(b) Draw the system curve on the accompanying pump curve.

(c) What is the pump efficiency at the operating point?

(d) What motor horsepower would you recommend?

(e) If the sand filter is overloaded by 25% above its design flow, what will be the depletion or accumulation in the clearwell during an 8-hr period?

WATER SUPPLY ENGINEERING–2

Due to the presence of organics, a water treatment plant receives water with excessive color (much higher than 100 CU). The water is taken through the following processes.

1) prechlorination
2) flash mixing
3) flocculation
4) sedimentation
5) filtration
6) storage
7) chlorination
8) distribution

(a) Briefly explain the purpose of each of the first seven steps.

(b) What problems might the plant have in meeting the EPA requirements for chlorinated organics?

(c) How are trihalomethanes (THMs) introduced into water supplies?

(d) How are THMs reduced?

(e) What are the problems inherent with THM treatments?

WATER SUPPLY ENGINEERING–3

A rapid mixing paddle device in a drinking water system has four flat plate rectangular paddles, each with an area of 2.81 ft^2. Each paddle is mounted to arms on a rotating spindle turning at 100 rpm. The maximum diameter of the mixing device is 3.75 ft. The plates are completely submerged in 55°F water, and each plate addresses the water flat-on. The mixing paddle unit is located in a 185.7 ft^3 tank through which 2 MGD of water flows.

(a) What power is required to drive the paddle mixer?

(b) What is the mean velocity gradient?

(c) Is the value of Gt_d (i.e., the product of the velocity gradient and the mean detention time) within the normal range for such mixing units?

WATER SUPPLY ENGINEERING–4

A water treatment plant receives 2 MGD of well water with a turbidity of 300 NTU.

(a) Design a treatment sequence.

(b) Draw a profile diagram showing the locations and elevations of the treatment processes.

(c) Size all settling basins in your design.

WATER SUPPLY ENGINEERING–5

In order to build an underground parking lot for a large office building, a 400-ft × 400 ft-square excavation site 35 ft deep needs to be made in a sandy soil (permeability = 190 ft/day) in a location where the water table is only 25 ft below the surface. An impervious clay stratum exists 120 ft down. In order for the soil to develop enough bearing strength to support the heavy excavation machinery, the water surface must be lowered 10 ft below the excavation floor. To accomplish this, it has been decided to drop one or more dewatering wells 1 ft in diameter in the middle of the excavation site.

(a) What amount of water must the well(s) withdraw to completely dewater the site to a depth of 10 ft below the excavation floor?

(b) Are dewatering wells a good method to use in this situation?

(c) What other methods could be used to provide a dry work site?

WATER SUPPLY ENGINEERING–6

The population of a small residential area with its own water supply system varies between 24,500 during the day and 35,000 at night. The area is relatively flat. The town is approximately rectangular in shape with length and width of 4.4 mi and 2.9 mi, respectively. The water district is under a mandate to provide all water customers with minimum and maximum water pressures of 50 psig and 80 psig, respectively. Disregard minor losses, and state all other assumptions.

(a) For what average demand should the water district design its water treatment facility?

(b) Considering hourly, daily, and seasonal variations, what is the maximum instantaneous demand the water district could expect to see?

(c) What additional requirements should the water district add for firefighting?

(d) In general, how would you recommend meeting the minimum and maximum water pressure requirements?

(e) What total volume of water storage would you recommend?

(f) Assuming that houses are uniformly distributed in the town, sketch a reasonable layout for the water distribution system, including all towers, mains, and submains.

(g) Design the water storage system, indicating the volume and elevation of each tank. Assume there are a total of 7 water tanks, evenly distributed.

Assume that the distribution system is made up of the following pipe lengths, types, and roughness coefficients.

type	diameter (in)	length (ft)	roughness coefficient
steel	16	4000	100
steel	12	4000	100
steel	6	1000	100
copper tube	$\frac{3}{4}$	80	130

(h) How often and at what time of day should the water tanks be refilled?

(i) Assuming that the water tank surface elevation is 80 ft above the ground, will the water pressure requirements be satisfied?

WATER SUPPLY ENGINEERING–7

The composition of water as delivered from an aquifer is as follows.

Ca^{++}	80 mg/ℓ
Cl^-	18 mg/ℓ
CO_2	15 mg/ℓ
Fe^{++}	3 mg/ℓ
HCO_3^-	336 mg/ℓ
Mg^{++}	30 mg/ℓ
SO_4^{--}	64 mg/ℓ
TDS	650 mg/ℓ

(a) What is the hardness?

(b) Is the water acidic or alkaline? Why?

(c) What is the acidity or alkalinity?

(d) What is the most likely color of this water?

(e) What is the pH of this water?

(f) Should this water be softened?

(g) What amounts of lime and soda ash are required to obtain a final hardness of 80 mg/ℓ?

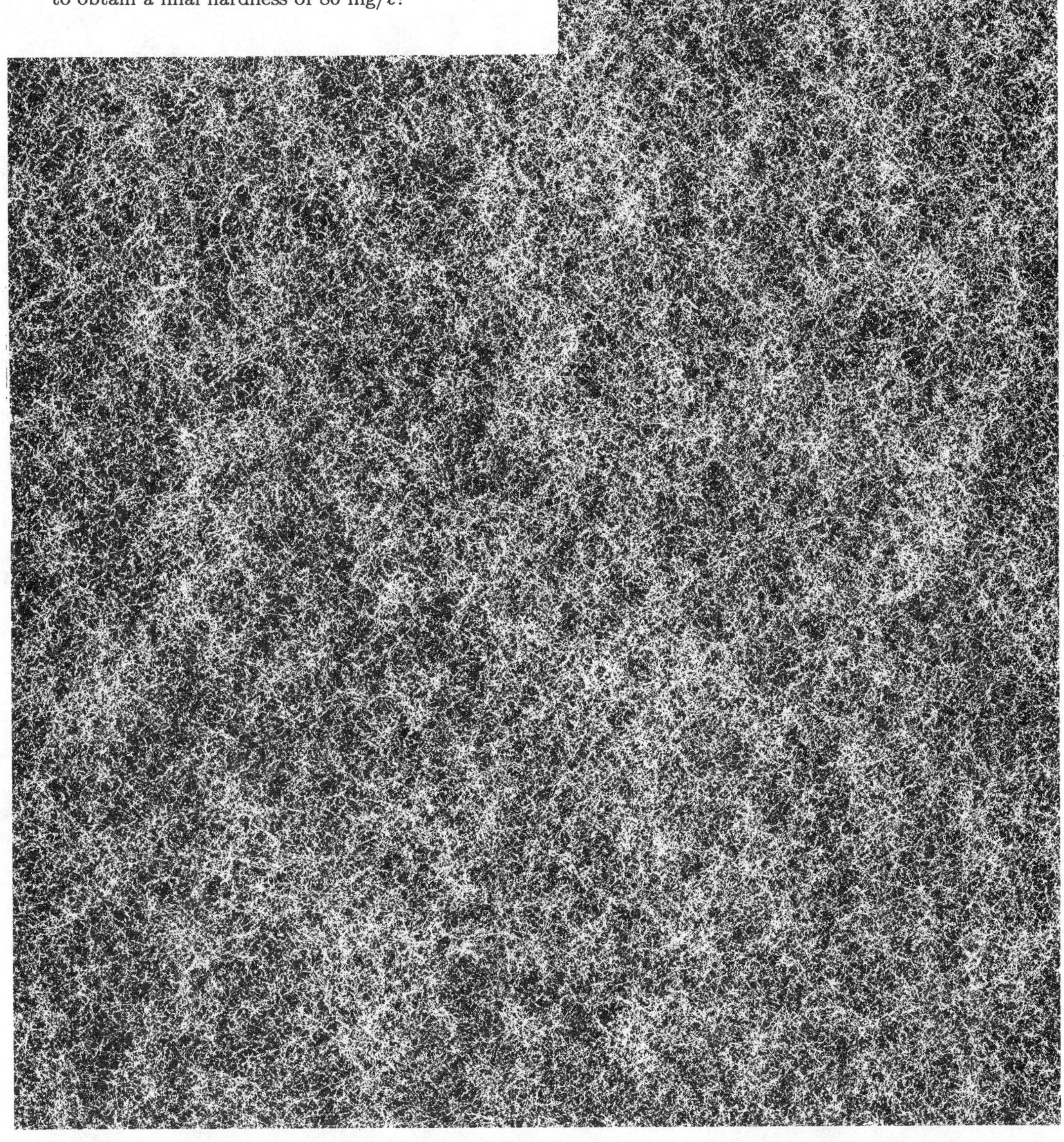

7 WASTEWATER ENGINEERING

WASTEWATER ENGINEERING–1

A wastewater treatment plant receives 4.5 MGD of 70% wastewater influent with the following characteristics.

- 950 lbm BOD/MG
- 3200 mg/ℓ MLSS
- 2 mg/ℓ dissolved oxygen
- 70% volatile solids

As part of its processing, the wastewater is aerated with diffusion aerators operating with the following characteristics.

BOD constant:	1.15 lbm O_2/lbm BOD removed
MLSS constant:	0.03 lbm O_2/lbm BOD removed
oxygen transfer efficiency:	15%
BOD removal efficiency:	85%
mechanical efficiency:	80%
outlet pressure:	22.7 lbf/in^2

What is the cost of 24 hr of constant aeration if the aerators have been sized with 100% excess capacity and electrical power costs $0.08/kW-hr?

WASTEWATER ENGINEERING–2

Three covered digesters are used to process 30,000 gal of waste sludge per day. Each digester has an internal diameter of 45 ft, vertical walls of unspecified height, and a 7.5 ft-high tapered section at the bottom. Prior to loading, the waste is thickened 5% and has the following characteristics.

BOD_5 loading:	12,000 lbm/day
volatile organics:	70%
temperature:	61°F
BOD reaction constant:	0.2/day

Use the following data in calculating methane production.

efficiency of waste utilization: E:	70%
yield coefficient, Y:	0.06 lbm/lbm
endogenous coefficient, R_d:	0.03/day
mean cell residence time, θ_c:	10 days

The average ambient temperature is 58°F, and the temperature of digestion is 91°F. The average heat loss through the vertical portion of the digester walls is 110,000 BTU/hr per digester. The overall coefficients of heat transfer for the floor and walls of the digester are 0.15 BTU/ft^2-°F-hr and 0.16 BTU/ft^2-°F-hr, respectively.

(a) How much heat (in BTUs) must be supplied to maintain the 91°F digestion temperature?

(b) Will the methane produced from the digestion process be sufficient to provide the energy required?

WASTEWATER ENGINEERING–3

An existing 24-in (inside diameter) unlined concrete sewer pipe currently serves a small town but is expected to be under capacity when the population peaks at 300,000. The existing line drops 100 ft as it travels 4000 ft to the first pumping station. A second pipe will be installed in parallel with the existing pipe. Use 250 gpcd (gal per capita per day) as an estimate of the future wastewater production.

(a) What diameter pipe should be installed so that the combined capacity will be adequate for the peak population?

(b) If both pipes are 24-in diameter, what will be the depth of flow in each pipe?

WASTEWATER ENGINEERING–4

A pretreatment plant processes waste from several local industries prior to discharging it into the municipal system. The following mix of influents must be handled simultaneously.

volume	source
13,000 gal/day	sanitary waste from employees
41,250 gal/day	hospital medical waste
5000 gal/day	plating plant chemical waste
15,000 gal/day	dairy animal waste

(a) What typical limits of BOD would be expected from each of these four sources?

(b) What other wastewater quality characteristics must be considered when designing the pretreatment process?

(c) Design a pretreatment process and briefly explain why each step is required.

WASTEWATER ENGINEERING–5

Design a stabilization pond to support a large national park campground. The population of the campground increases dramatically during the summer. During winter months, a skeleton crew remains to perform maintenance. The following characteristics must be considered in the pond design.

BOD loading (all sources):	250 mg/ℓ
average annual temperature:	70°F
latitude:	29°N
skeleton population:	300
peak summer population:	10,000 campers
hospital bed capacity:	100
laundromat:	20 washers
dining rooms:	500 seats
gas stations:	8 pumps

The dining room serves three meals a day. During the summer, the dining facility offers a choice of three sittings/servings per meal.

Use the following data in calculating the performance of the stabilization pond.

$$\begin{aligned}\text{BOD loading from all sources} &= 250\ \text{mg}/\ell\\ \text{BOD}_5\ \text{conversion} &= 90\%\\ \text{first-order BOD}_5\ \text{rate constant},\ k &= 0.25/\text{day at } 20^\circ\text{C}\\ \text{temperature coefficient},\ \theta &= 1.06 \text{ at } 20^\circ\text{C}\\ \text{summer pond temperature} &= 32^\circ\text{C } (90^\circ\text{F})\\ \text{winter pond temperature} &= 10^\circ\text{C } (50^\circ\text{F})\\ \text{maximum individual pond area} &= 10\ \text{ac}\\ \text{maximum pond depth} &= 2\ \text{ft}\end{aligned}$$

The product, kt, of the first order reaction constant, k, and the detention time, t, is 5.

WASTEWATER ENGINEERING–6

Specify five different methods for disposing of solid waste in open areas. For each, indicate the (a) advantages, (b) disadvantages, and (c) relative cost. Also specify (d) the approximate population size for which each method is appropriate.

WASTEWATER ENGINEERING–7

The present population of a large town is 500,000 and is expected to increase by 3% each year as it has in the past. The population currently disposes of 5.4 lbm of solid waste per capita day, but this amount is expected to increase at the rate of 1% per year as it has in the past.

Through a new recycling program, 3% of all solid waste can be recovered. A solid-waste disposal site has been in use (without recycling) for 16 years and was originally designed with a 20-year life. The site is roughly square in configuration with vertical side slopes. There is a 250-ft buffer zone around the site, and waste can be piled to an average height of 40 ft. Daily, intermittent, and final cover occupies 25% of the disposal site. The existing site has four years of capacity remaining. Solid waste is compacted to a density of 1000 lbm/yd^3.

(a) How many acres does the current site occupy?

(b) Specify the size (in acres) of a replacement site that would extend the useful life by an additional five years.

WASTEWATER ENGINEERING–8

One MGD of wastewater is treated in a lagoon/holding pond arrangement. The effluent from the ponds is subsequently processed by spraying into fields at the rate of 2.5 in-ac/ac-week. The static head available in the spray lines is 45 ft, but head losses due to friction through the piping system are 30 ft. Each spray nozzle covers a circular area 350 ft in diameter when operating at its rated capacity of 275 gal/min and 75 psig. Pumps increase the head above the static value. To ensure uniform coverage, only 25% of the field nozzles are in use at one time. Pump mechanical efficiency is 82%; motor electrical conversion efficiency is 92%.

(a) At rated capacity, what land area (in acres) can be in use at any given moment?

(b) How many nozzles (total) are needed?

(c) What horsepower motor(s) should be used to drive the pumps?

(d) What considerations should be taken into account when determining the application rate of effluent to the spray fields?

WASTEWATER ENGINEERING–9

A typical block in a 5 block × 8 block subdivision is illustrated. Streets are 24-feet wide. The subdivision is constructed on a 160-ac section of land. Individual building sites (parcels, lots, etc.) are 0.25 ac (90 ft × 120 ft) each, and building sites are grouped into blocks containing 12 sites as shown. There are 480 such sites planned. Each site will support, on the average, 4.1 residents. Infiltration is 200 gal/in-mi-day. Inflow is 2000 gal/in-mi-day. What flow must the wastewater pump station be able to handle?

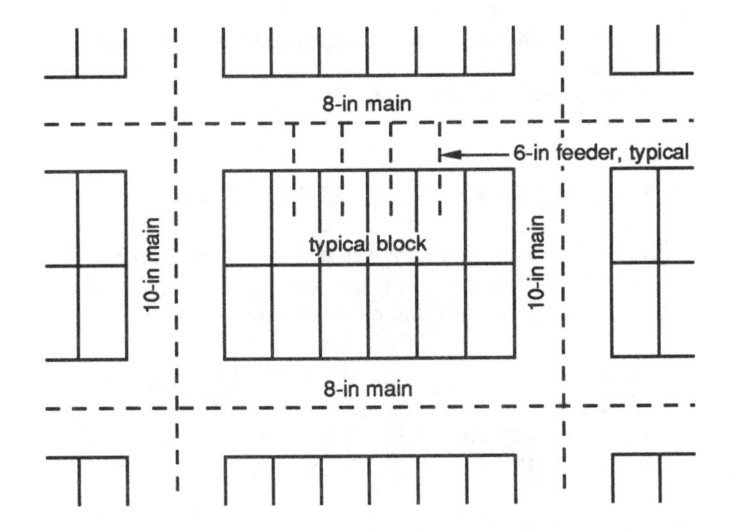

WASTEWATER ENGINEERING–10

A complete mix-activated sludge processing facility will be designed using the theory of kinetics and the following characteristics.

incoming flow, Q:	5 MGD
influent soluble BOD_5, S_0:	250 mg/ℓ
effluent BOD:	20 mg/ℓ
effluent suspended solids:	20 mg/ℓ
mixed liquor volatile suspended solids, MLVSS = 0.8 MLSS:	3500 mg/ℓ
return sludge suspended solids, SS:	10,000 mg/ℓ
return sludge volatile suspended solids, x_r:	8000 mg/ℓ
mean cell residence time, θ_c:	10 days
kinetic coefficient, y:	0.6 lbm VSS/lbm BOD_5
kinetic coefficient, k_d:	0.06/day
percent of effluent biodegradable suspended solids:	65%
ratio of BOD_5/BOD_L:	0.68

Calculate the following quantities.

(a) biological efficiency (in soluble BOD_5 removal)

(b) overall efficiency (in BOD_5 removal)

(c) biomass production (pounds per day)

(d) increase in mixed liquor suspended solids (pounds per day)

(e) sludge to be wasted (pounds per day)

(f) sludge-wasting rate (MGD)

(g) recirculation ratio

(h) hydraulic retention time (hours)

(i) specific substrate utilization rate (mg BOD_5/mg MLVSS-day)

(j) food-to-microorganism ratio (mg BOD_5/mg MLVSS-day)

(k) blower capacity (ft^3/min) if the oxygen transfer efficiency is 8% and 100% excess capacity is used

WASTEWATER ENGINEERING–11

Design an industrial wastewater treatment plant to process wastewater from a beet sugar processing facility. The facility will operate for three months during the fall and winter. The facility will have a total capacity of 1000 tons per day. The flows and characteristics of different sources within the plant are as follows.

source	unit volume (gal/ton)	daily flow (gal/day)	BOD (mg/ℓ)	TSS (mg/ℓ)	TDS (mg/ℓ)
flume/wash water	2200	2.2×10^6	200	800	780
process water	660	6.6×10^5	1230	1100	1120
lime drainage	75	7.5×10^4	1420	450	2850

(a) Suppose the flume/wash water will be recovered and reused. Design a sedimentation basin to treat the flume/wash water. Assume the detention time will be 2.0 hr and the water depth will be 10.0 ft.

(b) Evaluate the use of an aerobic lagoon to treat the facility's wastewater. Assume a detention time of 2.0 days, a working depth of 1.5 ft, and a target loading rate of 86 lbm BOD per acre-day.

(c) Briefly describe two alternative methods for treating the wastewater from this facility.

WASTEWATER ENGINEERING–12

A wastewater treatment plant will be designed according to *Ten-States' Standards* to process incoming wastewater with the following characteristics.

2.0 MGD flow (Q)
250 mg/ℓ BOD_5
180 mg/ℓ suspended solids (SS)

Treatment will consist of a circular trickling filter using synthetic plastic media, a grit chamber, rectangular primary clarifiers, and a secondary clarifier.

(a) Draw the process diagram showing each operation in sequence.

(b) Size the primary clarifier with an aspect ratio of 4:1. State your assumptions.

(c) Size the trickling filter. State your assumptions.

WASTEWATER ENGINEERING–13

Outline the general procedure and list the general equations you would use to design a three-cell aeration pond based on the following characteristics.

wastewater flow:	Q
wastewater organic strength:	S_0, BOD_5
wastewater nutrients:	P, NH_3-N or TKN
average winter wastewater temperature:	T_{ww}
average summer wastewater temperature:	T_{ws}
critical ambient winter air temperature:	T_{aw}
critical ambient summer air temperature:	T_{as}
substrate utilization coefficient:	k
theoretical yield coefficient:	$Y_{T,20°C}$
microbial decay coefficient:	$k_{d,20°C}$
mean cell residence time:	θ_c
temperature factor:	f
detention time in each cell:	t
depth of each cell:	D

(a) What are the cell volume and cell surface area?

(b) What are the winter and summer lagoon operating temperatures?

(c) How are k and k_d corrected for the winter and summer operating temperatures?

(d) What is the soluble BOD_5 from the first cell?

(e) What is the biomass concentration entering the first cell?

(f) What is the biomass concentration entering the second cell?

(g) What is the biomass concentration entering the third cell?

(h) Estimate the effluent BOD_5 before clarification, assuming the soluble BOD_5 from the third cell is S_e.

(i) Calculate the minimum mean cell residence time for nitrification. Assume μ_{max} and pH_{opt} are known.

(j) Calculate the oxygen requirements for each cell of the lagoon, including nitrification where appropriate.

(k) Determine the motor horsepower required to supply oxygen to each cell assuming a surface aerator delivers 1.8 lbf of oxygen per horsepower-hour.

(l) Determine the motor horsepower for complete mixing in each cell.

(m) Specify a maximum length:width aspect ratio for rectangular cells.

(n) Compare the nutrient requirements and supply in the first cell. Assume the first cell controls the nutrients requirements.

WASTEWATER ENGINEERING–14

Effluent from a primary treatment plant discharges from an 8-in-diameter pipe into an intermittent stream. The characteristics of the effluent are

velocity:	0.6 ft/sec (flowing full)
BOD_5:	150 mg/ℓ (20°C)
dissolved oxygen, DO:	1.0 mg/ℓ
temperature:	22°C

The characteristics of the intermittent stream are

flow:	2 ft^3/sec
velocity:	0.2 ft/sec
BOD_5:	4 mg/ℓ (20°C)
dissolved oxygen, DO:	10.0 mg/ℓ
summer temperature:	52°F
winter temperature:	45°F
deaeration coefficient, K_D:	0.1/day
reaeration coefficient, K_R:	0.25/day

(a) Find the effluent volumetric flow rate.

(b) Find the five-day BOD immediately after mixing.

(c) Find the dissolved oxygen immediately after mixing.

(d) Find the temperature in the summer immediately after mixing.

(e) Find the temperature in the winter immediately after mixing.

(f) Find the dissolved oxygen concentration at a point 20 mi downstream.

WASTEWATER ENGINEERING–15

A retirement community supports 2000 people. A single concrete sewer main connects the community to a local treatment plant.

(a) Determine the required diameter of the waste main assuming a minimum permissible slope. State all of your assumptions.

(b) Given that the waste main diameter is 8 in and the slope is 0.021 ft/ft, find the depth and velocity of flow.

WASTEWATER ENGINEERING–16

A large spill of a toxic liquid chemical has occurred. The following steps have already been taken.

1) The spill has been adequately contained by a makeshift dike.
2) The identity of the chemical has been determined and no other hazardous or toxic contaminants have been identified.
3) It has been determined that there are no immediate hazards due to fire or explosion.
4) It has been determined that there is no serious threat of contamination to municipal water supplies, major groundwater supplies, or surface water.

Specify the steps you would take to neutralize the chemical and clean up the site.

WASTEWATER ENGINEERING–17

Explain how you would design and use a lagoon as an equalizing basin to alter peak flow. Discuss the merits of such a method to equalize flow.

WASTEWATER ENGINEERING–18

(a) What is the difference (if any) between scouring and self-cleansing velocities? What are typical values?

(b) What minimum scouring velocity should be maintained given a roughly spherical particle with a diameter of 0.5 in?

WASTEWATER ENGINEERING–19

The following questions apply to sanitary landfills.

(a) What agencies in the U.S. control landfill design?

(b) What is the average density of shredded waste?

(c) What is the average density of baled and compacted waste?

(d) What is a typical depth of daily soil cover?

(e) What is a typical depth of intermediate soil cover?

(f) What is a typical depth of final soil cover?

(g) What is the maximum length of time that intermediate cover can be used?

(h) What is the purpose of a clay liner?

(i) What is the primary limitation of clay liners?

(j) What is the significance (as related to sanitary landfill design) of the acronym NIMBY?

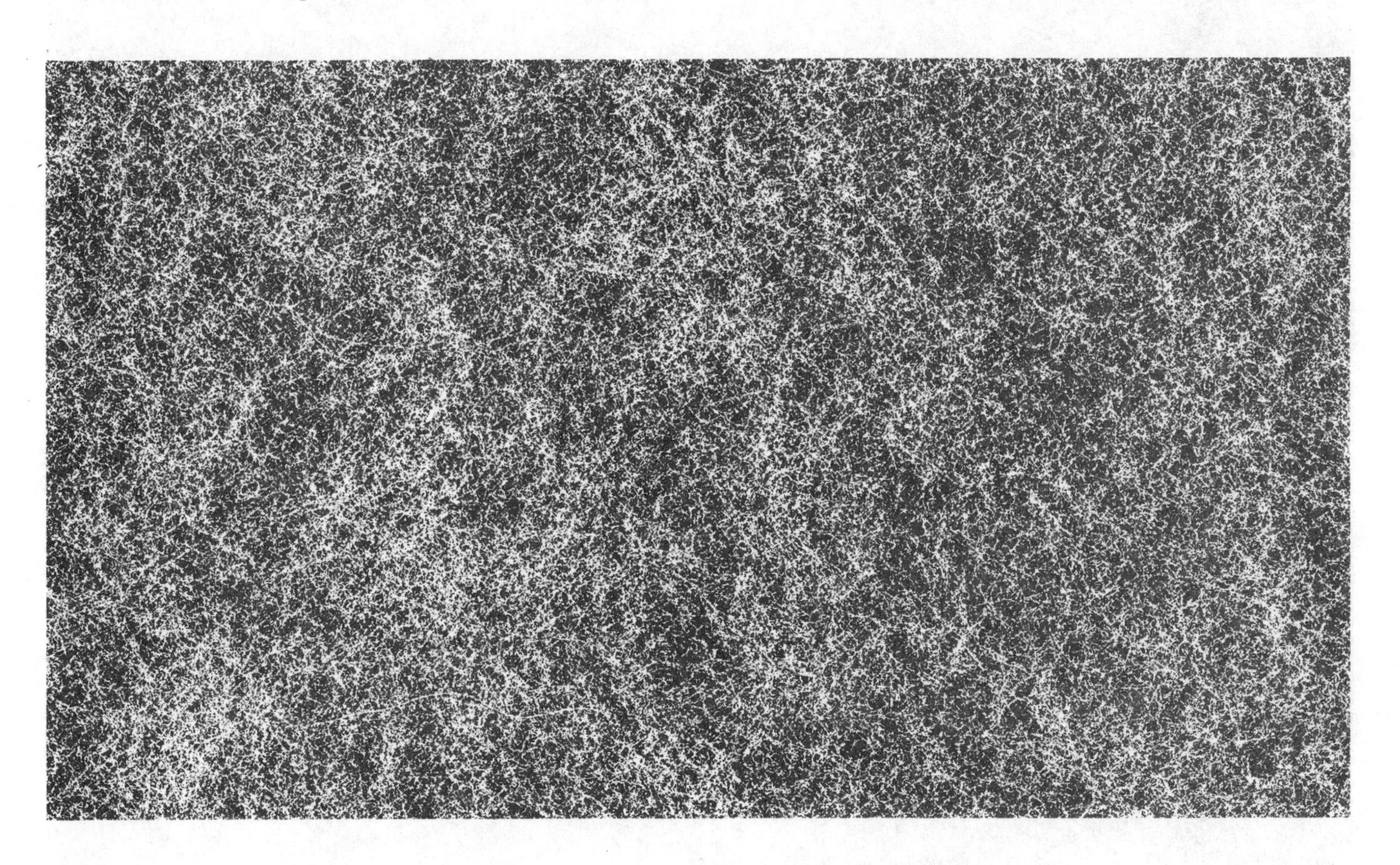

8 SOILS/SOIL MECHANICS

SOILS/SOIL MECHANICS–1

Soil is taken from a borrow site and used as fill at a construction site. A standard Proctor test indicates that the dry specific weight of a borrow soil is 105 lbf/ft^3 at an optimum moisture content of 20%. The borrow soil has a water content of 26%. The specific gravity of the soil solids is 2.65. Per contract, the fill must be compacted to 95% of the standard Proctor maximum, but the earthwork contractor complains that the fill pumps and ruts badly before reaching 95% compaction.

(a) Plot three points on the zero air void curve for moisture contents of 20%, 25%, and 30%.

(b) What is the maximum water content when the fill is compacted to 95% of the standard Proctor maximum?

(c) What is the maximum density without pumping or rutting at the original water content of 26%?

(d) Does the earthwork contractor have a valid argument?

(e) Briefly give your recommendations.

SOILS/SOIL MECHANICS–2

Samples of five different soils, A, B, C, D, and E, are taken. The properties of the five soils are given.

	sieve analysis: % passing			
soil	10	40	100	200
A	45	15	5	–
B	75	40	20	15
C	85	62	35	7
D	89	72	48	39
E	95	87	72	68

soil	w_l liquid limit	w_p plastic limit	I_p plasticity index
A	8	–	–
B	27	15	12
C	37	29	8
D	35	22	13
E	52	25	27

(a) What are the soils' classifications according to the AASHTO system?

(b) What are the soils' classifications according to the Unified Soils system?

(c) Which soil(s) would require undercutting if exposed after excavation and grading (i.e., would not be suitable for roadway base material)?

(d) Which soil(s) would not make suitable roadway embankment material?

(e) Which soil(s) would not make suitable borrow material for a subbase?

SOILS/SOIL MECHANICS–3

The cut and fill volumes (in cubic yards) of soil for 20 stations along a new highway are given. The soil has a shrinkage factor of 0.25. The contractual free haul distance is 600 ft.

station	cut	fill
1+00	–	1000
2+00	2000	–
3+00	1000	–
4+00	–	1500
5+00	500	–
6+00	1000	–
7+00	1500	–
8+00	–	500
9+00	–	1500
10+00	–	500
11+00	–	1000
12+00	2000	–
13+00	2500	–
14+00	1000	–
15+00	0	0
16+00	–	1000
17+00	–	1500
18+00	–	2000
19+00	–	2500
20+00	–	2000

(a) Draw the mass diagram.

(b) Show all balance points.

(c) Calculate all haul amounts and draw arrows to show all haul directions.

(d) What is the station yardage of overhaul between 1+00 and the first balance point?

(e) What is the definition of a bank yard?

(f) What methods can be used to protect the slope and control erosion from the sides of cuts made along the highway?

SOILS/SOIL MECHANICS–4

Two 20-ft-deep cuts are made at an angle of 45°, one in sandy soil (angle of internal friction = 30°, density = 120 lbf/ft^3), and another in homogeneous soft clay (specific weight = 110 lbf/ft^3; unconfined compressive strength = 1600 lbf/ft^2). Answer parts (a) through (d) for both soils.

(a) Find the slope stability factor of safety for the material alone.

(b) Find the factor of safety for translational or lateral block failure.

(c) Is either of the factors of safety (from parts (a) and (b)) adequate? Why or why not?

(d) If the slope is inadequate, what slope is required?

SOILS/SOIL MECHANICS–5

An 18-ft-high, 40-ft-wide braced cut is made in clay (cohesion = 450 lbf/ft^2; specific weight = 115 lbf/ft^3). The bottom of the cut is well above a hard clay layer.

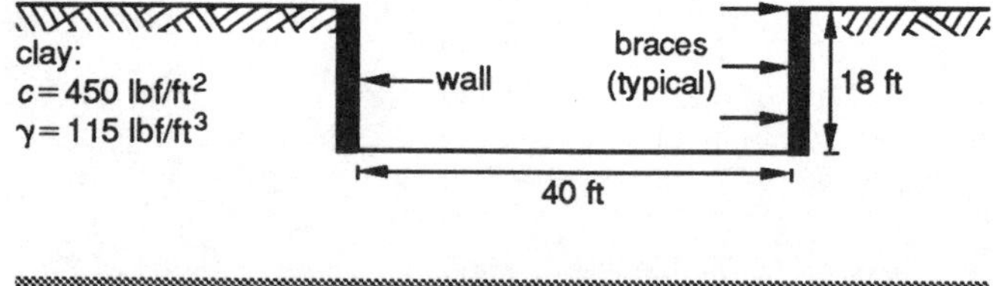

(a) What is the factor of safety against bottom heave?

(b) At what depth will the bottom heave?

(c) If the bottom of the cut is only 10 ft above a hard clay layer, what will be the factor of safety?

SOILS/SOIL MECHANICS–6

A 20-in-diameter concrete pipe with a crushing strength of 1000 lbf/ft^2 is buried 4 ft below grade. The trench is backfilled with granular soil having a dry specific weight of 120 lbf/ft^3. The pipe is loaded with an additional dead load of 100 lbf/ft^2. Determine if the pipe will break under H-20 loading for bedding classes B and D.

SOILS/SOIL MECHANICS–7

Two soils have been proposed for use as a landfill soil liner. Falling head permeability tests have been performed.

permeability test results

test parameter	soil A	soil B
sample length	15 cm	15 cm
sample area	40 cm^2	40 cm^2
initial head	100 cm	200 cm
final head	75 cm	190 cm
test time	300 min	6000 min
standpipe area	5 cm^2	2 cm^2

(a) What are the permeabilities of the two soils in units of ft/yr?

(b) Based on the permeabilities alone, make a recommendation as to which soil(s) should be used as a liner.

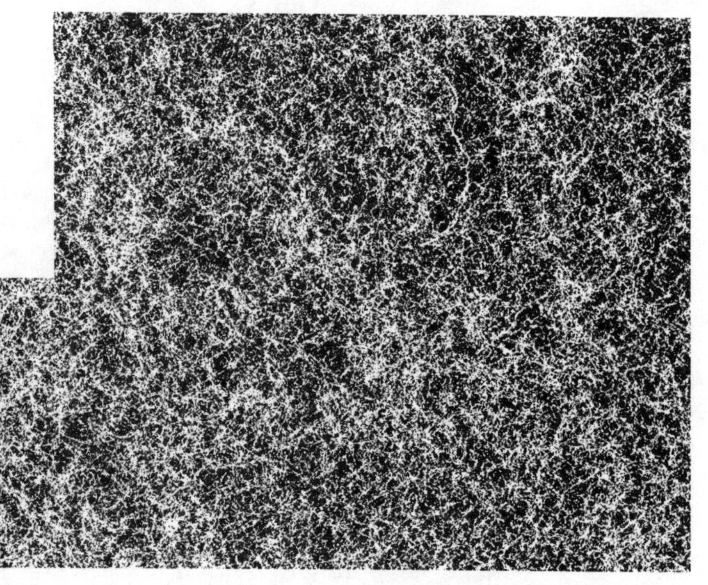

9 FOUNDATIONS

FOUNDATIONS–1

20-in-diameter concrete end-bearing piles placed every 12 ft (each way) carry a 24-in-thick, steel-reinforced concrete slab and a 1100 lbf/ft^2 live load. The piles penetrate 10 ft of recently placed granular fill (specific weight = 110 lbf/ft^3) and 24 ft of normally consolidated clay (specific weight = 95 lbf/ft^3; angle of internal friction = 30°; unconfined compressive strength = 900 lbf/ft^2). The pile ends bear on a dense sand layer ($\phi = 35°$; $\gamma = 12.5$ lbf/ft^3). Piles have been installed loosely through a steel casing for the first 10 ft.

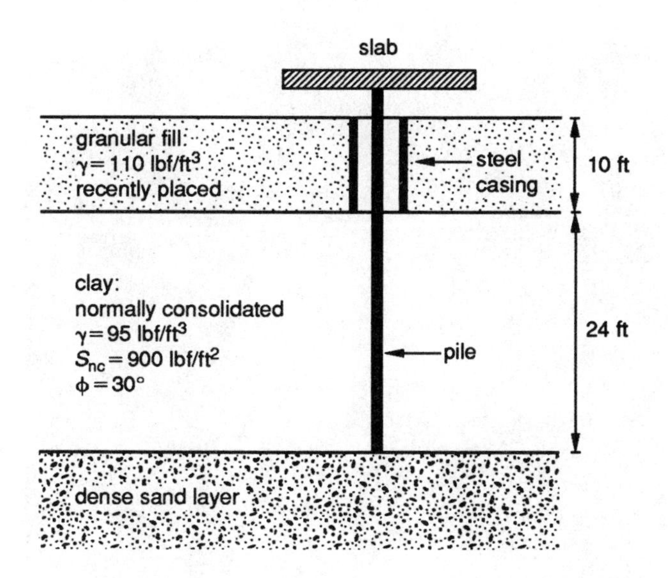

(a) Is the end capacity sufficient for this installation? Why or why not?

(b) Would your approach to determining the pile strength be any different if the clay layer was overconsolidated? If so, in what manner would your approach differ?

FOUNDATIONS–2

Two square footings in a structure must have the same settlement even though their loadings and sizes are different. Both footings are located at the same depth in a sand layer 2 ft above the water table. The first footing (A) is 6 ft square. (Refer to the illustration for more information.)

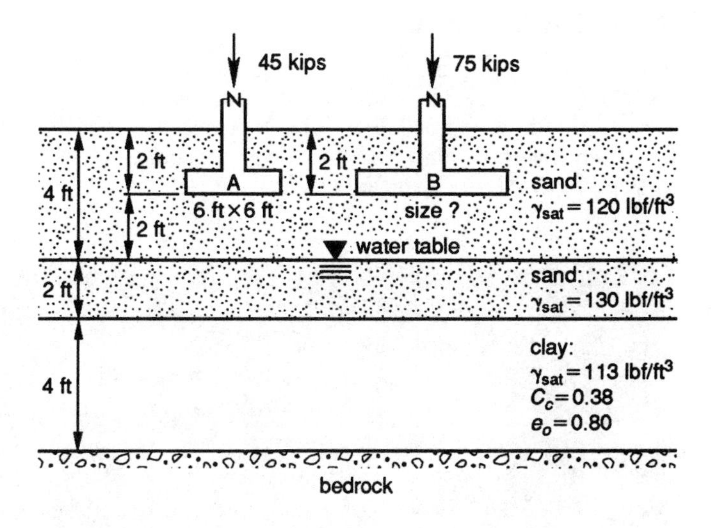

(a) What is the settlement of footing A?

(b) For a first approximation, assume a 2:1 (vertical:horizontal) pressure distribution and determine the size of footing B that will produce the same settlement as footing A.

FOUNDATIONS–3

A pipe support in a refinery consists of a single 20-ft-long, 20-in-diameter concrete pile. A lateral (horizontal to the ground) load is applied at the top of the pile. Six feet of the pile remain above ground. The modulus of elasticity of the concrete is 60,000 lbf/ft^2. The soil consists of sand (effective friction angle = 30°; specific weight = 120 lbf/ft^3; coefficient of modulus variation = 40,000 lbf/ft^3). Assume the water table is at the ground surface. Use a factor of safety of 2.5 to determine the maximum allowable lateral load. State the authority for your method.

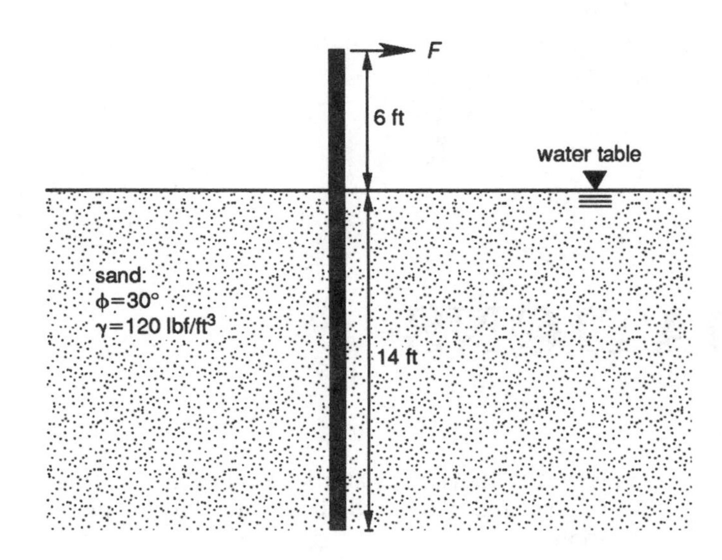

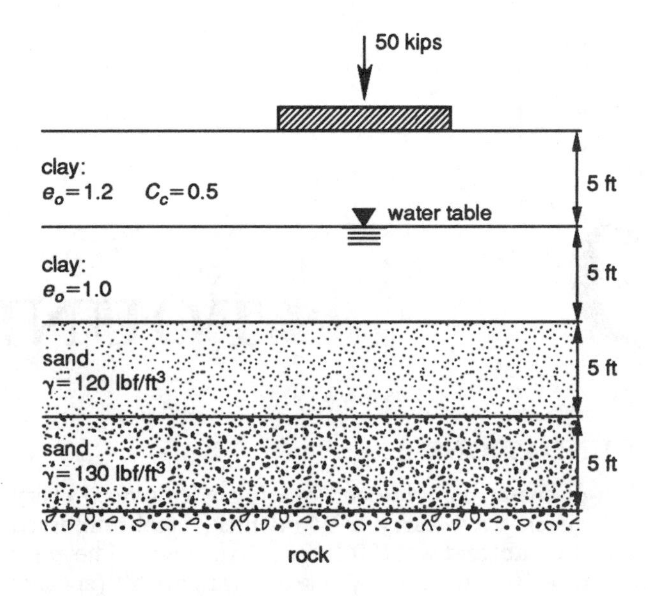

FOUNDATIONS–4

The bay spacing in a reinforced concrete frame is 20 ft. Each bay has a series of 6-ft × 6-ft footings that carry vertical loads of 50,000 lbf. The footings are located over two layers of clay and two layers of sand. (Refer to the illustration for more details.)

(a) Find the primary settlement under the center of the footing.

(b) Is this settlement tolerable? Why or why not?

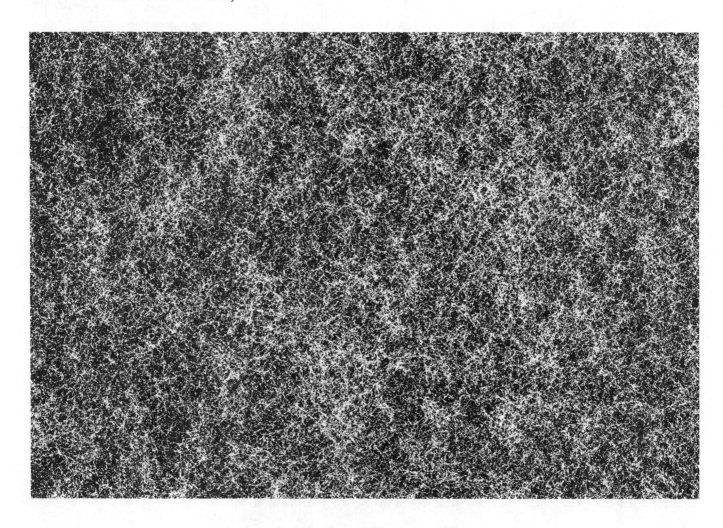

10 CONCRETE DESIGN

CONCRETE DESIGN–1

(a) What is an admixture?

(b) Give six examples of admixtures and explain the purpose of each.

CONCRETE DESIGN–2

A 1:1.6:2.6 (by weight) mixture of cement, sand, and coarse aggregate is produced with the following specifications.

cement:	specific gravity = 3.15 94 pounds per sack
sand:	specific gravity = 2.62 SSD
coarse aggregate:	specific gravity = 2.65 SSD
water:	5.8 gal per sack cement
entrained air:	3%

(a) How much cement, sand, coarse aggregate, and water are required to produce 1 yd^3 of concrete? Express all answers in pounds.

(b) If the sand absorbs 1.6% moisture and the coarse aggregate has 3.2% excess moisture (based on saturated surface dry conditions), what weight (in pounds) of water is needed?

(c) Discuss three techniques that can be used to assure proper curing of concrete pavement.

(d) Discuss two methods of installing transverse joints in concrete pavement.

CONCRETE DESIGN–3

A concrete mixture uses the following materials.

cement:	34 sacks specific gravity = 3.15 94 pounds per sack
fine aggregate:	6500 pounds (dry basis) specific gravity = 2.67 −2% moisture absorption from SSD
gravel:	11,500 pounds (dry basis) specific gravity = 2.64 +1.5% moisture excess from SSD
water:	142 gal (as delivered)
entrained air:	4%

(a) What is the concrete yield (in cubic yards)?

(b) What is the water-cement ratio in gallons per sack of cement?

(c) What is the cement factor in sacks per cubic yard?

CONCRETE DESIGN–4

A 16-in (gross dimension) square, tied column must carry 280-kip dead and 190-kip live loads. The dead load includes the column self-weight. The column is not exposed to any moments. Sidesway is prevented at the top, and slenderness effects are to be disregarded. The concrete compressive strength is 4000 lbf/in^2; the steel tensile yield strength is 60,000 lbf/in^2.

(a) Use the current ACI 318 or UBC to design the column. Specify longitudinal steel size and tie spacing.

(b) Sketch the column cross section.

CONCRETE DESIGN–5

A series of frames are spaced on 20-ft centers (measurement perpendicular to the page). The frames support a 110 lbf/ft^2 vertical downward dead load (which includes the concrete self-weight) and a 45 lbf/ft^2 vertical downward live load. The columns are rectangular, and the width dimension is fixed at 12 in (as shown) by architectural details. The material strengths are f'_c = 4000 lbf/in^2 for concrete and f_y = 60 $kips/in^2$ for steel.

(a) Use the current ACI 318 or UBC with ultimate strength design to determine the exterior column depth. (Do not use ACI moment coefficients.)

(b) Draw the design to scale showing dimensions, reinforcement placement, and other details.

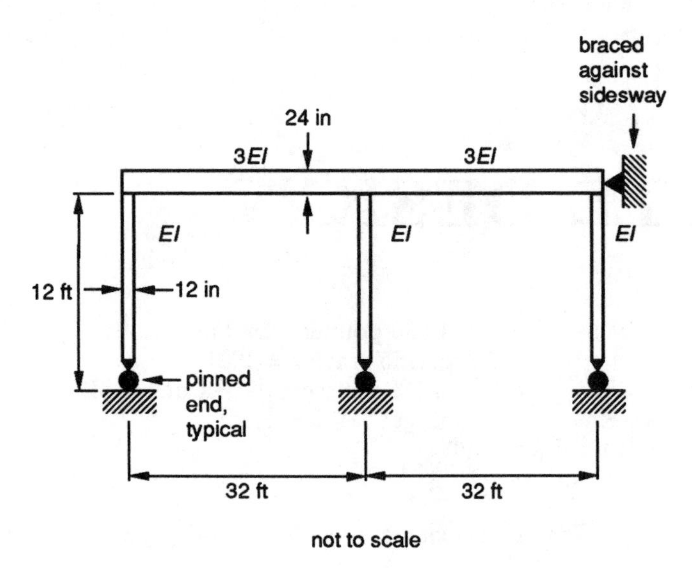

CONCRETE DESIGN–6

A simply supported exterior concrete beam has a length of 22 ft. The beam carries a live load of 3.3 kips/ft and a dead load of 1.3 kips/ft. The dead load does not include the beam weight. The beam width is 16 in. The concrete compressive strength is 3000 lbf/in^2; the steel yield strength is 36,000 lbf/in^2. Use the current ACI 318 or UBC and design the beam depth and steel reinforcement. No compressive reinforcement is to be used. Do not design shear reinforcement.

CONCRETE DESIGN–7

Determine the deflection of the concrete beam designed in the previous problem. Assume a depth to reinforcement, d, of 27.3 in.

CONCRETE DESIGN–8

A cantilever beam extends 8 ft from a rigid support and carries 100 lbf/ft dead load and 300 lbf/ft live load. The dead load does not include the concrete weight. The beam width is 10 in. 4000 lbf/in^2 concrete and 60 kips/in^2 steel will be used. #3 bar is available for stirrup and miscellaneous use. Design the beam.

(a) Specify the beam depth.

(b) Give the amount, type, and location of the longitudinal steel.

(c) Draw a cross section of the beam.

CONCRETE DESIGN–9

A combined footing supports two 18-in columns separated by a distance of 18 ft (center-to-center). The right column (column A) is located near a property line and cannot extend more than 2 ft to the right. There is no limitation on how far the footing can extend to the left of column B, nor is there any limitation on the footing width. The footing thickness is limited to 2 ft overall. Overburden loading and punching shear are to be neglected in the design. Use the following data.

column A:	80 kips dead load
	55 kips live load
column B:	110 kips dead load
	105 kips live load
concrete compressive strength:	3000 lbf/in^2
steel yield strength:	60,000 lbf/in^2
allowable soil bearing pressure:	3500 lbf/ft^2

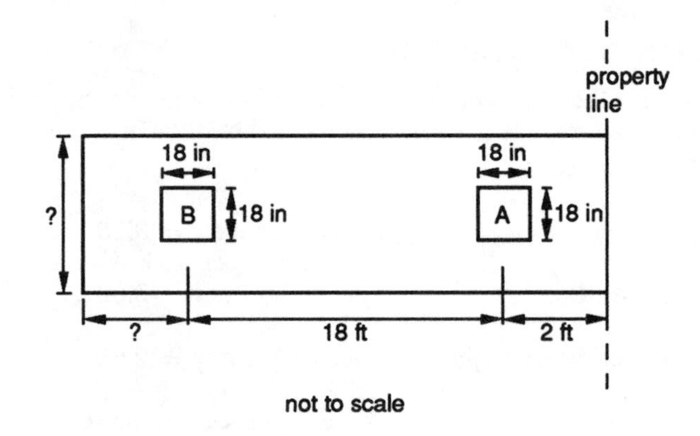

(a) Size the footing for a uniform pressure distribution.

(b) Draw the shear and moment diagrams for the footing.

(c) Design the longitudinal reinforcement for the maximum moment.

CONCRETE DESIGN–10

A 10-in × 10-in column carries a service-level axial load of 10,000 lbf and a moment of 5000 ft-lbf, and it rests on a square footing. The allowable soil pressure is 1500 lbf/ft^2. Ultimate or factored loads may be taken as 1.6 times service loads.

(a) Determine the size of the footing required to keep the soil pressure below the allowable value.

(b) Check for shear.

(c) Design the amount, spacing, and location of the steel reinforcement.

CONCRETE DESIGN–11

A water tank weighs 40,000 lbf when full. It is supported by four legs that, in turn, are supported by two footings with the dimensions illustrated. The footings rest on a 6-in-thick concrete slab. All concrete has a compressive strength of 3000 lbf/in^2. The allowable bearing pressure of the soil is 1300 lbf/ft^2.

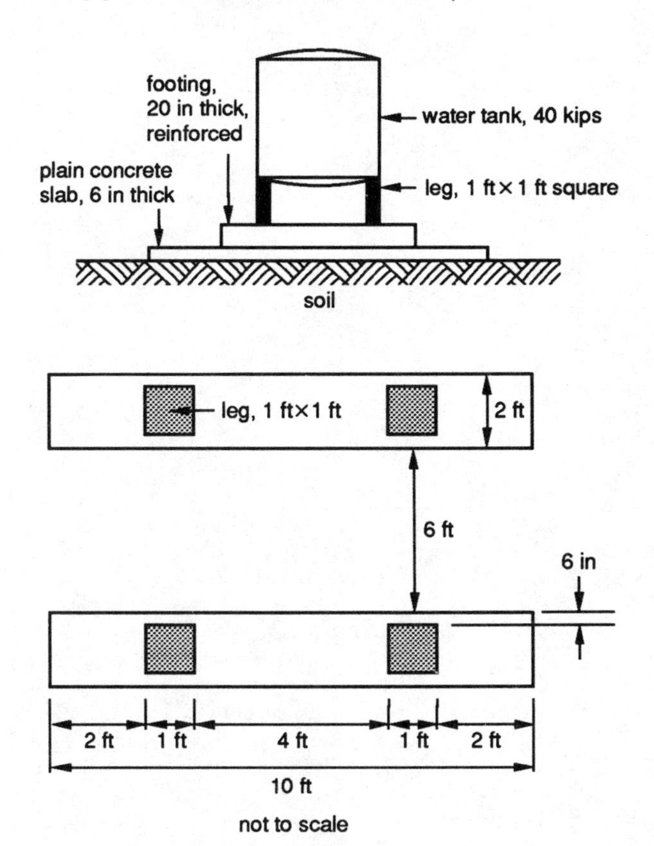

(a) Determine if the two footings are of sufficient size.

(b) Determine if the slab shear strength is sufficient.

CONCRETE DESIGN–12

A 32-ft-diameter tank is used to hold 55,000 gal of a water-based solution from a plating operation. The specific gravity of the solution is 1.0. The tank itself weighs 200,000 lbf. The tank is supported by a ringwall (circular) foundation, also 32 ft in diameter, that supports all of the tank and content weight. The soil in which the ringwall is located has a frost-line depth of 18 in and an allowable soil pressure of 1500 lbf/ft^2. Use the following material properties: $f'_c = 3000$ lbf/in^2, $f_y = 60$ kips/in^2. Do not design for seismic loading.

(a) Determine the required depth of the ringwall.

(b) Design the steel reinforcement. State your assumptions or the basis for your design.

(c) Draw a cross section of the wall you design and label all dimensions and steel bars.

CONCRETE DESIGN–13

A rib-slab system has the dimensions shown. Each beam uses three #9 longitudinal bars. The concrete has a compressive strength of 3000 lbf/in^2. The steel has a tensile yield strength of 60,000 lbf/in^2. Determine the ultimate moment capacity of this design per repeating unit.

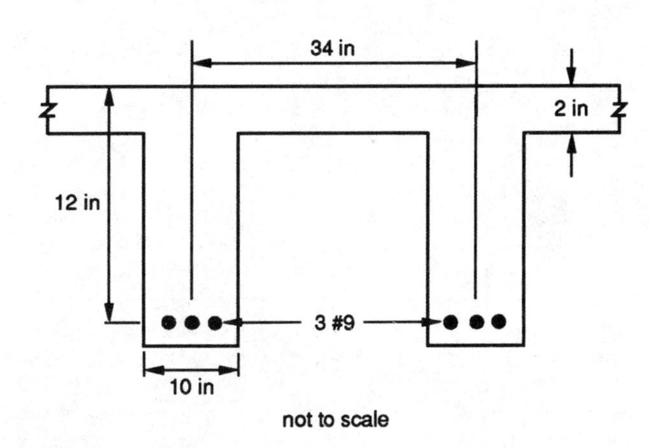

CONCRETE DESIGN–14

A 4-ft-wide, 30-ft-long simply supported beam is part of a roof section and is constructed as a double-tee, prestressed section as shown. The total superimposed dead load is 45 lbf/ft^2; the live load is 40 lbf/ft^2. The initial concrete compressive strength is 3500 lbf/in^2; the ultimate concrete compressive strength is 5000 lbf/in^2. 250-kips/in^2 tendons are used, and each web contains four tendons.

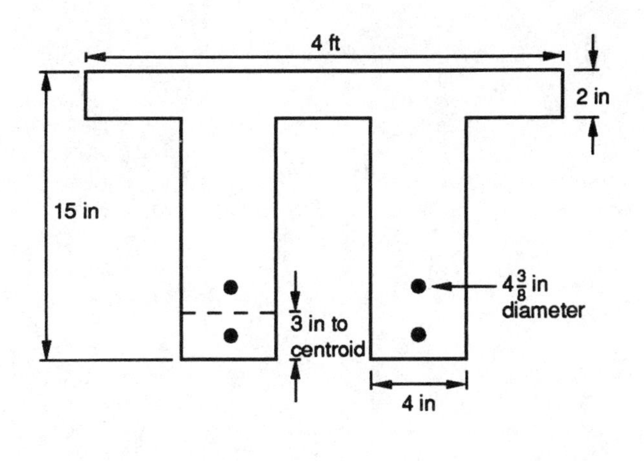

(a) Check all permissible concrete stresses immediately after prestress transfer.

(b) Check all permissible concrete stresses with service loading.

(c) What is the section's nominal moment strength?

11 STEEL DESIGN

STEEL DESIGN–1

A moment-resisting connection between a W18 × 46 beam and a W10 × 45 column is designed to resist a moment of 140 ft-kips and a 70-kip vertical load. All steel is A36. Use $\frac{7}{8}$-in A325 bolts.

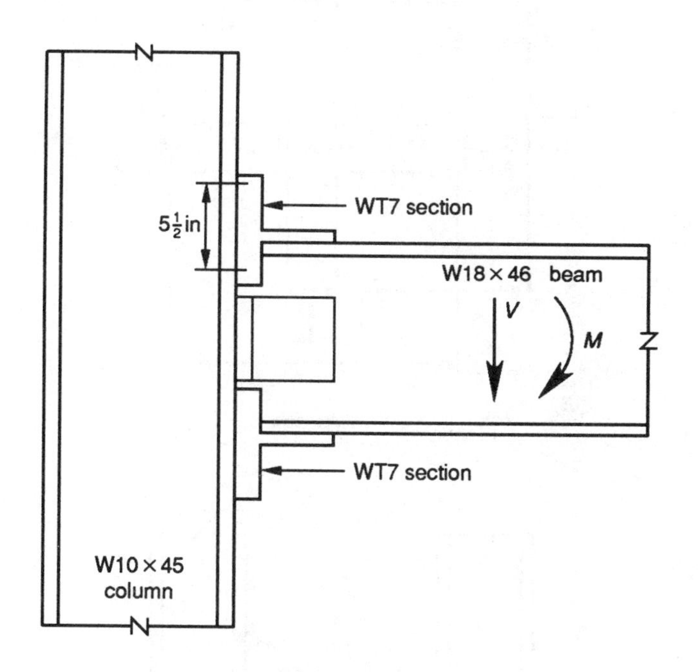

(a) How many bolts are needed to attach the WT7 section to the column flange?

(b) Specify the required flange thickness of the WT7 sections. Prying force is to be reduced to an insignificant amount

(c) If four $\frac{5}{8}$-in diameter A325 bolts are used in each WT section, what is the minimum beam size?

STEEL DESIGN–2

A 3400-pound sign is supported by three equally spaced cables from a W8 beam, the exact type to be determined. The beam, in turn, is supported by two side wires and a threaded tie rod. Assume all wire and rod connections are pinned.

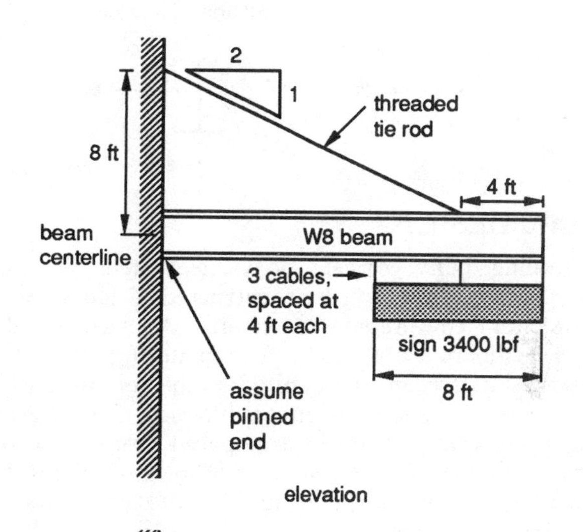

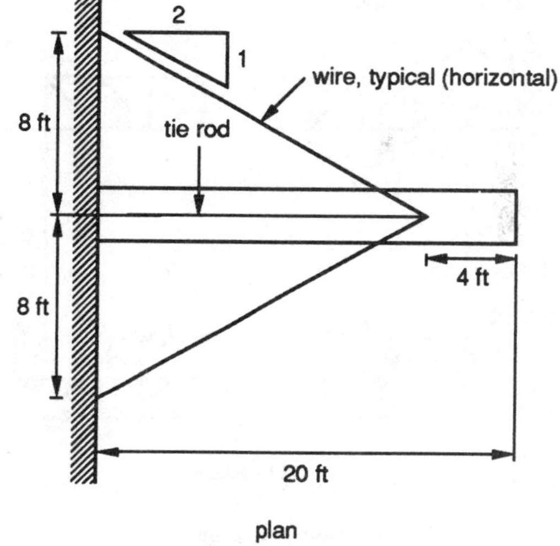

(a) What is the lightest W8 beam that can be used to support the sign?

(b) What are the reactions in the wires and rods?

STEEL DESIGN–3

Design the interior columns of the frame shown in the figure. The columns are braced continuously in the plane perpendicular to the frame. The girders are

W12 × 96. Use A36 steel. Left-to-right sidesway is uninhibited. Connections of columns to their supports are rigid.

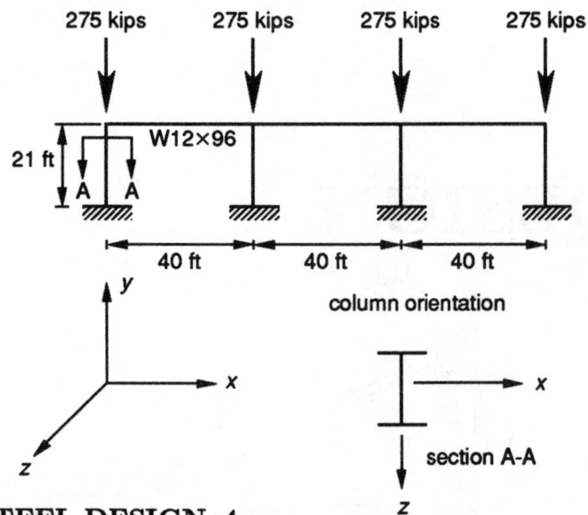

STEEL DESIGN–4

A framing bent consists of vertical steel W-shape columns with a truss girder constructed of double-angle tension and compression members. All steel is A36. Truss members are welded to the mounting plate (WT shapes) that are, in turn, bolted to the columns. (The centerlines of the truss members intersect the centroids of the bolt groups.) Bents are spaced every 22 ft and carry a uniformly distributed roof load of 70 lbf/ft^2, which includes an allowance for the truss girder itself.

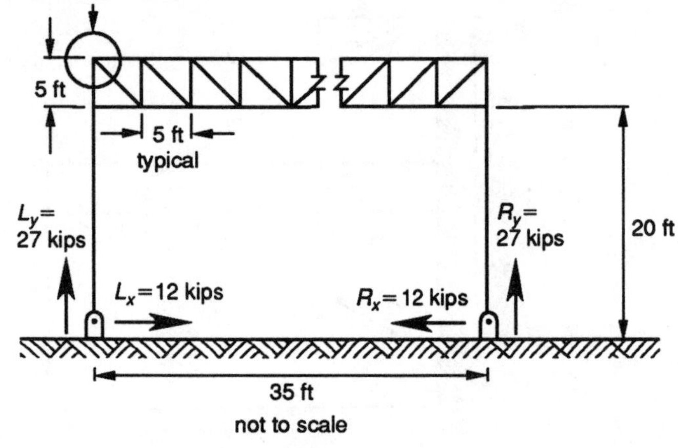

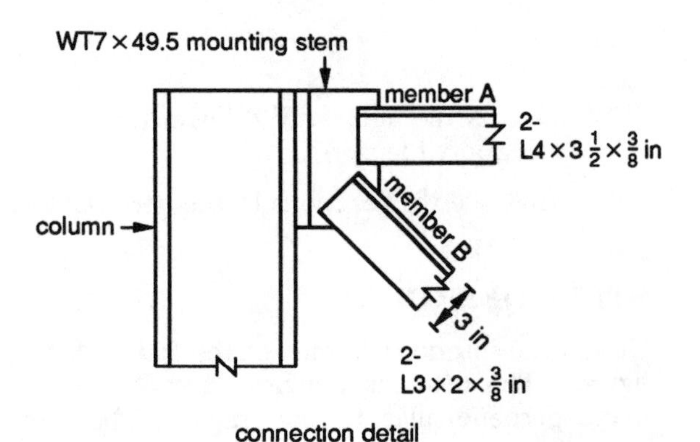

connection detail

(a) Use A325 bolts in a friction-type connection to design the bolted connections between the mounting stem and the column. Disregard prying action.

(b) Design the weld connection for member B. Assume E70 electrodes will be used.

STEEL DESIGN–5

A 22-ft-long W24 × 94 beam crosses at right angles and sits on an 18-ft-long W27 × 146 beam. The W27 × 146 beam is, in turn, supported on unstiffened L8×4×1 seat angles that are welded to the flanges of the columns. A 180-kip concentrated load is applied at midlength along the W24 × 94 beam. All beams are simply supported on seats that offer no moment resistance.

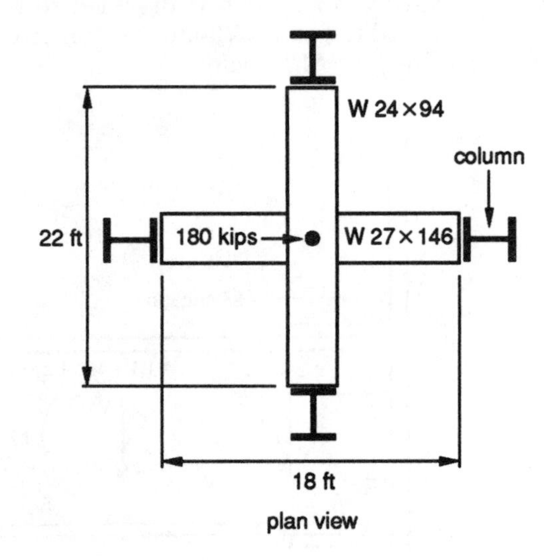

plan view

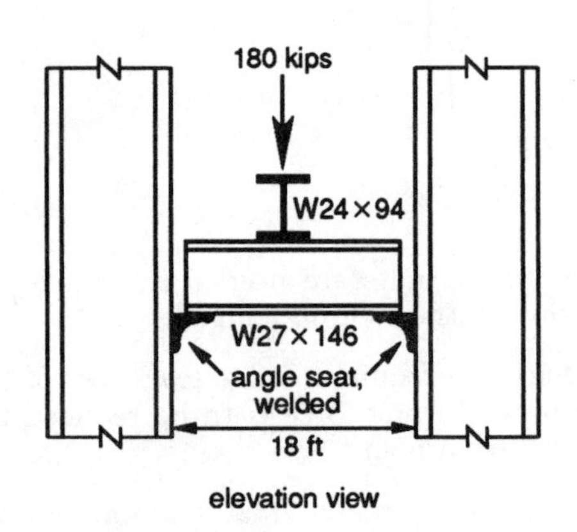

elevation view

(a) What fraction of the 180-kip load is carried by each beam?

(b) What are the maximum bending stresses in each beam?

(c) Evaluate the adequacy of the seat angles that support the W27 × 146 beam and design the welds. (The 4-in leg is horizontal.)

STEEL DESIGN–6

A horizontal roof is constructed with 36-ft girders supporting 32-ft joists. The girders are supported by 32-ft-high columns. All members are A36 steel. All horizontal members are continuous and can be assumed to be simply supported. The columns can be assumed to be pin-ended at both ends. The roof carries a live load of 0.02 kips/ft^2 and a dead load of 0.06 kips/ft^2. Wind and earthquake forces are carried by an external bracing system and should not be considered in this design.

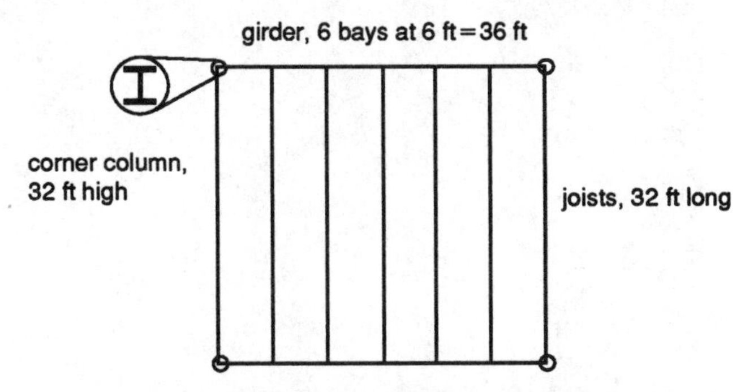

(a) Select the lightest W section for the joists such that the maximum deflection does not exceed 1/360th of the length.

(b) Select the lightest W section for the girders.

(c) Select an appropriate column.

STEEL DESIGN–7

A semibraced frame system is constructed and carries the loads shown. Special detailing is not provided. The columns are identical. The system is being evaluated for seismic forces according the Uniform Building Code.

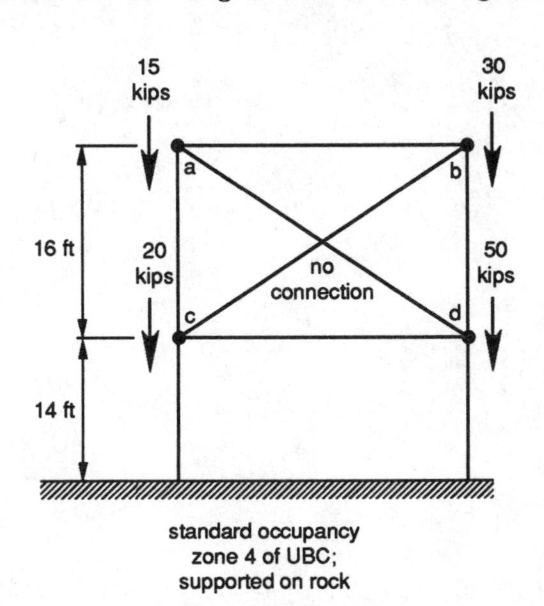

(a) What is the equation that gives the base shear?

(b) What is the total base shear?

(c) Calculate the lateral seismic forces at the floor levels.

(d) Calculate the maximum shear and moment from the lateral forces in the first story columns.

STEEL DESIGN–8

An exterior two-story, 24-ft-high steel column supports an exterior sign, a distributed roof load, and story loads. (Refer to the diagram for more details.) Select a column according to AISC procedures.

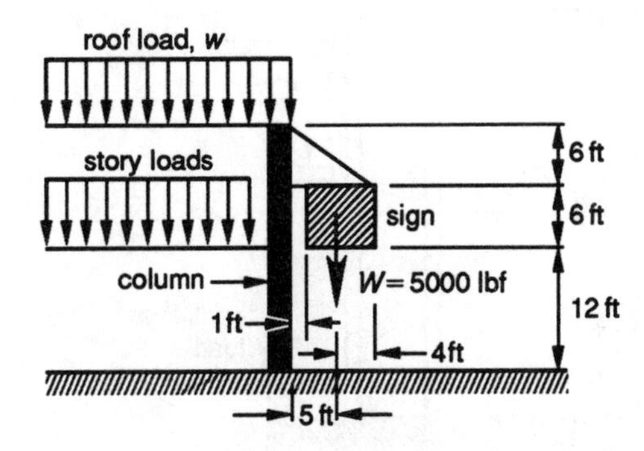

STEEL DESIGN–9

A single plate bracket is welded to the face of a column as shown. Determine the size of the fillet weld required. Use E70 electrodes and A36 steel.

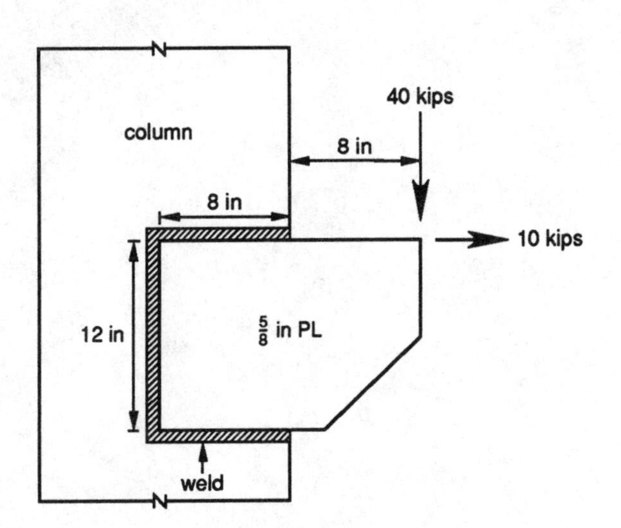

STEEL DESIGN–10

Design an A36 column to support the loads indicated.

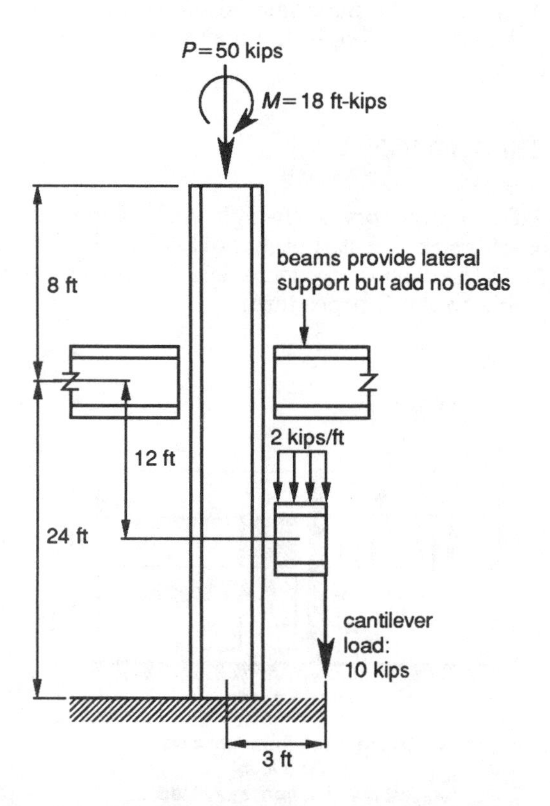

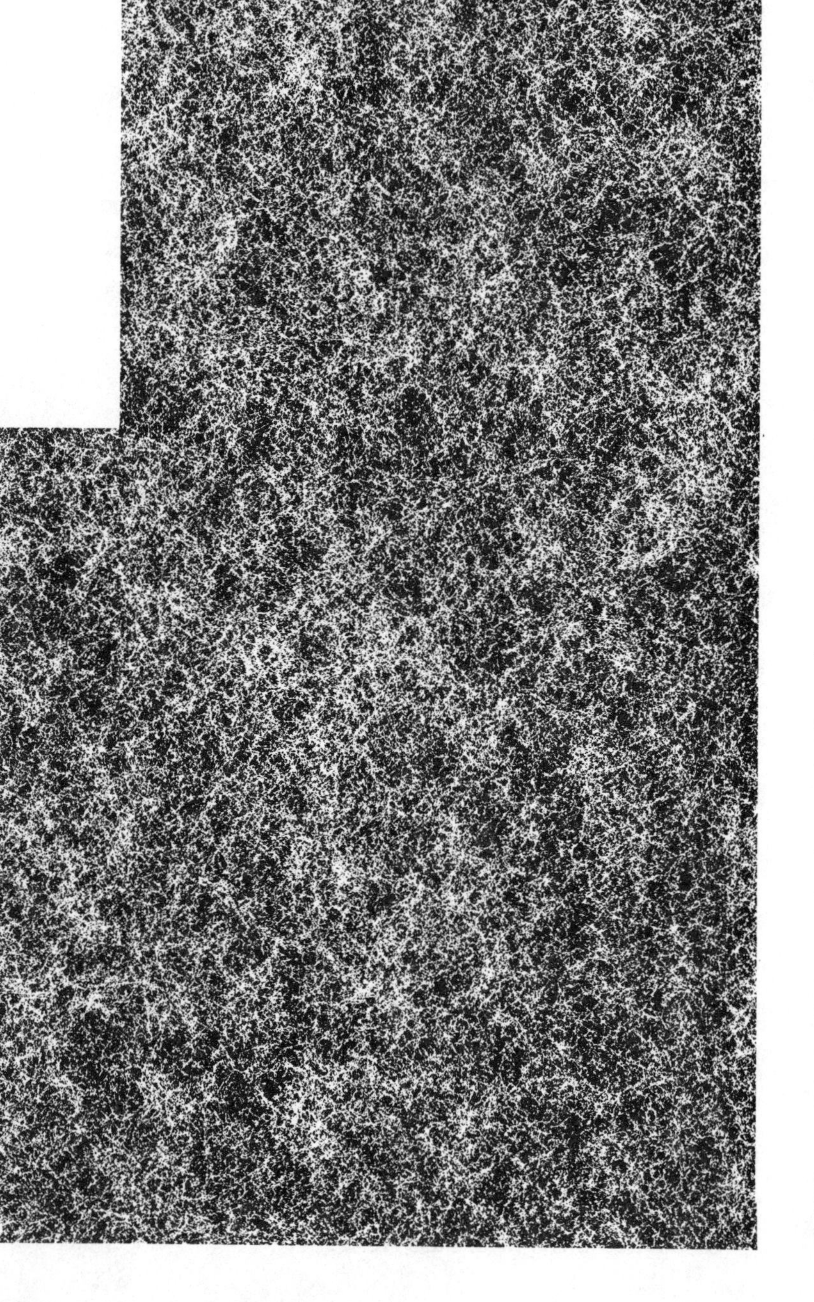

12 TIMBER DESIGN

TIMBER DESIGN–1

Lumber is used to form and shore up a cast-in-place, monolithic concrete slab during curing. The concrete has a specific weight of 150 lbf/ft^3, and the form lumber must also support a construction live load of 50 lbf/ft^2. The forms will be removed after seven days.

The 6-in-thick slab is poured on a layer of 1-in × 6-in (nominal) lumber (shown end-on in the figure). The lumber is, in turn, supported by joists. The joists sit on the ledger area of soffit beams with a minimum actual width of $15\frac{1}{2}$ in each. 4-in × 8-in (nominal) posts support the entire assembly. The 9-in × 12-in concrete beams are precast, and no form work is required for them.

To prevent cross-grain bending of the soffit beam, 4-in × 4-in headers are placed on top of the posts to extend the post support to the full width of the soffit beam.

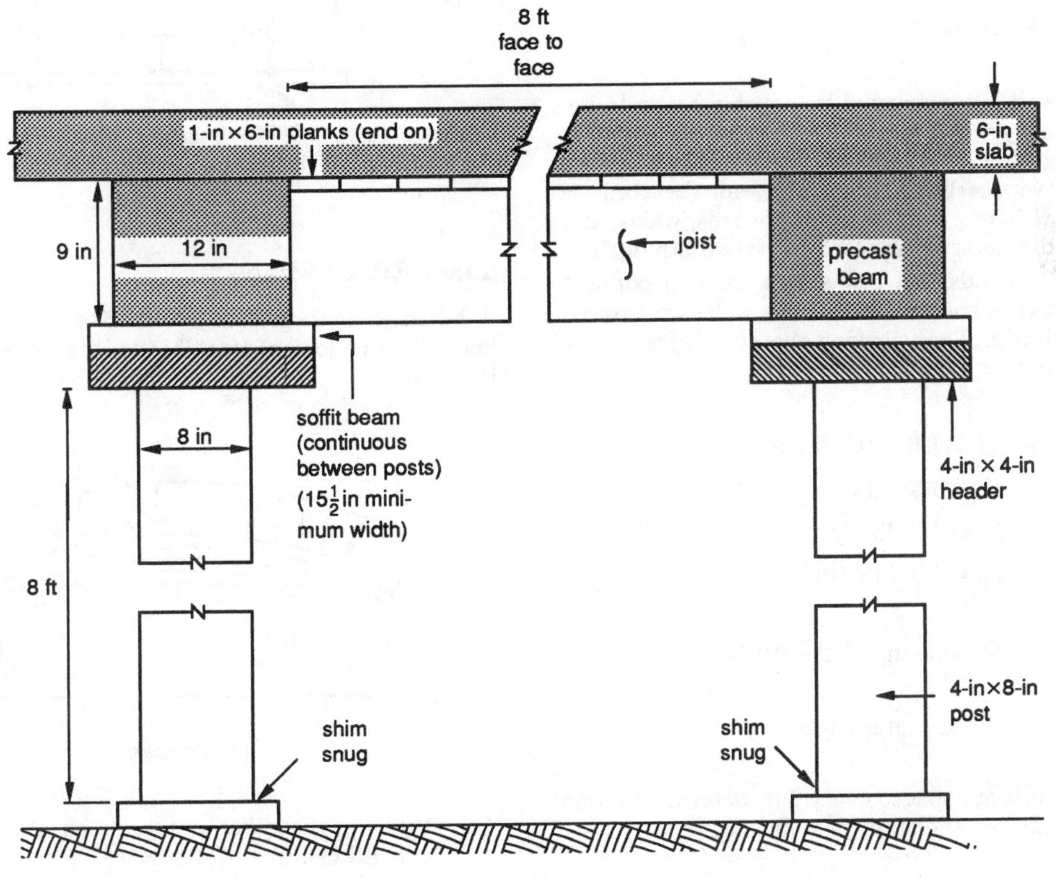

The concrete compressive strength is 3000 lbf/in^2; the reinforcing steel has a yield strength of 40,000 lbf/in^2. The following characteristics can be used for all lumber.

$$E = 1.6 \times 10^6 \text{ lbf/in}^2$$
$$F_b = 1200 \text{ lbf/in}^2$$
$$F_v = 125 \text{ lbf/in}^2$$
$$F_c = 1200 \text{ lbf/in}^2$$
$$F_{c\perp} = 400 \text{ lbf/in}^2$$

(a) Find the maximum spacing between joists as limited by the 1-in × 6-in lumber.

(b) Given a 2-in (nominal) joist material and joist spacing of 2 ft, 6 in, find the required joist size. (Joists can be doubled if additional joist capacity is required.)

(c) What is the maximum post spacing?

(d) If the posts are spaced every 5 ft and joist spacing is 2 ft-6 in, what soffit beam size is required?

TIMBER DESIGN–2

A retaining wall is constructed by pouring concrete between temporary wood forms. The concrete is at 70°F, has a setting time of $1\frac{1}{2}$ hours, and is placed at the rate of 2 ft (vertical) per hour. The wall sheathing is constructed of 1-in × 6-in (nominal) boards, wales, and studs. $\frac{1}{2}$-in-diameter threaded steel (yield strength = 36 kips/in^2) tie rods (tie backs) are used to connect the wales between the two walls. The wales are spaced 2 ft-6 in vertically. The following characteristics can be used for all lumber.

$$E = 1.6 \times 10^6 \text{ lbf/in}^2$$
$$F_b = 1700 \text{ lbf/in}^2$$
$$F_v = 100 \text{ lbf/in}^2$$
$$F_c = 1200 \text{ lbf/in}^2$$

(a) Determine the spacing of the studs.

(b) Determine the size of the studs.

(c) If the studs are spaced every 2 ft, determine if one tie in between the studs is sufficient.

(d) Determine the size of the wales.

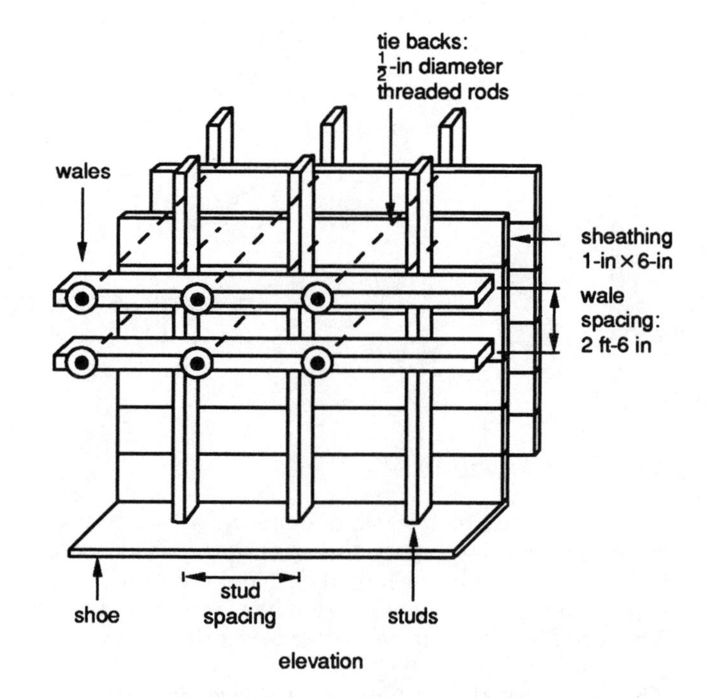

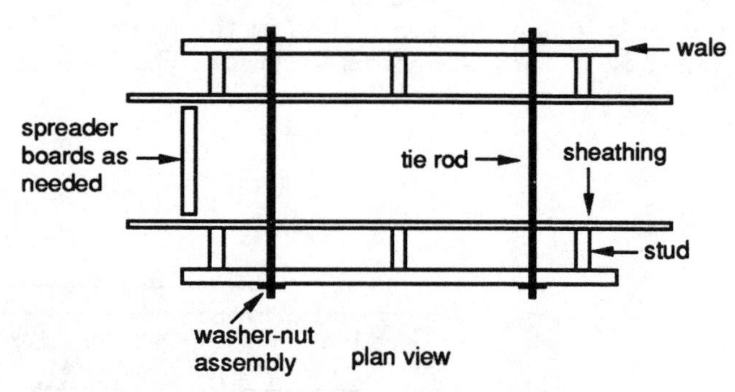

TIMBER DESIGN–3

A $6\frac{3}{4}$-in × 24-in glulam beam 25 ft long is uniformly loaded with a load of 1000 lbf/ft. Its radius of curvature is 100 ft.

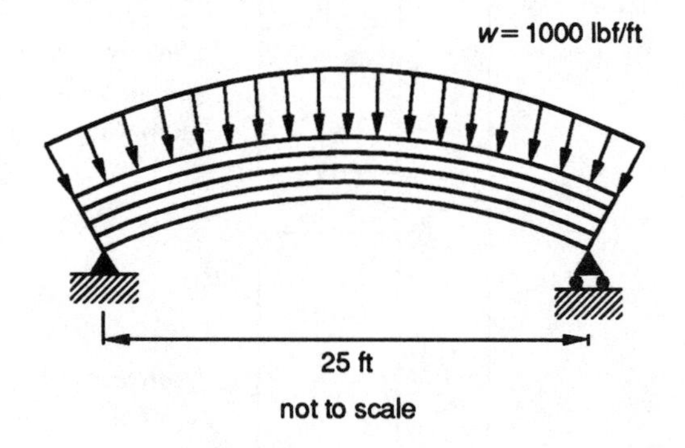

(a) What is the difference between a 22F-V3 and a 22F-E3 beam?

(b) What is the size factor for the beam described?

(c) What is the curvature factor for the beam?

(d) What is the maximum shear stress in the beam?

(e) What is the maximum bending stress in the beam?

(f) Given that the modulus of elasticity, E, is 1.6×10^6 lbf/in², what is the deflection?

(g) What moisture content would you expect the finished glulam beam to have?

TIMBER DESIGN–4

Design a 25-ft-high solid square timber member to serve as a signpost that carries the loads indicated. The column is set in a concrete pier at its base but is free to rotate and translate at the top. Assume the timber has the following properties.

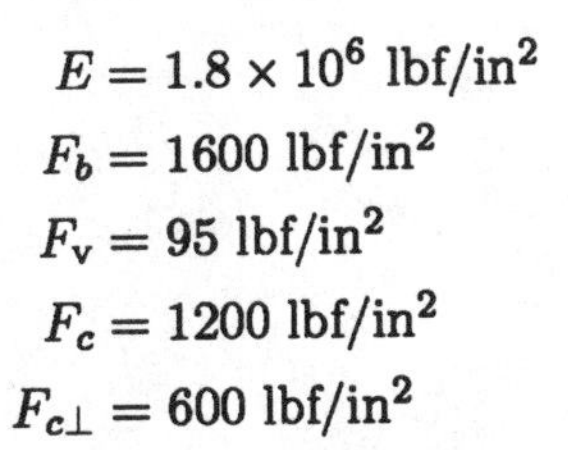

$$E = 1.8 \times 10^6 \text{ lbf/in}^2$$
$$F_b = 1600 \text{ lbf/in}^2$$
$$F_v = 95 \text{ lbf/in}^2$$
$$F_c = 1200 \text{ lbf/in}^2$$
$$F_{c\perp} = 600 \text{ lbf/in}^2$$

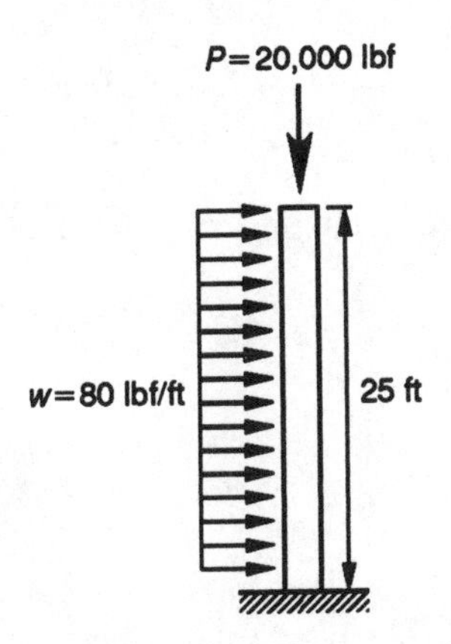

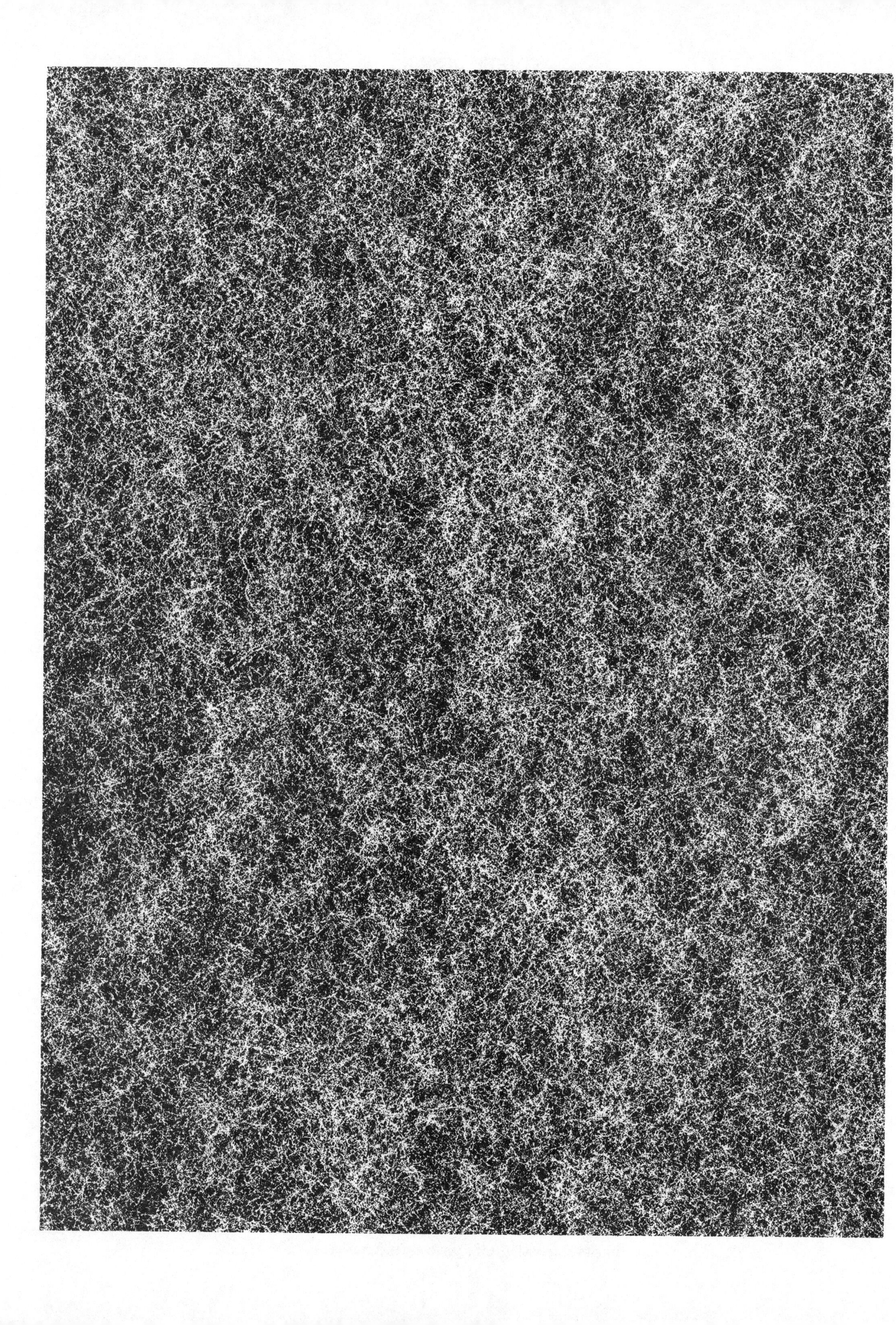

13 TRAFFIC

TRAFFIC–1

A suburban business district has a maximum daytime population of 120,000. Traffic passing through an intersection in the business district is controlled by a signal with a total cycle length of 60 sec. There are no bus stops in or near the intersection. The intersection is not part of any transit bus routes.

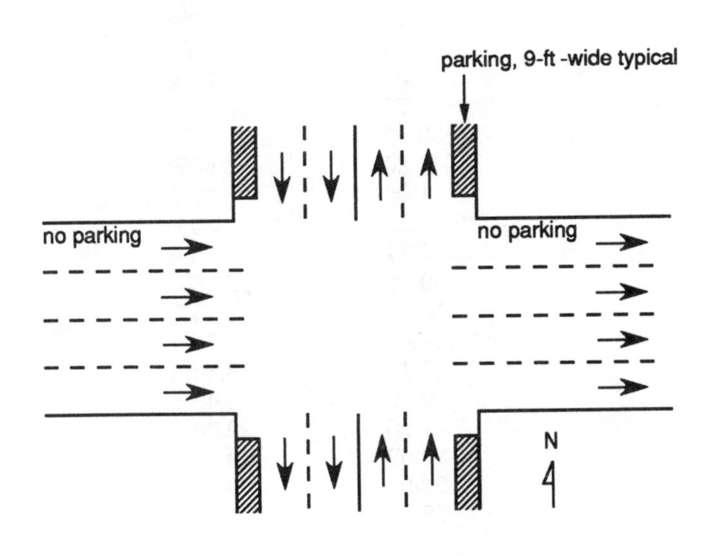

The following data have been collected on the nature of the traffic during a particular time of day.

	west-to-east	north-to-south and south-to-north
lane width	11 ft	12 ft
peak hour factor	0.80	0.70
percent right turns	0%	0%
percent left turns	20%	0%
percent trucks and local bus traffic	10%	7%
length of green and amber phases	32 sec	26 sec
length of all red phases	1 sec	1 sec

(a) Find the maximum service volume for level of service (LOS) C on the west-to-east approach to the intersection.

(b) Find the maximum service volume for LOS E on the west-to-east approach to the intersection.

(c) Find the maximum service volume for LOS B on the north-to-south approach to the intersection.

(d) Find the maximum service volume for LOS E on the south-to-north approach to the intersection.

TRAFFIC–2

List at least eight major factors that influence the capacity of a signalized intersection.

TRAFFIC–3

The average daily traffic (ADT) patterns for four different sets of traffic are shown. For each pattern, specify a type of interchange (e.g., cloverleaf, loop-connector, etc.) and briefly indicate your reason for choosing that type.

(a)

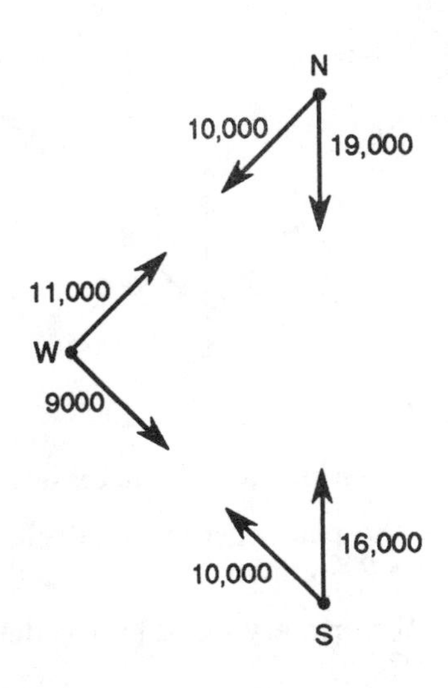

(b)

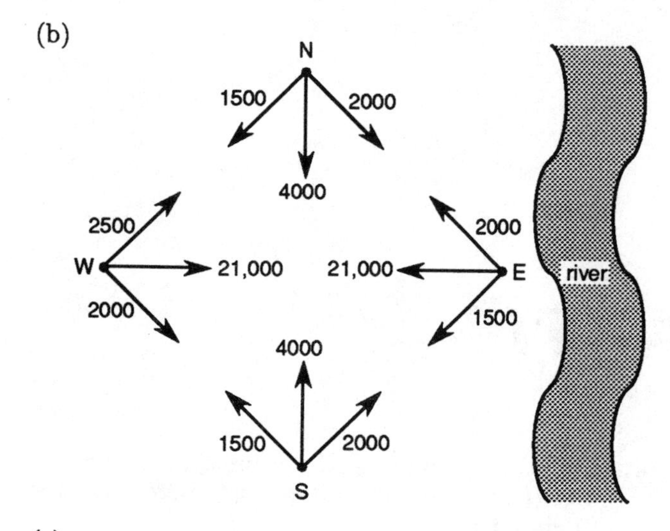

(c)

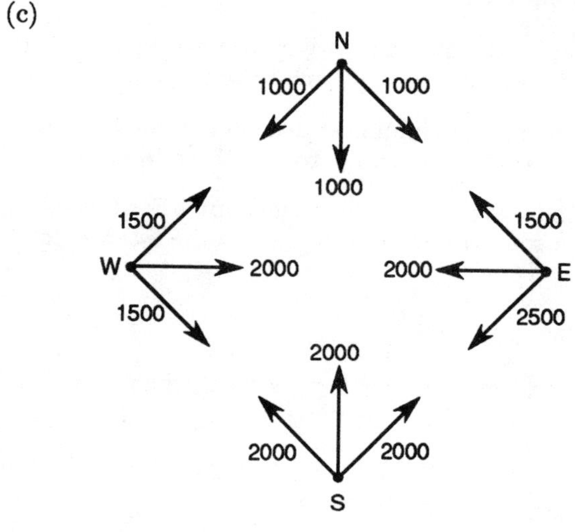

(d)

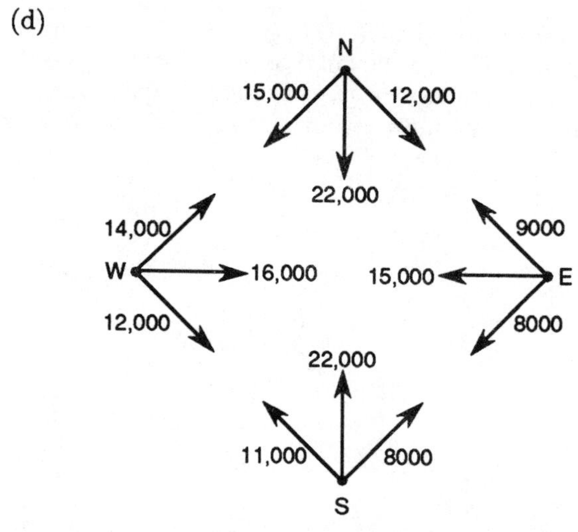

TRAFFIC–4

The following questions are not necessarily related.

(a) What is the passenger car equivalent (E_T) for a truck on a 2% grade?

(b) What is the primary variable that determines level of service?

(c) List five guardrail warrants. What values must these warrants achieve to justify the installation of guardrails?

(d) What is the primary variable used to rate the need for making future improvements to intersections?

(e) What dimensions should be given to a single-vehicle parking slot in a municipal parking structure?

(f) On the average (i.e., considering all other uses for space), what is the amount of space (in square feet) that should be allowed per parked vehicle in a municipal parking structure?

TRAFFIC–5

A rural two-lane highway was designed for 50-mph travel. The following traffic counts were taken over a one-year period.

annualized volume (millions of vehicles per year)	cumulative percentage
0	0%
0.11	10%
0.23	20%
0.39	30%
0.77	40%
0.99	50%
1.30	60%
1.96	70%
3.37	80%
7.81	90%
9.38	93.15%
11.50	95.43%
14.46	97.72%
17.15	98.86%
19.59	99.43%
21.55	99.66%
22.62	99.77%
23.98	99.89%
28.04	100.00%

(a) What is the 30th hour volume?

(b) What is the maximum hourly volume?

(c) What is the approximate average annual daily traffic (AADT)?

(d) Given a vehicle speed of 42 mph, what is the 30th hour density?

(e) What is the headway in each direction of flow under the conditions given in part (d) assuming a 50-50 directional split?

(f) Estimate the density when the road is at capacity.

(g) What is the space mean unit travel time?

(h) If traffic is expected to grow at the rate of 2% per year, what will be the percentage increase in traffic after four years?

TRAFFIC–6

The accident report of a highway patrol officer reads as follows.

> Responding to a radio message, I reached the scene of the accident at 18:34 hours and observed a stationary truck that had, apparently, just skidded into a parked car. There were no injuries. The road was dry. The truck left 180 ft of skidmarks up to the point of impact. The skid was down (south) on Scenic Drive, which has a uniform downward grade of −5.0% in the direction of travel. Based on the limited amount of damage to both vehicles, I estimate that the impact speed was 20 mph.

(a) Using the following coefficient of friction values, determine the initial velocity of the truck just prior to locking the brakes.

speed prior to skidding	coefficient of friction
20 mph	0.46
30 mph	0.38
40 mph	0.32
50 mph	0.30
60 mph	0.28

(b) How would your answer to (a) change if the accident report stated that the road was wet?

TRAFFIC–7

A new four-lane highway is being designed to cross an existing two-lane road. The intersection is in a wooded area and trees will have to be cleared to provide proper sight distance. The four-lane highway will be posted fo 50-mph travel; the two-lane road will be posted for 40-mph travel.

Draw a plan view of the intersection and show the areas that you would clear to provide sight distance. Label and dimension the sight triangles. Use current AASHTO specifications.

TRAFFIC–8

An airport is being redesigned to current FAA regulations.

(a) What is the value of a reasonable runway gradient?

(b) Draw two intersecting runways. What are the VFR and IFR maximum loads?

(c) What other factors must be considered when designing airport runway pavement sections?

TRAFFIC–9

The following questions are not necessarily related.

(a) What arrival type represents very good progression? How is arrival type determined?

(b) How many trips will be generated by a shopping center with 100,000 ft^2 of leasable space (1) per weekday, and (2) during the peak hour?

(c) What is the definition of the peak hour factor?

(d) What is the definition of (1) the critical gap, and (2) trip distribution?

(e) What are typical warrants and numerical criteria for signalizing an intersection?

TRAFFIC–10

A compacted asphalt pavement with a bulk specific gravity contains the following components. (Percentages are by weight of total mixture.)

asphalt cement:	specific gravity = 1.01 7% by weight
coarse aggregate:	specific gravity = 2.61 52% by weight
fine aggregate:	specific gravity = 2.71 34% by weight
mineral filler:	specific gravity = 2.70 7% by weight

The maximum specific gravity of a paving mixture sample (as measured according to ASTM D 2041) was 2.455.

The bulk specific gravity of the compacted paving mixture sample (as measured according to ASTM D 2726) was 2.360.

(a) What is the bulk specific gravity of the aggregate?

(b) What is the effective specific gravity of the aggregate?

(c) What would the maximum (zero voids) specific gravity of the paving mixture be if the asphalt mixture is increased to 8%?

(d) What is the asphalt absorption (by weight)?

(e) What is the effective asphalt content (by weight) of the paving mixture?

(f) What is the percent voids in the mineral aggregate (VMA)?

(g) What is the percent air voids in the compacted mixture?

(h) What is bleeding?

(i) What is (are) the primary cause(s) of bleeding in asphalt concrete?

(j) What are typical air void requirements for asphalt concrete?

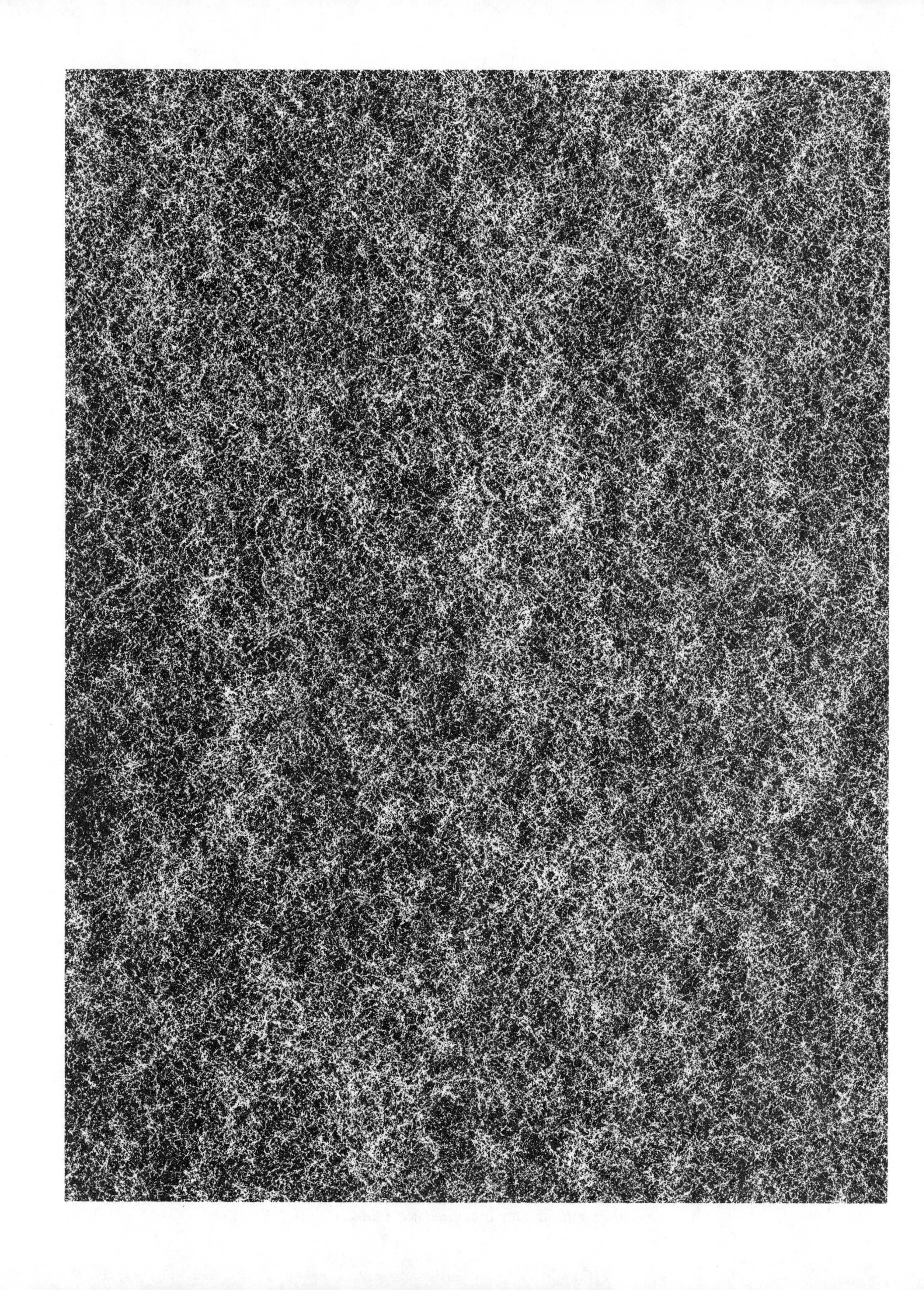

14 SURVEYING

SURVEYING–1

Two highway tangents intersect at station 60+50 as shown. The roadway follows a horizontal circular curve constructed between the two tangents, and it passes through point A. Other data are given in the drawing.

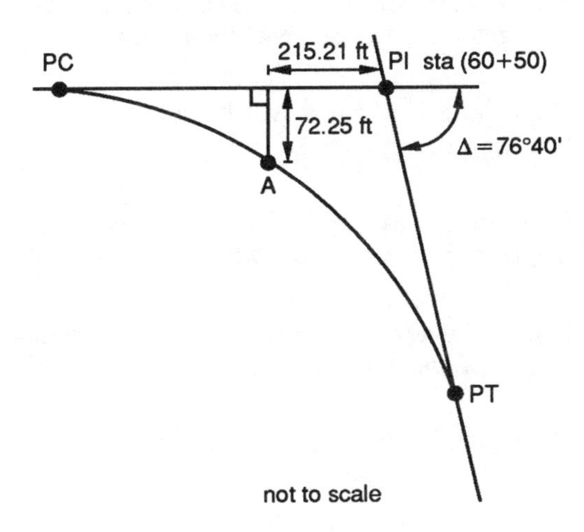

(a) What is the radius of the horizontal circular curve that passes through the PC, PT, and point A?

(b) Measuring from the intersection point, PI, what are the stations of the PC and PT?

(c) Assume the curve radius is 660 ft and the design speed is 40 mph. Use the current AASHTO standards to determine if the radius is acceptable.

SURVEYING–2

The PI of two tangents of a highway is located in a lake. Stations A and B are selected to replace the inaccessible PI.

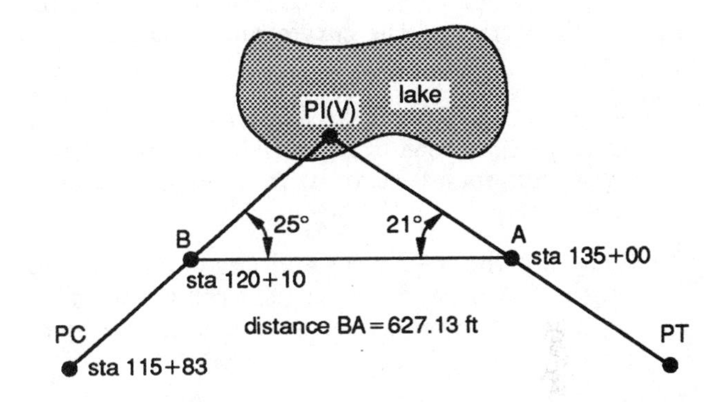

(a) What is the radius of the circular curve between the PC and PT?

(b) What is the degree of the curve?

(c) What is the length of the curve?

(d) What is the station of the PT?

(e) If the instrument is at the PC, what deflection angle should be used to locate staking station 116+00 on the curve?

(f) If the instrument is at the PC, what deflection angle should be used to locate staking station 117+00 on the curve?

(g) If the instrument is at station 116+00, what deflection angle should be used to locate staking station 118+00 on the curve?

SURVEYING–3

The bearings of two tangents connected by a horizontal circular curve are N50°E and S35°E, respectively. The tangents intersect at station 37+00. The curve radius is 800 ft.

(a) What is the length of the curve?

(b) What is the station of the PC?

(c) What is the station of the PT?

(d) What is the interior angle at the PI?

(e) What is the tangent distance from the PI to the PC?

(f) What is the long chord distance?

(g) What is the external distance?

(h) What is the degree of the curve (arc basis)?

(i) What is the degree of the curve (chord basis)?

(j) What is the chord length of a 100-ft arc (arc basis)?

SURVEYING–4

Numerous accidents have been occurring at several horizontal highway curves in your county. It is suspected that the geometries of the curves are contributing to the accidents.

(a) What are the geometric and trigonometric relationships that need to be considered when replacing a compound curve with a simple circular curve?

(b) What are the geometric and trigonometric relationships that need to be considered when replacing a simple circular curve with a simple spiral curve?

(c) What are the geometric and trigonometric relationships that need to be considered when replacing a broken-back curve with a simple circular curve?

SURVEYING–5

A road contains a vertical curve starting at station 60+00 and ending at station 68+00. The elevation at the beginning of the curve (BVC) is 562 ft. The grade prior to the BVC is −1.5%. Just after the end of the curve (EVC), the grade is +2.5%.

(a) What are the centerline elevations of the road every 50 ft from the BVC to the EVC?

(b) The road is constructed on marshy ground subject to flooding. At station 63+00, the road becomes elevated and is supported by five-pile bents every 50 ft. The piles are separated from each other by 10 ft (perpendicular to the road centerline) and are cut 3 ft below the finished road grade. A 4° horizontal circular curve to the right starts at station 64+50 and finishes at station 66+00. Superelevation on the curve is 12%. What are the cutoff elevations for each of the five piles in the four bents in the horizontal curve?

SURVEYING–6

The grade into a vertical sag curve is −2%. The curve length is 1400 ft. The grade out of the curve is +4%. The elevation and station of the grade intersection are 226.88 ft and 7+20, respectively. The curve goes through a flood plain where the 50-year flood elevation is 240 ft.

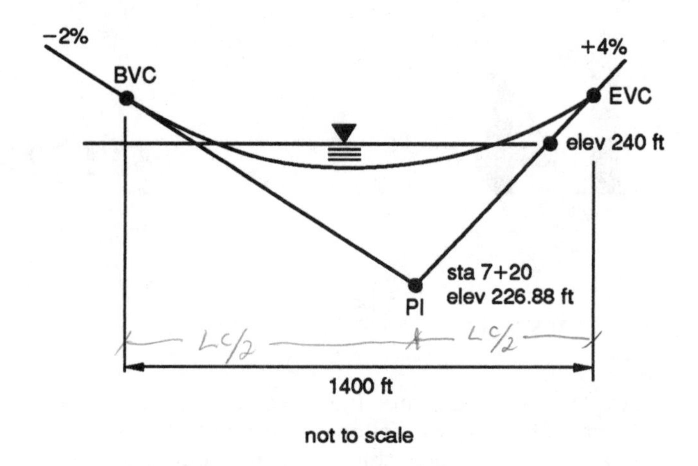

(a) Find the stations where the curve drops below and emerges from the 50-year flood plain.

(b) What length of curve will be submerged in a 50-year flood?

(c) How will increasing the length of curve change the length of the curve that will be submerged?

(d) How will decreasing the length of curve change the length of the curve that will be submerged?

(e) What can be done to the pavement so that flooding will not degrade it?

SURVEYING–7

A vertical curve connects a −1.3% tangent and a +1.8% tangent. The tangents intersect at station 74+00 and elevation 310 ft. A railroad line at grade crosses the curve at station 75+20 and elevation 314 ft.

(a) What is the length of the vertical curve?

(b) If the curve length is changed to exactly 1300 ft, what will be the difference in elevation between the railroad line and the roadway?

(c) The road is curbed and guttered. If the curve length is exactly 1300 ft, at what station and elevation would a single grate inlet be placed on the curve?

SURVEYING–8

During a survey, lengths and bearings are collected for legs of a traverse.

leg	bearing	distance
AB	N35.15°W	905.21 ft
BC	N81.28°E	1135.76 ft
CD	S7.19°E	1207.92 ft
DE	S15.25°W	800.25 ft
EF	N48.17°W	1100.85 ft
FA	N40.73°E	429.53 ft

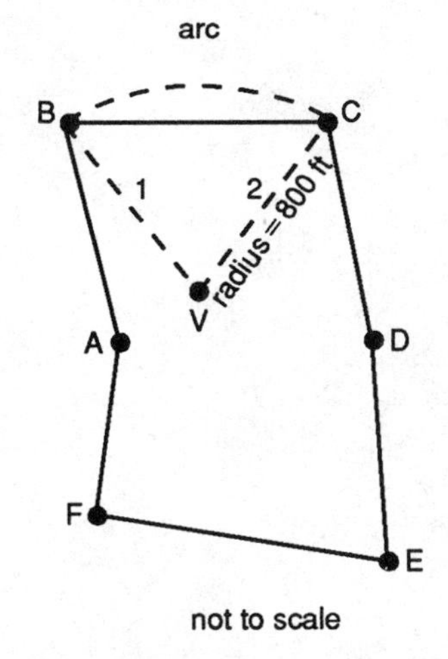

(a) What is the precision of the survey?

(b) Adjust the traverse by the compass rule.

(c) Using the adjusted traverse, what is the area of the figure ABCDEF?

(d) What is the area of the circular segment bounded by the chord BC?

(e) What is the length of the arc BC?

SURVEYING–9

The partial results of a survey of a traverse are shown in the following figure.

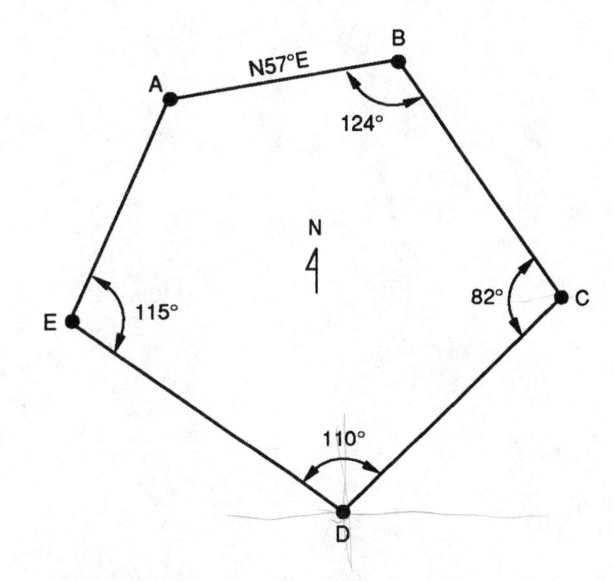

(a) What is the bearing of line BC?

(b) What is the bearing of line CD?

(c) What is the bearing of line DE?

(d) What is the bearing of line EA?

(e) What is the interior angle at A?

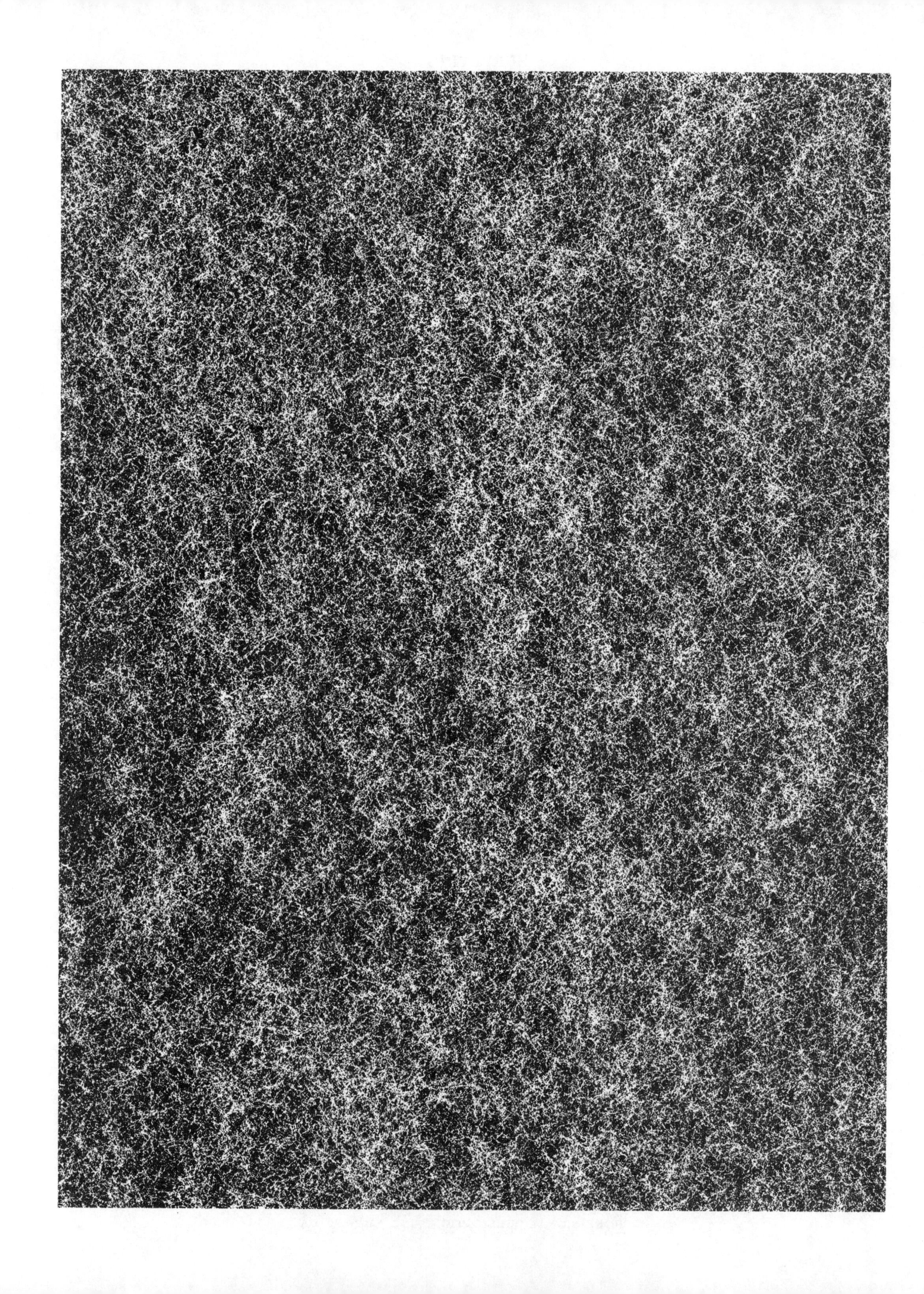

15 MASONRY

MASONRY–1

A short masonry column has a height of 10 ft and is constructed of concrete block with four #7 steel bars as shown. The column is subjected to a moment of 100 in-kips. Sidesway is permitted. Special inspection is used. What is the maximum axial load the column can support in addition to the applied moment?

concrete compressive strength:	3000 lbf/in^2
masonry compressive strength:	1350 lbf/in^2
allowable steel stress:	16,000 lbf/in^2
ratio of steel to masonry moduli of elasticity:	40

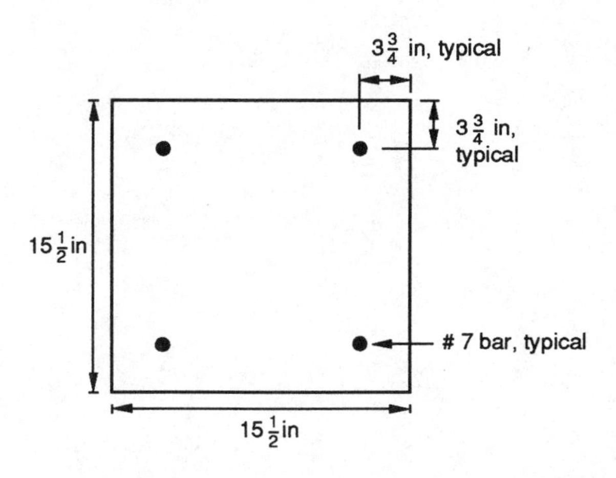

MASONRY–2

A square masonry column has a nominal 12-in outside dimension. It is constructed of concrete block (concrete compressive strength = 3000 lbf/in^2), mortar (masonry compressive strength = 1500 lbf/in^2), and four #8 bars (16,000 lbf/in^2 allowable stress). The modular ratio is 42. Special inspection is used.

(a) What is the maximum height this column can have?

(b) What is the maximum axial (noneccentric) load this column can support if the effective height equals 20 ft?

(c) What is the maximum eccentricity that would be permitted in this column if the axial load equals 40 kips?

MASONRY–3

A masonry column is eccentrically loaded by a steel beam as shown. Design the method of support and connection between the steel beam and column. Assume special inspection.

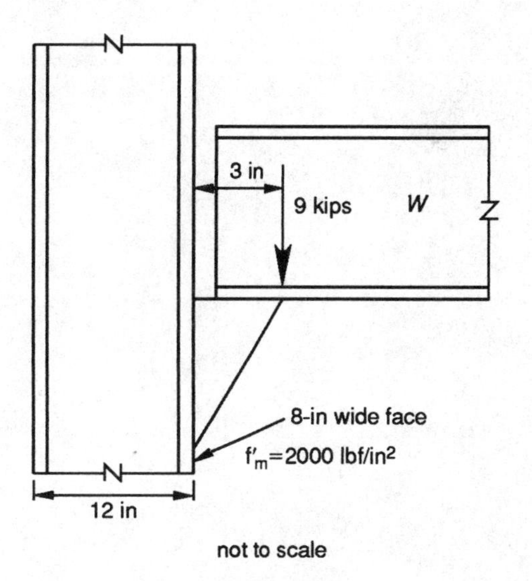

MASONRY–4

A two-story (20-ft-high) masonry wall is constructed of 8-in cinder blocks with total grouting. The concrete block compressive strength is 3000 lbf/in^2; the masonry compressive strength is 1500 lbf/in^2. Reinforcing steel (16,000 lbf/in^2 allowable stress) consists of vertical #5 bars spaced every 48 in. The modular ratio is 44. The wall is supported laterally by steel beams at the bottom, top, and midpoint as illustrated. The steel is rigid. Is the wall adequate with a 35 lbf/ft^2 wind load? State your assumptions.

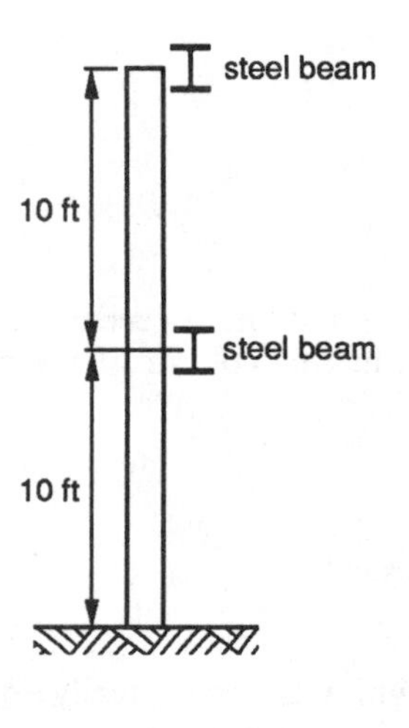

MASONRY–5

A 24-in-square brick column 32 ft high carries an axial load of 65 kips in addition to a moment of 45 ft-kips. The column is supported laterally only at its top and bottom, and the ends are free to rotate in the plane of bending. The brick compressive strength is 1800 lbf/in^2. Reinforcing steel (16,000 lbf/in^2 allowable stress) consists of four vertical #8 bars.

(a) What is the minimum cracked section?

(b) What is the maximum deflection?

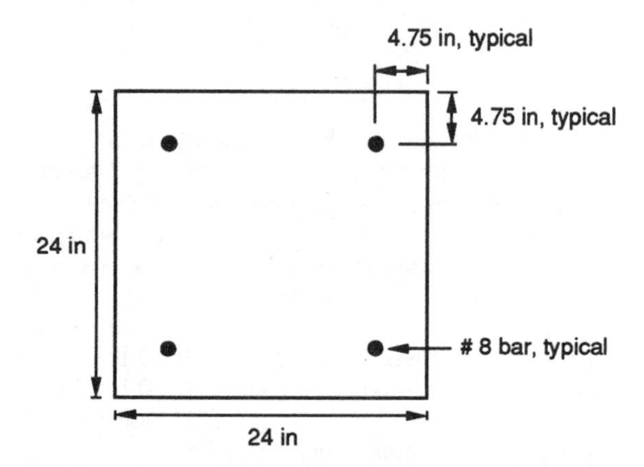

ENGINEERING ECONOMIC ANALYSIS

1. (a) Alternative A:

$$\begin{aligned} P &= \$7500 + \$15{,}000 + (\$6000)(P/A, 8\%, 30) \\ &\quad + (\$7500)(P/F, 8\%, 10) + (\$7500)(P/F, 8\%, 20) \\ &= \$7500 + \$15{,}000 + (\$6000)(11.2578) \\ &\quad + (\$7500)(0.4632 + 0.2145) \\ &= \$95{,}130 \end{aligned}$$

Alternative B:

$$\begin{aligned} P &= \$15{,}000 + \$15{,}000 + (\$4500)(P/A, 8\%, 30) \\ &\quad + (\$15{,}000)[(P/F, 8\%, 10) + (P/F, 8\%, 20)] \\ &= \$15{,}000 + \$15{,}000 + (\$4500)(11.2578) \\ &\quad + (\$15{,}000)(0.4632 + 0.2145) \\ &= \$90{,}826 \end{aligned}$$

Alternative C:

$$\begin{aligned} P &= \$30{,}000 + \$22{,}500 + (\$3000)(P/A, 8\%, 30) \\ &\quad + \$30{,}000[(P/F, 8\%, 10) + (P/F, 8\%, 20)] \\ &= \$30{,}000 + \$22{,}500 + (\$3000)(11.2578) \\ &\quad + (\$30{,}000)(0.4632 + 0.2145) \\ &= \$106{,}604 \end{aligned}$$

Choose Alternative B.

(b) Water pumped per year = V gal.

$$\begin{aligned} \text{A:}\quad P &= \$95{,}130 + \left(\frac{4.5}{100}V\right)(P/A, 8\%, 30) \\ &= \$95{,}130 + \left(\frac{4.5}{100}V\right)(11.2578) \\ &= \$95{,}130 + 0.5066V \end{aligned}$$

$$\begin{aligned} \text{B:}\quad P &= \$90{,}826 + \left(\frac{3.75}{100}V\right)(P/A, 8\%, 30) \\ &= \$90{,}826 + \left(\frac{3.75}{100}V\right)(11.2578) \\ &= \$90{,}826 + 0.4222V \end{aligned}$$

$$\begin{aligned} \text{C:}\quad P &= \$106{,}604 = \left(\frac{3.00}{100}V\right)(P/A, 8\%, 30) \\ &= \$106{,}604 + \left(\frac{3.00}{100}V\right)(11.2578) \\ &= \$106{,}604 + 0.3377V \end{aligned}$$

The selection is sensitive to the volume of water pumped. B is always superior to A, but C will become superior to B if $V > 186{,}722$ gal/yr. At $V = 186{,}722$ gal/yr,

$$P_B = P_C = \$169{,}660$$

If $V < 186{,}722$ gal/yr, choose B. If $V > 186{,}722$ gal/yr, choose C.

2. $t = \text{tax rate} = 46\%$

$$\begin{aligned} \text{total investment} &= \$55{,}000 + \$75{,}000 + \$85{,}000 \\ &= \$215{,}000 \end{aligned}$$

$$D_1 = (\$215{,}000)(0.25) = \$53{,}750$$
$$D_2 = (\$215{,}000)(0.38) = \$81{,}700$$
$$D_3 = (\$215{,}000)(0.37) = \$79{,}550$$

Production cost per year:

$$(4000)(\$60) = \$240{,}000$$
$$x = \text{price per item}$$
$$\text{revenue per item} = 4000x$$

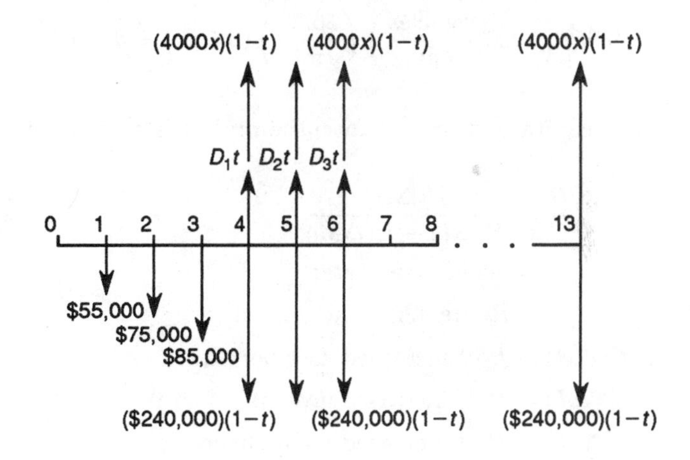

$$\begin{aligned} P = 0 &= -(\$55{,}000)(P/F, 18\%, 1) \\ &\quad - (\$75{,}000)(P/F, 18\%, 2) \\ &\quad - (\$85{,}000)(P/F, 18\%, 3) \\ &\quad - (\$240{,}000)(1 - 0.46) \\ &\quad \times (P/A, 18\%, 10)(P/F, 18\%, 3) \\ &\quad + (\$53{,}750)(0.46)(P/F, 18\%, 4) \\ &\quad + (\$81{,}700)(0.46)(P/F, 18\%, 5) \\ &\quad + (\$79{,}550)(0.46)(P/F, 18\%, 6) \\ &\quad + (4000x)(1 - 0.46)(P/A, 18\%, 10)(P/F, 18\%, 3) \end{aligned}$$

$$\begin{aligned}
P = 0 = &-(\$55{,}000)(0.8475) - (\$75{,}000)(0.7182) \\
&- (\$85{,}000)(0.6086) \\
&- (\$240{,}000)(1 - 0.46)(4.4941)(0.6086) \\
&+ (\$53{,}750)(0.46)(0.5158) \\
&+ (\$81{,}700)(0.46)(0.4371) \\
&+ (\$79{,}550)(0.46)(0.3704) \\
&+ (4000)(1 - 0.46)(4.4941)(0.6086)x \\
&5907.84x - 463{,}944.4 = 0
\end{aligned}$$

$$x = \boxed{\$78.53}$$

The amount to be added to the cost of each unit is

$$\$78.53 - \$60 = \boxed{\$18.53}$$

3. Route 420:

annual operating costs =

$$(365)\left(1200\,\frac{\text{vehicles}}{\text{day}}\right)\left(\frac{\$0.22}{\text{mi}}\right)(9.2\text{ mi}) = \$886{,}512$$

Route 422:

annual operating costs =

$$(365)\left(1200\,\frac{\text{vehicles}}{\text{day}}\right)\left(\frac{\$0.22}{\text{mi}}\right)(6.5\text{ mi}) = \$626{,}340$$

Use capitalized cost because the project has an infinite life.

$$\begin{aligned}
\frac{B}{C} &= \frac{P(\Delta B)}{P(\Delta I) + P(\Delta M)} \\
&= \text{benefit-cost ratio} \\
\Delta &= \text{Route 422} - \text{Route 420} \\
P(\Delta B) &= P \text{ of increased user benefits} \\
P(\Delta I) &= P \text{ of increased investment costs} \\
P(\Delta M) &= P \text{ of increased maintenance costs}
\end{aligned}$$

Calculate users' benefits derived from using Route 422 instead of Route 420.

value of time saved =

$$\begin{aligned}
&(365\text{ days})\left(1200\,\frac{\text{vehicles}}{\text{day}}\right)\left[\left(\frac{9.2\text{ mi}}{50\text{ mph}}\right)\left(\frac{60\text{ min}}{\text{hr}}\right)\right. \\
&\quad \left.-\left(\frac{6.5\text{ mi}}{50\text{ mph}}\right)\left(\frac{60\text{ min}}{\text{hr}}\right)\right]\left(\frac{\$0.30}{\text{min}}\right) \\
&\quad = \$425{,}736 \text{ per year}
\end{aligned}$$

operating costs saved =

$$\$886{,}512 - \$626{,}340 = \$260{,}172 \text{ per year}$$

$$\begin{aligned}
P(\Delta B) &= \frac{\$425{,}736 + \$260{,}172}{0.10} \\
&= \$6{,}859{,}080 \\
P(\Delta I) &= \$1{,}200{,}000 - 0 \\
&= \$1{,}200{,}000
\end{aligned}$$

ΔM is the decrease in the cost of resurfacing. It is $\$260{,}000 - \$300{,}000 = -\$40{,}000$, and it occurs once every 15 years.

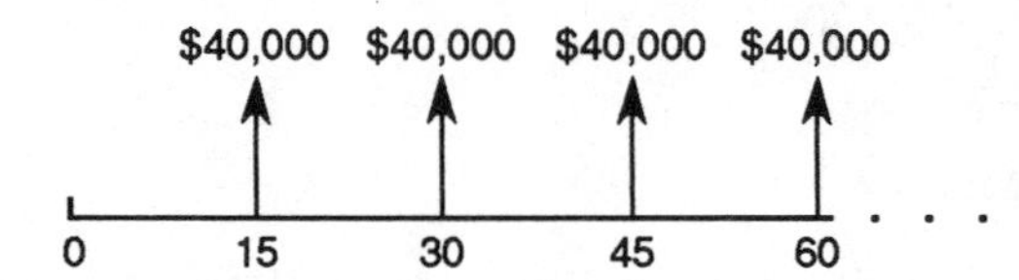

$$\begin{aligned}
P(\Delta M) &= -(\$40{,}000)(P/F, 10\%, 15) \\
&\quad - (\$40{,}000)(P/F, 10\%, 30) \\
&\quad - (\$40{,}000)(P/F, 10\%, 45) - \cdots \\
&= -(\$40{,}000)(0.2394 + 0.0573 + 0.0137 + \cdots) \\
&= -\$12{,}416 \\
\frac{B}{C} &= \frac{\$6{,}859{,}080}{\$1{,}200{,}000 - \$12{,}416} \\
&= 5.78 > 1.0
\end{aligned}$$

$$\boxed{\text{Choose Route 422.}}$$

4. Daily MSW volume (first year):

$$\begin{aligned}
&= \left(\frac{5\text{ lbm}}{\text{person}}\right)(100{,}000\text{ persons}) \\
&= 500{,}000\text{ lbm/day}
\end{aligned}$$

Total MSW volume (first year):

$$(365\text{ days})\left(\frac{500{,}000\,\frac{\text{lbm}}{\text{day}}}{2000\,\frac{\text{lbm}}{\text{ton}}}\right) = 91{,}250\text{ tons}$$

MSW composition	fraction recoverable	price/ton
combustibles	$(0.50)(0.60) = 0.30$	$(0.5)(\$45) = \$22.50$
ferrous materials	$(0.08)(0.90) = 0.072$	$(0.5)(\$80) = \$40$
glass	$(0.15)(0.80) = 0.12$	\$20
aluminum	$(0.05)(0.70) = 0.035$	\$200

Because the MSW volume increases 5% annually and prices increase in varying degrees, effective growth rates, i, must be calculated for each type of recoverable.

$$\begin{aligned}
\text{combustibles: } i_C &= (1+0.05)(1+0.04)-1 \\
&= 0.092 \\
\text{ferrous materials: } i_F &= (1+0.05)(1+0.08)-1 \\
&= 0.134 \\
\text{glass: } i_G &= (1+0.05)(1+0.02)-1 \\
&= 0.071 \\
\text{aluminum: } i_A &= (1+0.05)(1+0.12)-1 \\
&= 0.176
\end{aligned}$$

(a) First year's revenues:

combustibles: (91,250)(0.30)($22.5) =	$615,937
ferrous materials: (91,250)(0.072)($40) =	$262,800
glass: (91,250)(0.12)($20) =	$219,000
aluminum:(91,250)(0.035)($200) =	$638,750
	$1,736,487

Total revenues = **$1,736,487**

(b) Fifth year's revenues:

$$\begin{aligned}
&(\$615{,}937)(1+i_C)^4 + (\$262{,}800)(1+i_F)^4 \\
&\quad + (\$219{,}000)(i+i_G)^4 + (\$638{,}750)(1+i_A)^4 \\
&= (\$615{,}937)(1+0.092)^4 + (\$262{,}800)(1+0.134)^4 \\
&\quad + (\$219{,}000)(i+0.071)^4 + (\$638{,}750)(1+0.176)^4
\end{aligned}$$

$= \boxed{\$2{,}820{,}259}$

(c) Tenth year's revenues:

$$\begin{aligned}
&(\$615{,}937)(1+i_C)^9 + (\$262{,}800)(1+i_F)^9 \\
&\quad + (\$219{,}000)(i+i_G)^9 + (\$638{,}750)(1+i_A)^9 \\
&= (\$615{,}937)(1+0.092)^9 + (\$262{,}800)(1+0.134)^9 \\
&\quad + (\$219{,}000)(i+0.071)^9 + (\$638{,}750)(1+0.176)^9
\end{aligned}$$

$= \boxed{\$5{,}328{,}872}$

5.

salvage = $1000
1 2 3 . . . 8
$1250 $1250 $1250 $1250
$500 $600 $700
$1750 $1850 $1950
$500 + (7)($100) = $1200
$10,000

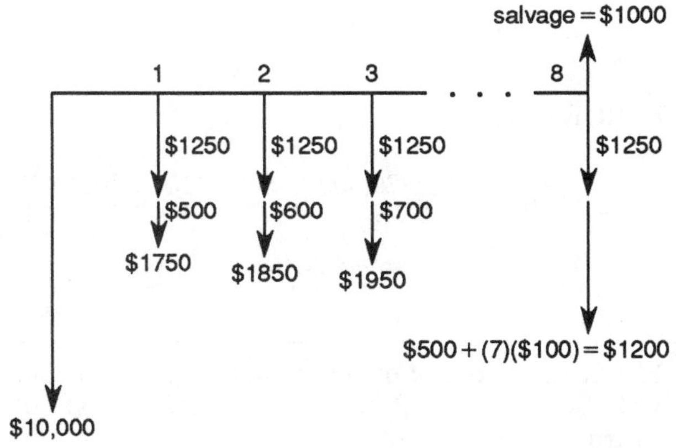

(a)
$$\begin{aligned}
A_8 &= (\$10{,}000)(A/P, 10\%, 8) + \$1250 + \$500 \\
&\quad + (\$100)(A/G, 10\%, 8) \\
&\quad - (\$1000)(A/F, 10\%, 8) \\
&= (\$10{,}000)(0.1874) + \$1250 + \$500 \\
&\quad + (\$100)(3.0045) - (\$1000)(0.0874) \\
&= \boxed{\$3837.05}
\end{aligned}$$

(b)
$$\begin{aligned}
P &= (A_8)(P/A, i\%, n) \\
&= -(\$3837.05)(P/A, 10\%, 8) \\
&= -(\$3837.05)(5.3349) \\
&= \boxed{-\$20{,}470.28}
\end{aligned}$$

(c)
$$\begin{aligned}
A_4 &= (\$10{,}000)(A/P, 10\%, 4) + \$1750 \\
&\quad + (\$100)(A/G, 10\%, 4) \\
&\quad - (\$3000)(A/F, 10\%, 4) \\
&= (\$10{,}000)(0.3155) + \$1750 \\
&\quad + (\$100)(1.3812) - (\$3000)(0.2155) \\
&= \boxed{\$4396.62}
\end{aligned}$$

(d)
$$P = -\$10{,}000 - \frac{\$1750}{1.1} - \frac{\$1850}{(1.1)^2} + \frac{\$6500}{(1.1)^2}$$

$= \boxed{-\$7747.93}$

(e) Calculate EUAC for 7 years.

$$\begin{aligned}
A_7 &= (\$10{,}000)(A/P, 10\%, 7) + (\$100) \\
&\quad \times (A/G, 10\%, 7) + \$1750 - (\$1500)(A/F, 10\%, 7) \\
&= (\$10{,}000)(0.2054) + (\$100)(2.6216) \\
&\quad + \$1750 - (\$1500)(0.1054) \\
&= \$3908.06
\end{aligned}$$

A_8 was calculated in (a) as $3837.05.

Because $A_8 < A_7$, the most economical life is 8 years.

(f)

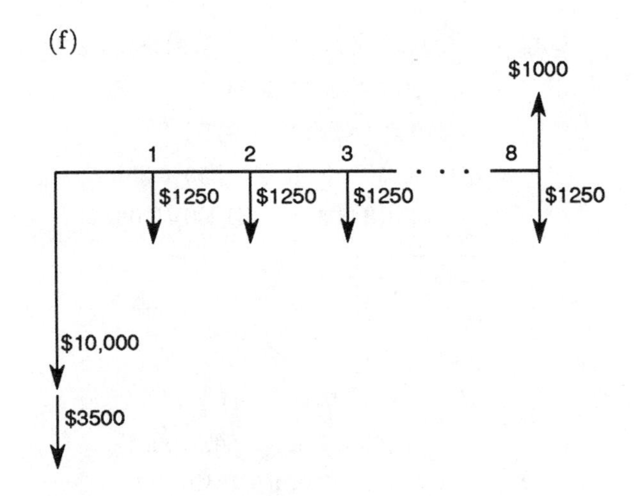

$$\begin{aligned} A_8 &= (\$13{,}500)(A/P, 10\%, 8) + \$1250 \\ &\quad - (\$1000)(A/F, 10\%, 8) \\ &= (\$13{,}500)(0.1874) + \$1250 - (\$1000)(0.0874) \\ &= \boxed{\$3692.50} \end{aligned}$$

(g) $P = (A_8)(P/A, i\%, n)$

From (f), $A_8 = \$3692.50$.

$$\begin{aligned} P &= -(\$3692.50)(P/A, 10\%, 8) \\ &= -(\$3692.50)(5.3349) \\ &= \boxed{-\$19{,}699.12} \end{aligned}$$

(h) Calculate A_7 and compare with A_8 as calculated in (f).

$$\begin{aligned} A_7 &= (\$13{,}500)(A/P, 10\%, 7) + \$1250 \\ &\quad - (\$1500)(A/F, 10\%, 7) \\ &= (\$13{,}500)(0.2054) + \$1250 - (\$1500)(0.1054) \\ &= \boxed{\$3864.80} \end{aligned}$$

Since $A_7 > A_8$, keep the vehicle for 8 years.

6. (a)

$$\begin{aligned} \text{profit} &= \$2{,}400{,}000 - \$2{,}000{,}000 \\ &= \$400{,}000 \\ \text{tax} &= (\$400{,}000)(0.45) = \$180{,}000 \\ \text{after-tax profit} &= \$400{,}000 - \$180{,}000 \\ &= \$220{,}000 \\ \text{after-tax ROR} &= \frac{\$220{,}000}{\$2{,}000{,}000} \\ &= \boxed{11\%} \end{aligned}$$

(b)

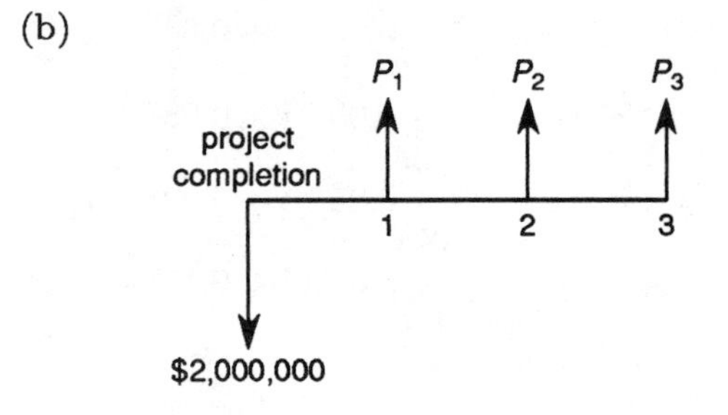

The effective rate, i', considering interest, i, and inflation, e, is

$$\begin{aligned} i' &= i + e + ie \\ &= 0.10 + 0.05 + (0.10)(0.05) \\ &= 0.155 \end{aligned}$$

$$\begin{aligned} P_1 &= \text{first payment} \\ &= (0.50)(\$2{,}400{,}000)(1.155) \\ &= \$1{,}386{,}000 \\ P_2 &= \text{second payment} \\ &= (0.25)(\$2{,}400{,}000)(1.155)^2 \\ &= \$800{,}415 \\ P_3 &= \text{third payment} \\ &= (0.25)(\$2{,}400{,}000)(1.155)^3 \\ &= \$924{,}479 \end{aligned}$$

The ROR is the rate that results in a zero present worth.

$$\begin{aligned} P &= -\$2{,}000{,}000 + \frac{P_1}{1+\text{ROR}} + \frac{P_2}{(1+\text{ROR})^2} \\ &\quad + \frac{P_3}{(1+\text{ROR})^3} \\ &= 0 \end{aligned}$$

Try ROR = 25%.

$$\begin{aligned} -\$2{,}000{,}000 + \frac{\$1{,}386{,}000}{1.25} + \frac{\$800{,}415}{(1.25)^2} + \frac{\$924{,}479}{(1.25)^3} \\ = \$94{,}399 \end{aligned}$$

Try ROR = 30%.

$$\begin{aligned} -\$2{,}000{,}000 + \frac{\$1{,}386{,}000}{1.30} + \frac{\$800{,}415}{(1.30)^2} + \frac{\$924{,}479}{(1.30)^3} \\ = -\$39{,}436 \end{aligned}$$

The before-tax rate of return is between 25% and 30%; by further trial and error, before-tax rate of return = 28.5%.

(c)

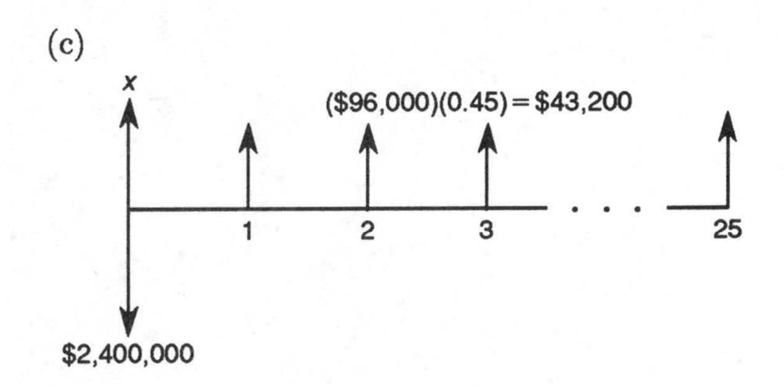

$$D = \frac{\$2{,}400{,}000}{25} = \$96{,}000$$

$$(x - \$2{,}400{,}000) + (\$96{,}000)(0.45)(P/A, 15\%, 25) = 0$$

$$(x - \$2{,}400{,}000) + (\$96{,}000)(0.45)(6.4641) = 0$$

$$x = \$2{,}120{,}750 \text{ (after tax)}$$

The before-tax present worth of revenues is

$$\frac{\$2{,}120{,}750}{1 - 0.45} = \boxed{\$3{,}855{,}909}$$

FLUID MECHANICS

1. (a) With the energy-datum zero level at point C, the energy equation between this point and the reservoir level is

$$h + 30\text{ ft} + 15\text{ ft} = \frac{v_{BC}^2}{2g} + \frac{p_C}{\gamma} + h_f$$

h is the water depth in the reservoir, and $p_C = 0$, assuming that the system discharges to the atmosphere. The local losses have been neglected. The friction losses are

$$h_f = f_{AB}\left(\frac{L_{AB}}{D_{AB}}\right)\left(\frac{v_{AB}^2}{2g}\right) + f_{BC}\left(\frac{L_{BC}}{D_{BC}}\right)\left(\frac{v_{BC}^2}{2g}\right)$$

The roughness for PVC is $\epsilon = 5 \times 10^{-6}$ ft, so

$$\frac{\epsilon}{D_{AB}} = \frac{5 \times 10^{-6}\text{ ft}}{\left(\frac{8\text{ in}}{12\,\frac{\text{in}}{\text{ft}}}\right)} = 7.5 \times 10^{-6}$$

$$\frac{\epsilon}{D_{BC}} = \frac{5 \times 10^{-6}\text{ ft}}{\left(\frac{6\text{ in}}{12\,\frac{\text{in}}{\text{ft}}}\right)} = 1.0 \times 10^{-5}$$

From the Moody diagram, the pipes are smooth for most of the Reynolds number range. As a first approximation, choose

$$f_{AB} = f_{BC} = 0.013$$

The friction losses are

$$\begin{aligned} h_f &= (0.013)\left[\frac{1000\text{ ft}}{\left(\frac{8\text{ in}}{12\,\frac{\text{in}}{\text{ft}}}\right)}\right]\left[\frac{v_{AB}^2}{(2)\left(32.2\,\frac{\text{ft}}{\text{sec}^2}\right)}\right] \\ &\quad + (0.013)\left[\frac{800\text{ ft}}{\left(\frac{6\text{ in}}{12\,\frac{\text{in}}{\text{ft}}}\right)}\right]\left[\frac{v_{BC}^2}{(2)\left(32.2\,\frac{\text{ft}}{\text{sec}^2}\right)}\right] \\ &= 0.303v_{AB}^2 + 0.323v_{BC}^2 \end{aligned}$$

The continuity equation relates the velocities v_{AB} and v_{BC}.

$$\begin{aligned} \pi\left(\frac{D_{AB}^2}{4}\right)v_{AB} &= \pi\left(\frac{D_{BC}^2}{4}\right)v_{BC} \\ v_{AB} &= \left(\frac{D_{BC}}{D_{AB}}\right)^2 v_{BC} \\ &= \left(\frac{6\text{ in}}{8\text{ in}}\right)^2 v_{BC} \\ &= 0.563v_{BC} \end{aligned}$$

Substituting this into the friction loss equation,

$$\begin{aligned} h_f &= (0.303)(0.563v_{BC})^2 + 0.323v_{BC}^2 \\ &= 0.419v_{BC}^2 \end{aligned}$$

The energy equation becomes

$$h + 45\text{ ft} = \frac{v_{BC}^2}{(2)\left(32.2\,\frac{\text{ft}}{\text{sec}^2}\right)} + 0.419v_{BC}^2$$

$$h = 0.435v_{BC}^2 - 45\text{ ft} \qquad \text{[Eq. 1]}$$

A second equation of h and v_{BC} is obtained from the required 40 lbf/in^2 pressure at point A. The energy equation between the reservoir level and A is

$$h + 45\text{ ft} = 45\text{ ft} + \frac{v_{AB}^2}{2g} + \frac{p_A}{\gamma}$$

Using the continuity equation and the known values produces

$$\begin{aligned} h &= \frac{(0.563v_{BC})^2}{(2)\left(32.2\,\frac{\text{ft}}{\text{sec}^2}\right)} + \frac{\left(40\,\frac{\text{lbf}}{\text{in}^2}\right)\left(144\,\frac{\text{in}^2}{\text{ft}^2}\right)}{62.4\,\frac{\text{lbf}}{\text{ft}^2}} \\ &= 0.0049v_{BC}^2 + 92.3\text{ ft} \qquad \text{[Eq. 2]} \end{aligned}$$

Solving Eqs. 1 and 2 simultaneously yields

$$h = 93.9\text{ ft}$$

$$v_{BC} = 17.9\text{ ft/sec}$$

From the continuity equation,

$$v_{AB} = 0.563\,V_{BC} = (0.563)\left(17.9\,\frac{\text{ft}}{\text{sec}}\right) = 10.1\text{ ft/sec}$$

In order to check the friction parameters, the Reynolds numbers are needed. The water viscosity at 50°F is $\nu = 1.41 \times 10^{-5}$ ft^2/sec.

$$(N_{Re})_{AB} = \frac{v_{AB}D_{AB}}{\nu} = \frac{\left(10.1\ \frac{\text{ft}}{\text{sec}}\right)\left(\frac{8\ \text{in}}{12\ \frac{\text{in}}{\text{ft}}}\right)}{1.41 \times 10^{-5}\ \frac{\text{ft}^2}{\text{sec}}}$$

$$= 4.8 \times 10^5$$

$$(N_{Re})_{BC} = \frac{v_{BC}D_{BC}}{\nu} = \frac{\left(17.9\ \frac{\text{ft}}{\text{sec}}\right)\left(\frac{6\ \text{in}}{12\ \frac{\text{in}}{\text{ft}}}\right)}{1.41 \times 10^{-5}\ \frac{\text{ft}^2}{\text{sec}}}$$

$$= 6.3 \times 10^5$$

From the Moody diagram, the friction factors chosen are correct. The minor losses should also be checked.

$$\text{exit loss} \approx (0.5)\left(\frac{v_{AB}^2}{2g}\right)$$

$$= (0.5)\left[\frac{\left(10.1\ \frac{\text{ft}}{\text{sec}}\right)^2}{(2)\left(32.2\ \frac{\text{ft}}{\text{sec}^2}\right)}\right]$$

$$= 0.8\ \text{ft}$$

$$\text{contraction loss} \approx (0.185)\left(\frac{v_{BC}^2}{2g}\right)$$

$$= (0.185)\left[\frac{\left(17.9\ \frac{\text{ft}}{\text{sec}}\right)^2}{(2)\left(32.2\ \frac{\text{ft}}{\text{sec}^2}\right)}\right]$$

$$= 0.9\ \text{ft}$$

The losses are small compared to the friction losses but should be added to the water depth in the reservoir.

$$h = 93.9\ \text{ft} + 0.8\ \text{ft} + 0.9\ \text{ft}$$

$$= \boxed{95.6\ \text{ft}}$$

(b) The energy equation between the reservoir and point C is

$$70\ \text{ft} + 45\ \text{ft} = \frac{v_{BC}^2}{2g} + \frac{p_C}{\gamma} + f_{AB}\left(\frac{L_{AB}}{D_{AB}}\right)\left(\frac{v_{AB}^2}{2g}\right) + f_{BC}\left(\frac{L_{BC}}{D_{BC}}\right)\left(\frac{v_{BC}^2}{2g}\right)$$

From part (a),

$$v_{AB} = 0.563 v_{BC}$$

Assuming $f_{AB} = 0.016$ and $f_{BC} = 0.015$ yields

$$\frac{v_{BC}^2}{(2)\left(32.2\ \frac{\text{ft}}{\text{sec}^2}\right)} + \frac{\left(40\ \frac{\text{lbf}}{\text{in}^2}\right)\left(144\ \frac{\text{in}^2}{\text{ft}^2}\right)}{62.4\ \frac{\text{lbf}}{\text{ft}^3}}$$

$$+ (0.016)\left[\frac{1000\ \text{ft}}{\left(\frac{8\ \text{in}}{12\ \frac{\text{in}}{\text{ft}}}\right)}\right]\left[\frac{(0.563 v_{BC})^2}{(2)\left(32.2\ \frac{\text{ft}}{\text{sec}^2}\right)}\right]$$

$$+ (0.015)\left[\frac{800\ \text{ft}}{\left(\frac{6\ \text{in}}{12\ \frac{\text{in}}{\text{ft}}}\right)}\right]\left[\frac{v_{BC}^2}{(2)\left(32.2\ \frac{\text{ft}}{\text{sec}^2}\right)}\right] = 115\ \text{ft}$$

Solving for the velocity,

$$v_{BC} = 6.7\ \text{ft/sec}$$

From the continuity equation,

$$v_{AB} = (0.563)\left(6.7\ \frac{\text{ft}}{\text{sec}}\right) = 3.8\ \text{ft/sec}$$

In order to check the friction factors, calculate the Reynolds numbers.

$$(N_{Re})_{AB} = \frac{v_{AB}D_{AB}}{\nu} = \frac{\left(3.8\ \frac{\text{ft}}{\text{sec}}\right)\left(\frac{8\ \text{in}}{12\ \frac{\text{in}}{\text{ft}}}\right)}{1.41 \times 10^{-5}\ \frac{\text{ft}^2}{\text{sec}}}$$

$$= 1.8 \times 10^5$$

$$(N_{Re})_{BC} = \frac{v_{BC}D_{BC}}{\nu} = \frac{\left(6.7\ \frac{\text{ft}}{\text{sec}}\right)\left(\frac{6\ \text{in}}{12\ \frac{\text{in}}{\text{ft}}}\right)}{1.41 \times 10^{-5}\ \frac{\text{ft}^2}{\text{sec}}}$$

$$= 2.4 \times 10^5$$

Referring to the Moody diagram, the friction factor is correct. The flow through the system is

$$Q = \pi\left(\frac{D_{BC}^2}{4}\right)v_{BC}$$

$$= \pi\left[\frac{\left(\dfrac{6\text{ in}}{12\,\dfrac{\text{in}}{\text{ft}}}\right)^2}{4}\right]\left(6.7\,\frac{\text{ft}}{\text{sec}}\right)$$

$$= \boxed{1.3\text{ ft}^3/\text{sec (590 gal/min)}}$$

(c) Pipeline AC must carry the flow not provided by pipeline ABC.

$$1900\,\frac{\text{gal}}{\text{min}} - 590\,\frac{\text{gal}}{\text{min}} = 1310\text{ gal/min}\left(2.9\text{ ft}^3/\text{sec}\right)$$

Try $D_{AC} = 10$ in. Using the roughness dimension in (a), the relative roughness is

$$\frac{\epsilon}{D_{AC}} = \frac{5\times10^{-6}\text{ ft}}{\dfrac{10\text{ in}}{12\,\dfrac{\text{in}}{\text{ft}}}} = 6\times10^{-6}$$

From the Moody diagram, choose an initial value for the friction factor $f_{AC} = 0.014$. The energy equation along pipeline AC is

$$70\text{ ft} + 45\text{ ft} = \frac{v_{AC}^2}{2g} + \frac{p_C}{\gamma} + f_{AC}\left(\frac{L_{AC}}{D_{AC}}\right)\left(\frac{v_{AC}^2}{2g}\right)$$

$$115\text{ ft} = \frac{v_{AC}^2}{(2)\left(32.2\,\dfrac{\text{ft}}{\text{sec}^2}\right)} + \frac{\left(40\,\dfrac{\text{lbf}}{\text{in}^2}\right)\left(144\,\dfrac{\text{in}^2}{\text{ft}^2}\right)}{62.4\,\dfrac{\text{lbf}}{\text{ft}^3}}$$

$$+ (0.014)\left(\frac{1800\text{ ft}}{\dfrac{10\text{ in}}{12\,\dfrac{\text{in}}{\text{ft}}}}\right)\left[\frac{v_{AC}^2}{(2)\left(32.2\,\dfrac{\text{ft}}{\text{sec}^2}\right)}\right]$$

Solving for the velocity yields

$$v_{AC} = 6.8\text{ ft/sec}$$

The diameter can be calculated from the continuity equation.

$$Q = \pi\left(\frac{D_{AC}^2}{4}\right)v_{AC}$$

$$2.9\,\frac{\text{ft}^3}{\text{sec}} = \pi\left(\frac{D_{AC}^2}{4}\right)\left(6.8\,\frac{\text{ft}}{\text{sec}}\right)$$

Solving for D_{AC} yields

$$D_{AC} = 0.74\text{ ft}$$

$$D_{AC} = (0.74\text{ ft})\left(12\,\frac{\text{in}}{\text{ft}}\right) = 8.8\text{ in}$$

This is close to the guessed value. The final step is to calculate the Reynolds number to check the friction factor.

$$(N_{Re})_{AC} = \frac{v_{AC}D_{AC}}{\nu}$$

$$= \frac{\left(6.8\,\dfrac{\text{ft}}{\text{sec}}\right)\left(\dfrac{10\text{ in}}{12\,\dfrac{\text{in}}{\text{ft}}}\right)}{1.41\times10^{-5}\,\dfrac{\text{ft}^2}{\text{sec}}}$$

$$= 4.0\times10^5$$

The value for this number in the Moody diagram is close to the initial choice. Therefore, the new PVC pipe should have a standard-size diameter of

$$D_{AC} = \boxed{10\text{ in}}$$

2. (a) The maximum static head at C occurs in the absence of flow through the system. In this case, the friction losses are zero and the energy equation between the tank and point C is

$$734\text{ ft} = z_C + \frac{p_C}{\gamma}$$

$$= 624\text{ ft} + \frac{p_c}{\gamma}$$

Solving for the static head p_C/γ yields

$$\frac{p_C}{\gamma} = \boxed{110\text{ ft}}$$

(b) The maximum flow takes place when C is at its minimum allowable pressure of 22 psig.

For path ABC, the continuity equation is

$$\pi\left(\frac{D_{AB}^2}{4}\right)(v_{AB}) = \pi\left(\frac{D_{BC}^2}{4}\right)v_{BC}$$

$$v_{AB} = \left(\frac{D_{BC}}{D_{AB}}\right)^2 v_{BC}$$

$$= \left(\frac{6\text{ in}}{10\text{ in}}\right)^2 v_{BC}$$

$$= 0.36v_{BC}$$

The energy equation is

$$734 \text{ ft} = 624 \text{ ft} + \frac{p_C}{\gamma} + \frac{v_{BC}^2}{2g} + f_{AB}\left(\frac{L_{AB}}{D_{AB}}\right)\left(\frac{v_{AB}^2}{2g}\right) + f_{BC}\left(\frac{L_{BC}}{D_{BC}}\right)\left(\frac{v_{BC}^2}{2g}\right)$$

Introducing the continuity equation and simplifying,

$$110 \text{ ft} = \frac{p_C}{\gamma} + \left(\frac{v_{BC}^2}{2g}\right)\left[1 + (0.36)^2 f_{AB}\left(\frac{L_{AB}}{D_{AB}}\right) + f_{BC}\left(\frac{L_{BC}}{D_{BC}}\right)\right]$$

For a 5- to 10-year-old cast iron pipe, the specific roughness is $\epsilon = 0.001$ ft.

$$\frac{\epsilon}{D_{AB}} = \frac{0.001 \text{ ft}}{\frac{10 \text{ in}}{12 \frac{\text{in}}{\text{ft}}}} = 0.0012$$

$$\frac{\epsilon}{D_{BC}} = \frac{0.001 \text{ ft}}{\frac{6 \text{ in}}{12 \frac{\text{in}}{\text{ft}}}} = 0.002$$

From the Moody diagram, initial estimates for the friction factors are

$$f_{AB} = 0.021$$
$$f_{BC} = 0.024$$

Substituting all known values in the energy equation yields

$$110 \text{ ft} = \frac{\left(22 \frac{\text{lbf}}{\text{in}^2}\right)\left(144 \frac{\text{in}^2}{\text{ft}^2}\right)}{62.4 \frac{\text{lbf}}{\text{ft}^3}} + \left[\frac{v_{BC}^2}{(2)\left(32.2 \frac{\text{ft}}{\text{sec}^2}\right)}\right] \times \left[1 + (0.36)^2(0.021)\left(\frac{500 \text{ ft}}{\frac{10 \text{ in}}{12 \frac{\text{in}}{\text{ft}}}}\right) + (0.024)\left(\frac{500 \text{ ft}}{\frac{6 \text{ in}}{12 \frac{\text{in}}{\text{ft}}}}\right)\right]$$

Solving for the velocity v_{BC} yields

$$v_{BC} = 12.0 \text{ ft/sec}$$

From the continuity equation,

$$v_{AB} = 0.36\, v_{BC} = (0.36)\left(12.0 \frac{\text{ft}}{\text{sec}}\right) = 4.3 \text{ ft/sec}$$

The friction factors are checked by calculating the Reynolds numbers.

$$(N_{Re})_{AB} = \frac{v_{AB} D_{AB}}{\nu} = \frac{\left(4.3 \frac{\text{ft}}{\text{sec}}\right)\left(\frac{10 \text{ in}}{12 \frac{\text{in}}{\text{ft}}}\right)}{1.41 \times 10^{-5} \frac{\text{ft}^2}{\text{sec}}} = 2.5 \times 10^5$$

$$(N_{Re})_{BC} = \frac{v_{BC} D_{BC}}{\nu} = \frac{\left(12.0 \frac{\text{ft}}{\text{sec}}\right)\left(\frac{6 \text{ in}}{12 \frac{\text{in}}{\text{ft}}}\right)}{1.41 \times 10^{-5} \frac{\text{ft}^2}{\text{sec}}} = 4.3 \times 10^5$$

The Moody diagram indicates that the friction factors remain unchanged with these values. The flow through branch ABC is

$$Q_{ABC} = \pi\left(\frac{D_{AB}^2}{4}\right)v_{AB} = \pi\left[\frac{\left(\frac{10 \text{ in}}{12 \frac{\text{in}}{\text{ft}}}\right)^2}{4}\right]\left(4.3 \frac{\text{ft}}{\text{sec}}\right) = 2.3 \text{ ft}^3/\text{sec} \text{ (1030 gal/min)}$$

The continuity equation for branch ADC yields

$$v_{AD} = \left(\frac{D_{DC}}{D_{AD}}\right)^2 v_{DC} = \left(\frac{8 \text{ in}}{12 \text{ in}}\right)^2 v_{DC} = 0.44 v_{DC}$$

The energy equation is

$$734 \text{ ft} = 624 \text{ ft} + \frac{p_c}{\gamma} + \frac{v_{DC}^2}{2g} + f_{AD}\left(\frac{L_{AD}}{D_{AD}}\right)\left(\frac{v_{AD}^2}{2g}\right) + f_{DC}\left(\frac{L_{DC}}{D_{DC}}\right)\left(\frac{v_{DC}^2}{2g}\right)$$

After introducing the continuity equation and collecting terms, this becomes

$$110\ \text{ft} = \frac{p_c}{\gamma} + \left(\frac{v_{DC}^2}{2g}\right) \times \left[1 + (0.44 f_{AD})\left(\frac{L_{AD}}{D_{AD}}\right) + f_{DC}\left(\frac{L_{DC}}{D_{DC}}\right)\right]$$

The relative roughnesses are

$$\frac{\epsilon}{D_{AD}} = \frac{0.001\ \text{ft}}{\frac{12\ \text{in}}{12\ \frac{\text{in}}{\text{ft}}}} = 0.001$$

$$\frac{\epsilon}{D_{DC}} = \frac{0.001\ \text{ft}}{\frac{10\ \text{in}}{12\ \frac{\text{in}}{\text{ft}}}} = 0.0015$$

From the Moody diagram, estimate initial values for the friction factors.

$$f_{AD} = 0.020$$
$$f_{DC} = 0.022$$

Replacing all known values in the energy equation produces

$$110\ \text{ft} = \frac{\left(22\ \frac{\text{lbf}}{\text{in}^2}\right)\left(144\ \frac{\text{in}^2}{\text{ft}^2}\right)}{62.4\ \frac{\text{lbf}}{\text{ft}^3}} + \left[\frac{v_{DC}^2}{(2)\left(32.2\ \frac{\text{ft}}{\text{sec}^2}\right)}\right] \times \left[1 + (0.44)(0.020)\left(\frac{700\ \text{ft}}{\frac{12\ \text{in}}{12\ \frac{\text{in}}{\text{ft}}}}\right) + (0.022)\left(\frac{400\ \text{ft}}{\frac{10\ \text{in}}{12\ \frac{\text{in}}{\text{ft}}}}\right)\right]$$

Solving for the velocity,

$$v_{DC} = 15.0\ \text{ft/sec}$$

From the continuity equation,

$$v_{AD} = 0.44 v_{DC} = (0.44)\left(15.0\ \frac{\text{ft}}{\text{sec}}\right) = 6.6\ \text{ft/sec}$$

The Reynolds numbers for this branch are

$$(N_{Re})_{AD} = \frac{v_{AD} D_{AD}}{\nu} = \frac{\left(6.6\ \frac{\text{ft}}{\text{sec}}\right)\left(\frac{12\ \text{in}}{12\ \frac{\text{in}}{\text{ft}}}\right)}{1.41 \times 10^{-5}\ \frac{\text{ft}^2}{\text{sec}}} = 4.7 \times 10^5$$

$$(N_{Re})_{DC} = \frac{v_{DC} D_{DC}}{\nu} = \frac{\left(15.0\ \frac{\text{ft}}{\text{sec}}\right)\left(\frac{8\ \text{in}}{12\ \frac{\text{in}}{\text{ft}}}\right)}{1.41 \times 10^{-5}\ \frac{\text{ft}^2}{\text{sec}}} = 7.1 \times 10^5$$

The friction values obtained from the Moody diagram are close to the initial estimates. The flow in branch ADC is

$$Q_{ADC} = \pi\left(\frac{D_{AD}^2}{4}\right) v_{AD} = \pi\left[\frac{\left(\frac{12\ \text{in}}{12\ \frac{\text{in}}{\text{ft}}}\right)^2}{4}\right]\left(6.6\ \frac{\text{ft}}{\text{sec}}\right) = 5.2\ \text{ft}^3/\text{sec}\ (2330\ \text{gal/min})$$

The total flow in the hydrant is the sum of both components.

$$Q = Q_{ABC} + Q_{ADC} = 1030\ \frac{\text{gal}}{\text{min}} + 2330\ \frac{\text{gal}}{\text{min}} = \boxed{3360\ \text{gal/min}}$$

3. (a) Neglecting the velocity heads because of the pipe lengths involved, the energy equation between the tank and point D is

$$1160\ \text{ft} + 122\ \text{ft} = 890\ \text{ft} + \frac{p_D}{\gamma} + (3.022)\left(\frac{v_{AB}^{1.85} L_{AB}}{C_{AB}^{1.85} D_{AB}^{1.167}}\right) + (3.022)\left(\frac{v_{BC}^{1.85} L_{BC}}{C_{BC}^{1.85} D_{BC}^{1.167}}\right) + (3.022)\left(\frac{v_{CD}^{1.85} L_{CD}}{C_{CD}^{1.85} D_{CD}^{1.167}}\right)$$

The continuity equation for each pipe produces the following flows and velocities.

$$Q_{CD} = 70\,\frac{\text{gal}}{\text{min}} = 0.156\ \text{ft}^3/\text{sec}$$

$$\text{v}_{CD} = \frac{Q_{CD}}{\pi\left(\frac{D_{CD}^2}{4}\right)} = \frac{0.156\,\frac{\text{ft}^3}{\text{sec}}}{\pi\left[\frac{(0.3333\ \text{ft})^2}{4}\right]} = 1.8\ \text{ft/sec}$$

$$\begin{aligned} Q_{BC} &= 130\,\frac{\text{gal}}{\text{min}} + Q_{CD} \\ &= 130\,\frac{\text{gal}}{\text{min}} + 70\,\frac{\text{gal}}{\text{min}} \\ &= 200\ \text{gal/min}\ (0.446\ \text{ft}^3/\text{sec}) \end{aligned}$$

$$\text{v}_{BC} = \frac{Q_{BC}}{\pi\left(\frac{D_{BC}^2}{4}\right)} = \frac{0.446\,\frac{\text{ft}^3}{\text{sec}}}{\pi\left[\frac{(0.50\ \text{ft})^2}{4}\right]} = 2.3\ \text{ft/sec}$$

$$\begin{aligned} Q_{AB} &= 140\,\frac{\text{gal}}{\text{min}} + Q_{BC} \\ &= 140\,\frac{\text{gal}}{\text{min}} + 200\,\frac{\text{gal}}{\text{min}} \\ &= 340\ \text{gal/min}\ (0.758\ \text{ft}^3/\text{sec}) \end{aligned}$$

$$\text{v}_{AB} = \frac{Q_{AB}}{\pi\left(\frac{D_{AB}^2}{4}\right)} = \frac{0.758\,\frac{\text{ft}^3}{\text{sec}}}{\pi\left[\frac{(0.5\ \text{ft})^2}{4}\right]} = 3.9\ \text{ft/sec}$$

Replacing the known values in the energy equation yields

$$392\ \text{ft} = \frac{p_D}{\gamma} + \left[\frac{3.022}{(150)^{1.85}}\right]\left[\frac{\left(3.9\,\frac{\text{ft}}{\text{sec}}\right)^{1.85}(16{,}000\ \text{ft})}{(0.5\ \text{ft})^{1.167}} + \frac{\left(2.3\,\frac{\text{ft}}{\text{sec}}\right)^{1.85}(14{,}000\ \text{ft})}{(0.5\ \text{ft})^{1.167}} + \frac{\left(1.8\,\frac{\text{ft}}{\text{sec}}\right)^{1.85}(8000\ \text{ft})}{(0.3333\ \text{ft})^{1.167}}\right]$$

Solving for the pressure head,

$$\frac{p_D}{\gamma} = 202.9\ \text{ft}$$

The pressure is

$$\begin{aligned} p_D &= (202.9\ \text{ft})\left(62.4\,\frac{\text{lbf}}{\text{ft}^3}\right) \\ &= 12{,}660\,\frac{\text{lbf}}{\text{ft}^2} \\ &= \boxed{88\ \text{lbf/in}^2\ \text{(psig)}} \end{aligned}$$

(b) If this pressure is insufficient for the demand at point D, most likely the pipes do not have enough capacity or the water level in the tank is too low.

(c) Assuming that the peak demand is 1.5 times the average, the flows and velocities through the pipe are

$$Q_{CD} = (1.5)\left(0.156\,\frac{\text{ft}^3}{\text{sec}}\right) = 0.234\ \text{ft}^3/\text{sec}$$

$$\text{v}_{CD} = (1.5)\left(1.8\,\frac{\text{ft}}{\text{sec}}\right) = 2.7\ \text{ft/sec}$$

$$Q_{BC} = (1.5)\left(0.446\,\frac{\text{ft}^3}{\text{sec}}\right) = 0.669\ \text{ft}^3/\text{sec}$$

$$\text{v}_{BC} = (1.5)\left(2.3\,\frac{\text{ft}}{\text{sec}}\right) = 3.4\ \text{ft/sec}$$

$$Q_{AB} = (1.5)\left(0.758\,\frac{\text{ft}^3}{\text{sec}}\right) = 1.137\ \text{ft}^3/\text{sec}$$

$$\text{v}_{AB} = (1.5)\left(3.9\,\frac{\text{ft}}{\text{sec}}\right) = 5.8\ \text{ft/sec}$$

With these new values, the energy equation is

$$392\ \text{ft} = \frac{p_D}{\gamma} + \left[\frac{3.022}{(150)^{1.85}}\right]\left[\frac{\left(5.8\,\frac{\text{ft}}{\text{sec}}\right)^{1.85}(16{,}000\ \text{ft})}{(0.5\ \text{ft})^{1.167}} + \frac{\left(3.4\,\frac{\text{ft}}{\text{sec}}\right)^{1.85}(14{,}000\ \text{ft})}{(0.5\ \text{ft})^{1.167}} + \frac{\left(2.7\,\frac{\text{ft}}{\text{sec}}\right)^{1.85}(8000\ \text{ft})}{(0.3333\ \text{ft})^{1.167}}\right]$$

The new pressure head is

$$\frac{p_D}{\gamma} = -10.1\ \text{ft}$$

The pressure at D is

$$\begin{aligned} p_D &= (10.1\ \text{ft})\left(62.4\,\frac{\text{lbf}}{\text{ft}^3}\right) \\ &= -630.2\,\frac{\text{lbf}}{\text{ft}^2} \\ &= \boxed{-4.4\ \text{lbf/in}^2\ \text{(psig)}} \end{aligned}$$

The negative pressure indicates a problem at this flow volume.

(d) For this configuration, it appears that the pipes are too small for the required pressure. Nevertheless, changing the existing pipeline would be costly. Possible alternatives include increasing the water level in the tank, pressurizing the tank, adding a pump somewhere in the pipeline, or replacing only a section of the pipeline. The solution chosen must be adopted based on cost.

4. Assume the flow directions shown in the figure.

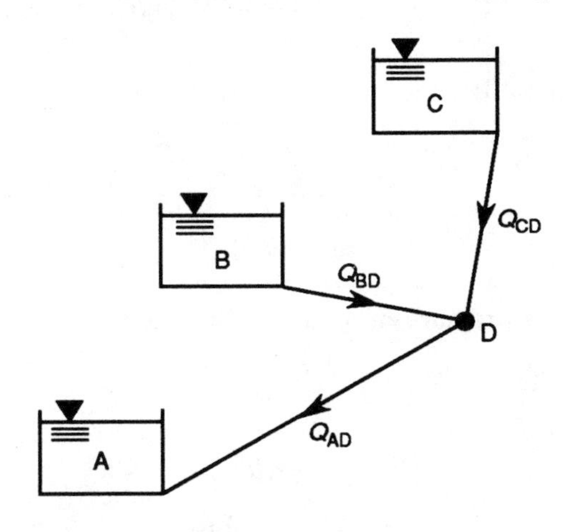

(a) The energy equations between the reservoirs and point D are

$$\text{CD:}\quad 1350\text{ ft} = 1200\text{ ft} + \frac{p_D}{\gamma} + h_{f,\text{CD}}$$

$$\text{BD:}\quad 1290\text{ ft} = 1200\text{ ft} + \frac{p_D}{\gamma} + h_{f,\text{BD}}$$

$$\text{AB:}\quad 1190\text{ ft} = 1200\text{ ft} + \frac{p_D}{\gamma} + h_{f,\text{AB}}$$

The velocity heads and the local losses have been neglected. The friction losses are calculated according to the Hazen-William equation as

$$h_f = (4.73)\left(\frac{Q^{1.85}L}{C^{1.85}D^{4.867}}\right)$$

The friction losses are

$$h_{f,\text{CD}} = (4.73)\left[\frac{Q_{\text{CD}}^{1.85}(10{,}000\text{ ft})}{(150)^{1.85}(0.8333\text{ ft})^{4.867}}\right] = 10.82Q_{\text{CD}}^{1.85}$$

$$h_{f,\text{BD}} = (4.73)\left[\frac{Q_{\text{BD}}^{1.85}(4000\text{ ft})}{(150)^{1.85}(0.6667\text{ ft})^{4.867}}\right] = 12.82Q_{\text{BD}}^{1.85}$$

$$h_{f,\text{AD}} = (4.73)\left[\frac{Q_{\text{AD}}^{1.85}(5000\text{ ft})}{(150)^{1.85}(1\text{ ft})^{4.867}}\right] = 2.23Q_{\text{AD}}^{1.85}$$

From the preceding equations, the flows are

$$Q_{\text{CD}} = 0.276h_{f,\text{CD}}^{0.54}$$

$$Q_{\text{BC}} = 0.252h_{f,\text{BC}}^{0.54}$$

$$Q_{\text{AD}} = 0.649h_{f,\text{AD}}^{0.54}$$

Rewrite the energy equations as

$$h_{f,\text{CD}} = 150\text{ ft} - \frac{p_D}{\gamma}$$

$$h_{f,\text{BC}} = 90\text{ ft} - \frac{p_D}{\gamma}$$

$$h_{f,\text{AD}} = 10\text{ ft} + \frac{p_D}{\gamma}$$

Rewrite the continuity equation at point D as

$$Q_{\text{CD}} + Q_{\text{BD}} - Q_{\text{AD}} = 0$$

This produces a system of seven unknowns: three flows, three head losses, and one pressure head at D. These equations are nonlinear and can be best solved iteratively using the following table.

$\frac{p_D}{\gamma}$ (assumed) (ft)	$h_{f,\text{CD}}$ (ft)	Q_{CD} (ft^3/sec)	$h_{f,\text{BC}}$ (ft)
60.0	90.0	3.13	30.0
50.0	100.0	3.22	40.0
40.0	110.0	3.49	50.0
42.2	107.8	3.46	47.8

Q_{BC} (ft^3/sec)	$h_{f,\text{AD}}$ (ft)	Q_{AD} (ft^3/sec)	$Q_{\text{CD}} + Q_{\text{BC}} - Q_{\text{AD}}$ (ft^3/sec)
1.58	70.0	6.43	−1.73
1.85	60.0	5.92	−0.76
2.08	50.0	5.37	0.20
2.03	52.2	5.49	−0.003

In the table, a value of p_D/γ is assumed. Replacing this value in the energy equations yields the head losses in each pipe. The flows are calculated from the energy loss expressions. Finally, the flows are checked for continuity. The last line in the table uses an estimate of p_D/γ interpolated from the previous two; the continuity closure error is very small, so the choice is correct and the flows are

$$Q_{\text{CD}} = 3.46\ \frac{\text{ft}^3}{\text{sec}} = \boxed{1553\text{ gal/min into D}}$$

$$Q_{\text{BC}} = 2.03\ \frac{\text{ft}^3}{\text{sec}} = \boxed{911\text{ gal/min into D}}$$

$$Q_{\text{AD}} = 5.49\ \frac{\text{ft}^3}{\text{sec}} = \boxed{2464\text{ gal/min out of D}}$$

(b) The table includes the following head losses.

$$h_{f,\mathrm{CD}} = \boxed{107.8 \text{ ft}}$$

$$h_{f,\mathrm{BD}} = \boxed{47.8 \text{ ft}}$$

$$h_{f,\mathrm{AD}} = \boxed{52.2 \text{ ft}}$$

5. The Hardy Cross method is used to solve the flow loop. The friction losses are given by

$$h_f = KQ^{1.85}$$
$$K = (10.6)\left(\frac{L}{C^{1.85}D^{4.867}}\right)$$

Q is in MGD, and L and D are in feet. Using the data provided produces the K values shown in the following computational tables.

The initial estimates for the directions and magnitudes of the flows are

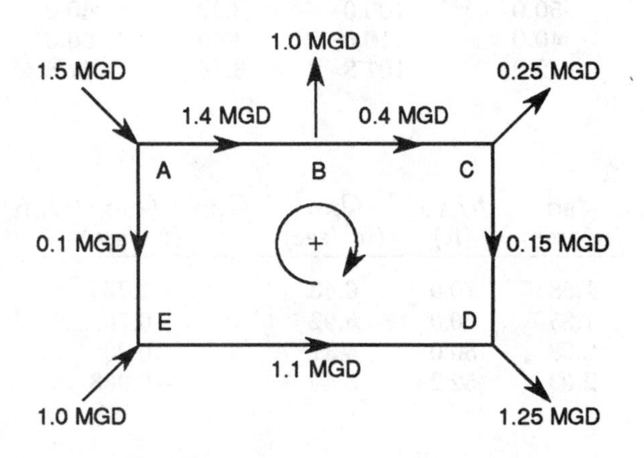

The method converges after two iterations as shown in the tables.

Iteration 1:

pipe	K	Q	h_f	$\frac{h_f}{Q}$	corrected Q
AB	6.53	1.40	12.17	8.70	1.26
BC	8.17	0.40	1.50	3.75	0.26
CD	52.86	0.15	1.58	10.54	0.01
DE	4.31	−1.10	−5.14	4.67	−1.24
EA	61.67	−0.10	−0.87	8.71	−0.24
			9.25	36.36	

correction −0.14

Iteration 2:

pipe	K	Q	h_f	$\frac{h_f}{Q}$	corrected Q
AB	6.53	1.26	10.06	7.96	1.13
BC	8.17	0.26	0.69	2.62	0.13
CD	52.86	0.01	0.02	1.28	−0.12
DE	4.31	−1.24	−6.39	5.16	−1.37
EA	61.67	−0.24	−4.31	18.17	−0.37
			0.06	35.19	

correction 0.00

The last table contains the solution to the network.

(a) $Q_{\mathrm{AB}} = \boxed{1.13 \text{ MGD (from A to B)}}$

(b) $\boxed{\text{Water flows from A to B.}}$

(c) The pressure drop is equal to the friction losses.

$$\frac{\Delta p_{\mathrm{AB}}}{\gamma} = K_{\mathrm{AB}}Q_{\mathrm{AB}}^{1.85}$$
$$= (6.53)(1.13 \text{ MGD})^{1.85} = 8.19 \text{ ft}$$
$$\Delta p_{\mathrm{AB}} = (8.19 \text{ ft})\left(62.4 \frac{\text{lbf}}{\text{ft}^3}\right)$$
$$= 511 \text{ lbf/ft}^2$$
$$= \boxed{3.5 \text{ lbf/in}^2 \text{ (psig)}}$$

(d) $Q_{\mathrm{BC}} = \boxed{0.13 \text{ MGD (from B to C)}}$

(e) $Q_{\mathrm{AE}} = \boxed{0.37 \text{ MGD (from A to E)}}$

(f) $Q_{\mathrm{ED}} = \boxed{1.37 \text{ MGD (from E to D)}}$

(g) $Q_{\mathrm{CD}} = \boxed{0.12 \text{ MGD (from C to D)}}$

(h)
$$\frac{\Delta p_{\mathrm{CD}}}{\gamma} = K_{\mathrm{CD}}Q_{\mathrm{CD}}^{1.85}$$
$$= (52.86)(0.12)^{1.85}$$
$$= 1.05 \text{ ft}$$
$$\Delta p_{\mathrm{CD}} = (1.05 \text{ ft})\left(62.4 \frac{\text{lbf}}{\text{ft}^3}\right)$$
$$= 65.5 \frac{\text{lbf}}{\text{ft}^2}$$
$$= \boxed{0.46 \text{ lbf/in}^2 \text{ (psig)}}$$

(i) **Flow goes from C to D.**

(j) The speed of wave propagation assuming a rigid conduit is

$$c = \sqrt{\frac{E}{\gamma}}$$

E is the bulk modulus of elasticity for the fluid, and γ is the specific weight. For water at 50°F,

$$E = 305{,}000 \text{ lbf/in}^2$$

$$\gamma = 62.4 \text{ lbf/ft}^3$$

The wave speed is

$$c = \sqrt{\frac{\left(305{,}000 \, \frac{\text{lbf}}{\text{in}^2}\right)\left(144 \, \frac{\text{in}^2}{\text{ft}^2}\right)\left(32.2 \, \frac{\text{ft}}{\text{sec}^2}\right)}{62.4 \, \frac{\text{lbf}}{\text{ft}^3}}}$$

$$= 4761 \text{ ft/sec}$$

The pressure surge can be minimized with a very slow closure of the valve (see Gupta, R.S., *Hydrology and Hydraulic Systems*, Prentice Hall, 1989, p. 575). If t_c is the closure time, this objective can be achieved ensuring that

$$t_c > \frac{20L}{c}$$

L is the pipe length.

$$t_c > \frac{(20)(2000 \text{ ft})}{4761 \, \frac{\text{ft}}{\text{sec}}} = \boxed{8.4 \text{ sec}}$$

6. (a) Call Q_1 the flow through branch AD and Q_2 the flow through ABCD. The energy equations in branches AD and ABCD are

AD:

$$\frac{p_A}{\gamma} = \frac{p_D}{\gamma} + f_{\text{AD}}\left(\frac{L_{\text{AD}}}{D_{\text{AD}}}\right) \times \left(\frac{Q_1^2}{\left(32.2 \, \frac{\text{ft}}{\text{sec}^2}\right)\left[\pi\left(\frac{D_{\text{AD}}^2}{4}\right)\right]^2}\right)$$

ABCD:

$$\frac{p_A}{\gamma} = \frac{p_D}{\gamma} + f_{\text{AC}}\left(\frac{L_{\text{AC}}}{D_{\text{AC}}}\right) \times \left(\frac{Q_2^2}{\left(32.2 \, \frac{\text{ft}}{\text{sec}^2}\right)\left[\pi\left(\frac{D_{\text{AC}}^2}{4}\right)\right]^2}\right) + f_{\text{CD}}\left(\frac{L_{\text{CD}}}{D_{\text{CD}}}\right)\left(\frac{Q_2^2}{\left(32.2 \, \frac{\text{ft}}{\text{sec}^2}\right)\left[\pi\left(\frac{D_{\text{CD}}^2}{4}\right)\right]^2}\right)$$

For 5-year-old cast iron pipes, the specific roughness is $\epsilon = 0.001$ ft. The relative roughnesses are

$$\frac{\epsilon}{D_{\text{AD}}} = \frac{0.001 \text{ ft}}{0.8333 \text{ ft}} = 0.0012$$

$$\frac{\epsilon}{D_{\text{AC}}} = \frac{0.001 \text{ ft}}{0.6667 \text{ ft}} = 0.0015$$

$$\frac{\epsilon}{D_{\text{CD}}} = \frac{0.001 \text{ ft}}{1 \text{ ft}} = 0.001$$

Using the Moody diagram, initially assume that $f_{\text{AD}} = f_{\text{DC}} = 0.021$ and $f_{\text{AC}} = 0.023$. Given that $p_A - p_D = 10 \text{ lbf/in}^2 = 1440 \text{ lbf/ft}^2$, the energy equations become

AD:

$$\frac{1440 \, \frac{\text{lbf}}{\text{ft}^2}}{62.4 \, \frac{\text{lbf}}{\text{ft}^3}} = (0.021)\left(\frac{2000 \text{ ft}}{0.8333 \text{ ft}}\right) \times \left[\frac{Q_1^2}{(2)\left(32.2 \, \frac{\text{ft}}{\text{sec}^2}\right)\left(\pi\left[\frac{(0.8333 \text{ ft})^2}{4}\right]\right)^2}\right]$$

ABCD:

$$\frac{1440 \, \frac{\text{lbf}}{\text{ft}^2}}{62.4 \, \frac{\text{lbf}}{\text{ft}^3}} = \left[\frac{Q_2^2}{(2)\left(32.2 \, \frac{\text{ft}}{\text{sec}^2}\right)}\right] \times \left[(0.023)\left(\frac{2100 \text{ ft}}{0.6667 \text{ ft}}\right)\left[\frac{1}{\left(\pi\left[\frac{(0.6667 \text{ ft})^2}{4}\right]\right)^2}\right] + (0.021)\left(\frac{900 \text{ ft}}{1 \text{ ft}}\right)\frac{1}{\left(\pi\left[\frac{(1 \text{ ft})^2}{4}\right]\right)^2}\right]$$

The resulting system of equations is

$$23.08 = 2.631Q_1^2$$
$$23.08 = 9.709Q_2^2$$

Solving for Q_1 and Q_2 yields

$$Q_1 = 2.96 \text{ ft}^3/\text{sec}$$
$$Q_2 = 1.54 \text{ ft}^3/\text{sec}$$

The velocities in the pipes are

$$\text{v}_{\text{AD}} = \frac{Q_1}{\pi\left(\frac{D_{\text{AD}}^2}{4}\right)} = \frac{2.96 \frac{\text{ft}^3}{\text{sec}}}{\pi\left[\frac{(0.8333 \text{ ft})^2}{4}\right]}$$
$$= 5.4 \text{ ft/sec}$$

$$\text{v}_{\text{AC}} = \frac{Q_2}{\pi\left(\frac{D_{\text{AC}}^2}{4}\right)} = \frac{1.54 \frac{\text{ft}^3}{\text{sec}}}{\pi\left[\frac{(0.6667 \text{ ft})^2}{4}\right]}$$
$$= 4.4 \text{ ft/sec}$$

$$\text{v}_{\text{CD}} = \frac{Q_3}{\pi\left(\frac{D_{\text{CD}}^2}{4}\right)} = \frac{1.54 \frac{\text{ft}^3}{\text{sec}}}{\pi\left[\frac{(1 \text{ ft})^2}{4}\right]}$$
$$= 2.0 \text{ ft/sec}$$

The Reynolds numbers are

$$(N_{\text{Re}})_{\text{AD}} = \frac{\text{v}_{\text{AD}} D_{\text{AD}}}{\nu} = \frac{\left(5.4 \frac{\text{ft}}{\text{sec}}\right)(0.8333 \text{ ft})}{1.41 \times 10^{-5} \frac{\text{ft}^2}{\text{sec}}}$$
$$= 3.2 \times 10^5$$

$$(N_{\text{Re}})_{\text{AC}} = \frac{\text{v}_{\text{AC}} D_{\text{AC}}}{\nu} = \frac{\left(4.4 \frac{\text{ft}}{\text{sec}}\right)(0.6667 \text{ ft})}{1.41 \times 10^{-5} \frac{\text{ft}^2}{\text{sec}}}$$
$$= 2.1 \times 10^5$$

$$(N_{\text{Re}})_{\text{CD}} = \frac{\text{v}_{\text{CD}} D_{\text{CD}}}{\nu} = \frac{\left(2.0 \frac{\text{ft}}{\text{sec}}\right)(1 \text{ ft})}{1.41 \times 10^{-5} \frac{\text{ft}^2}{\text{sec}}}$$
$$= 1.4 \times 10^5$$

With these values, the revised friction factors from the Moody diagram are

$$f_{\text{AD}} = 0.021$$
$$f_{\text{AC}} = 0.023$$
$$f_{\text{CD}} = 0.0215$$

Inserting these new values in the energy equations and recalculating the flows produces no substantial change.

$$Q_1 = 2.96 \text{ ft}^3/\text{sec}$$
$$Q_2 = 1.54 \text{ ft}^3/\text{sec}$$

The flow through the pump is

$$Q_{\text{pump}} = Q_1 + Q_2$$
$$= 2.96 \frac{\text{ft}^3}{\text{sec}} + 1.54 \frac{\text{ft}^3}{\text{sec}}$$
$$= \boxed{4.5 \text{ ft}^3/\text{sec (2020 gal/min)}}$$

(b)
$$\text{v}_{\text{AD}} = \frac{Q_1}{\pi\left(\frac{D_{\text{AD}}^2}{4}\right)} = \frac{2.96 \frac{\text{ft}^3}{\text{sec}}}{\pi\left[\frac{(0.8333 \text{ ft})^2}{4}\right]}$$
$$= \boxed{5.5 \text{ ft/sec}}$$

(c) From the problem statement,

$$p_A - p_D = \boxed{10 \text{ lbf/in}^2 \text{ (psig)}}$$

(d) The pressure drop is the same regardless of the path, so

$$p_A - p_D = \boxed{10 \text{ lbf/in}^2 \text{ (psig)}}$$

(e) Assuming that outlet D discharges to the atmosphere, the head provided by the pump is equal to the pressure drop.

$$h_{\text{pump}} = \frac{\left(10 \frac{\text{lbf}}{\text{in}^2}\right)\left(144 \frac{\text{in}^2}{\text{ft}^2}\right)}{62.4 \frac{\text{lbf}}{\text{ft}^3}}$$
$$= \boxed{23.1 \text{ ft}}$$

7. (a) The energy equation between the reservoirs is

$$12 \text{ ft} = f\left(\frac{\text{L}}{\text{D}}\right)\left[\frac{Q^2}{2g\left(\pi\frac{D^2}{4}\right)^2}\right]$$
$$+ 2K\left[\frac{Q^2}{2g\left(\pi\frac{D^2}{4}\right)^2}\right]$$

The last term represents the local losses caused by the two elbows. The loss coefficient for a single elbow is $K = 0.9$.

$$Q = \frac{6 \text{ MGD}}{0.64632 \dfrac{\text{MGD}}{\dfrac{\text{ft}^3}{\text{sec}}}} = 9.3 \text{ ft}^3/\text{sec}$$

For the maximum flow of $Q = 6$ MGD $= 9.3$ ft^3/sec and assuming initially that $f = 0.02$, the energy equation becomes

$$12 \text{ ft} = (0.02)\left(\frac{240 \text{ ft}}{D}\right)\left[\frac{\left(9.3 \dfrac{\text{ft}^3}{\text{sec}}\right)^2}{(2)\left(32.2 \dfrac{\text{ft}}{\text{sec}^2}\right)\left(\pi \dfrac{D^2}{4}\right)^2}\right]$$

$$+ (2)(0.9)\left[\frac{\left(9.3 \dfrac{\text{ft}^3}{\text{sec}^2}\right)^2}{(2)\left(32.2 \dfrac{\text{ft}}{\text{sec}^2}\right)\left(\pi \dfrac{D^2}{4}\right)^2}\right]$$

$$12 = \frac{10.45}{D^5} + \frac{3.90}{D^4}$$

Solve this nonlinear equation by trial and error for D.

$$D \approx 1.04 \text{ ft } (12.5 \text{ in})$$

To check the friction factor, the following quantities are needed. Assume that the physical properties of the sludge are similar to water.

$$\text{v} = \frac{Q}{\pi \dfrac{D^2}{4}} = \frac{9.3 \dfrac{\text{ft}^3}{\text{sec}}}{\pi \left(\dfrac{1.04 \text{ ft}}{4}\right)} = 10.9 \text{ ft/sec}$$

$$\text{N}_{\text{Re}} = \frac{\text{v}D}{\nu} = \frac{\left(10.9 \dfrac{\text{ft}}{\text{sec}}\right)(1.04 \text{ ft})}{1.54 \times 10^{-5} \dfrac{\text{ft}^2}{\text{sec}}} = 7.4 \times 10^5$$

The viscosity is selected for the worst case of $T = 45°$F. The roughness dimension for ductile iron is $\epsilon = 0.0008$ ft; therefore,

$$\frac{\epsilon}{D} = 0.0008 \text{ ft}/1.04 \text{ ft} = 0.0008$$

The Moody diagram produces $f = 0.019$, which is close to the value used.

Choose a standard-size pipe of

$$\boxed{D = 14 \text{ in}}$$

(b) The system must perform adequately under the worst possible conditions, that is, maximum flow and cold weather. A slightly oversized pipe would compensate for increased roughness due to the corrosion caused by the sludge.

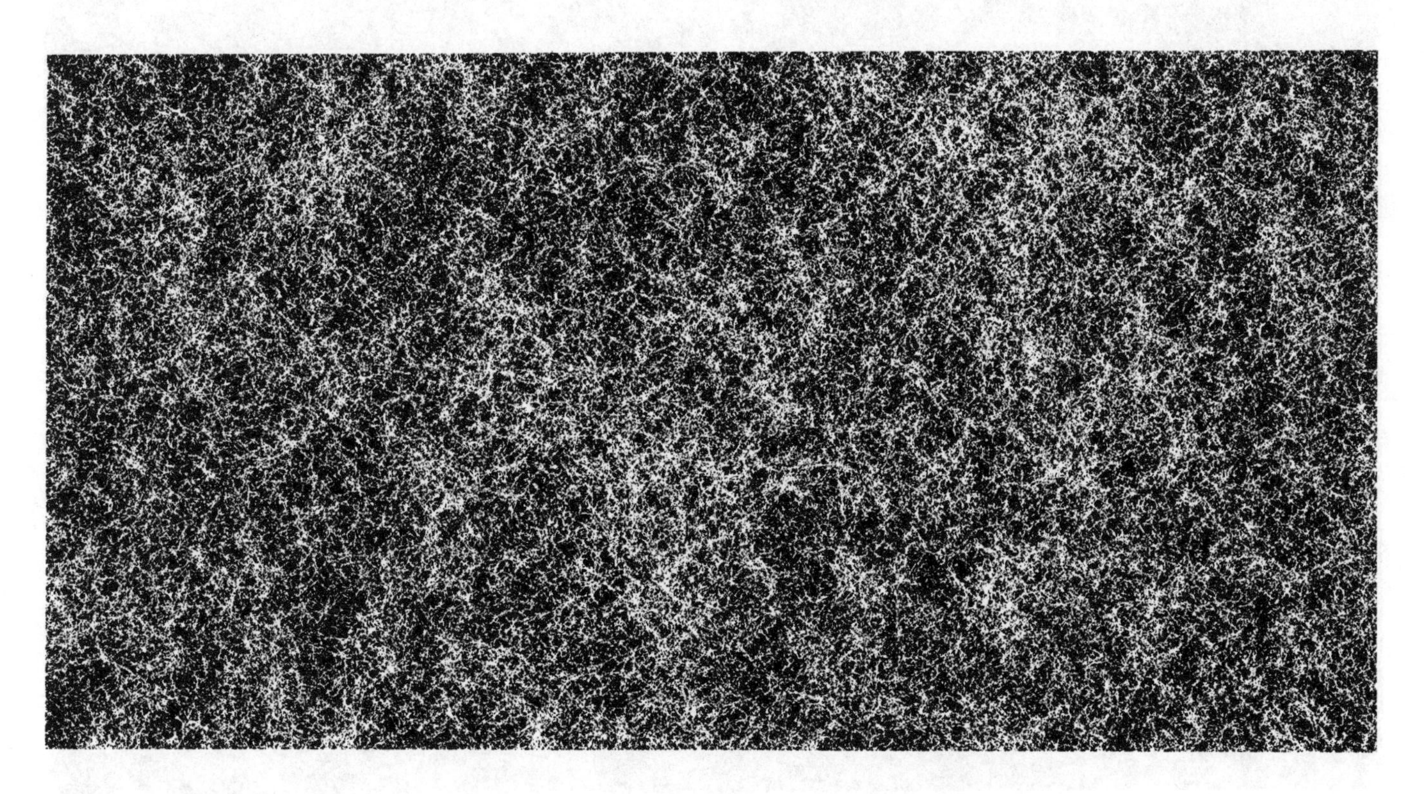

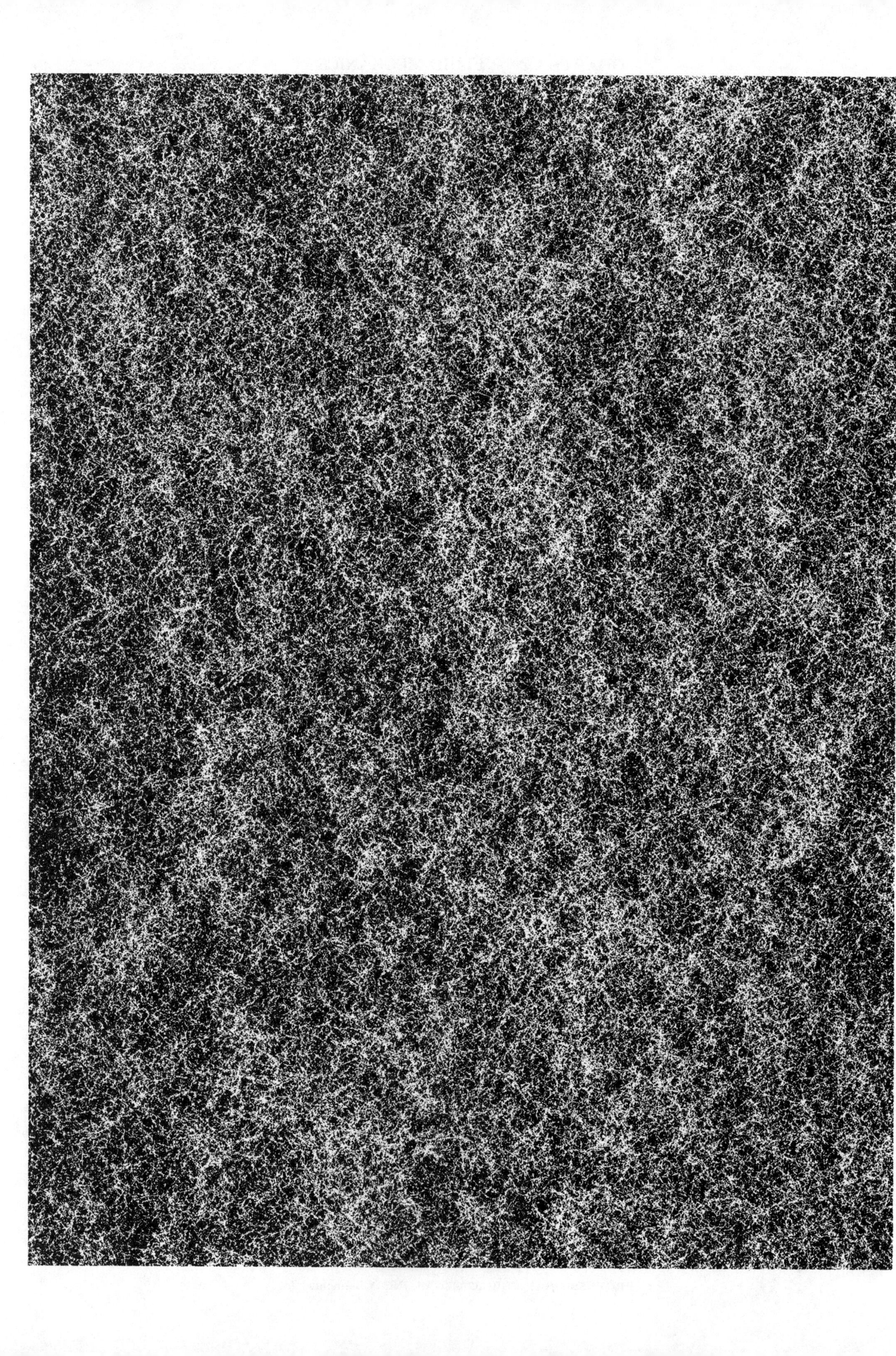

HYDRAULIC MACHINES

1. (a) Neglecting the velocity head, the energy equation is

$$70\text{ ft} + h_A = 150\text{ ft} + f_{AB}\left(\frac{L_{AB}}{D_{AB}}\right)\left(\frac{Q^2}{2g\left[\pi\left(\frac{D_{AB}^2}{4}\right)\right]^2}\right) + f_{BC}\left(\frac{L_{BC}}{D_{BC}}\right)\left(\frac{Q^2}{2g\left[\pi\left(\frac{D_{BC}^2}{4}\right)\right]^2}\right)$$

Assume initially that $f_{AB} = f_{BC} = 0.02$. The energy equation becomes

$$h_A = 80\text{ ft} + \left[(0.02)\left(\frac{8000\text{ ft}}{0.6667\text{ ft}}\right)\left[\frac{1}{\left(\pi\left[\frac{(0.6667\text{ ft})^2}{4}\right]\right)^2}\right] + (0.02)\left(\frac{4000\text{ ft}}{0.5\text{ ft}}\right)\left[\frac{1}{\left(\pi\left[\frac{(0.5\text{ ft})^2}{4}\right]\right)^2}\right]\right] \times \left[\frac{Q^2}{(2)\left(32.2\ \frac{\text{ft}}{\text{sec}^2}\right)}\right]$$

The equation of the system curve is

$$h = 80 + 95.02Q^2 \quad (Q \text{ in cfs})$$

The operating point is the intersection of this curve with the pump curve, as shown in the following diagram.

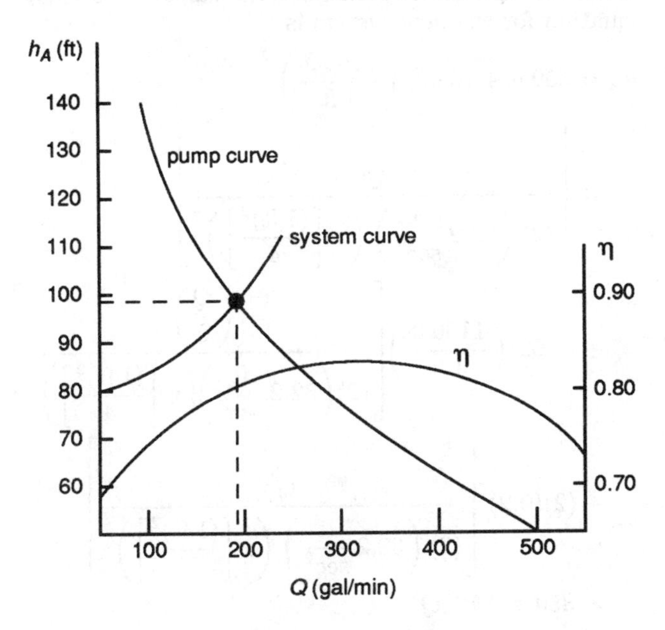

The operating point is at $Q = 190$ gal/min (0.42 ft^3/sec), $h_A = 97$ ft. In order to check the friction factors, the velocities and Reynolds numbers are calculated.

$$v_{AB} = \frac{Q}{\pi\left(\frac{D_{AB}^2}{4}\right)} = \frac{0.42\ \frac{\text{ft}^3}{\text{sec}}}{\pi\left[\frac{(0.6667\text{ ft})^2}{4}\right]} = 1.2\text{ ft/sec}$$

$$(N_{Re})_{AB} = \frac{v_{AB}D_{AB}}{\nu} = \frac{\left(1.2\ \frac{\text{ft}}{\text{sec}}\right)(0.6667\text{ ft})}{1.41\times 10^{-5}\ \frac{\text{ft}^2}{\text{sec}}} = 5.7\times 10^4$$

$$v_{BC} = \frac{Q}{\pi\left(\frac{D_{BC}^2}{4}\right)} = \frac{0.42\ \frac{\text{ft}^3}{\text{sec}}}{\pi\left[\frac{(0.5\text{ ft})^2}{4}\right]} = 2.1\text{ ft/sec}$$

$$(N_{Re})_{BC} = \frac{v_{BC}D_{BC}}{\nu} = \frac{\left(2.1\ \frac{\text{ft}}{\text{sec}}\right)(0.5\text{ ft})}{1.41\times 10^{-5}\ \frac{\text{ft}^2}{\text{sec}}} = 7.4\times 10^4$$

For PVC pipe, the specific roughness is $\epsilon = 5\times 10^{-6}$ ft, so the relative roughness is

$$\frac{\epsilon}{D_{AB}} = \frac{5\times 10^{-6}\text{ ft}}{0.6667\text{ ft}} = 0.0000075$$

$$\frac{\epsilon}{D_{BC}} = \frac{5\times 10^{-6}\text{ ft}}{0.5\text{ ft}} = 0.00001$$

From the Moody diagram, $f_{AB} = 0.020$ and $f_{BC} = 0.019$, which are close to the initial estimate. The flow through the system is

$$Q = \boxed{190\text{ gal/min}}$$

(b) Neglecting the velocity heads, the energy equation between the upper tank and point B is

$$150\text{ ft} = 90\text{ ft} + \frac{p_B}{\gamma} + f_{AB}\left(\frac{L_{AB}}{D_{AB}}\right)\left(\frac{Q^2}{2g\left[\pi\left(\frac{D_{AB}^2}{4}\right)\right]^2}\right)$$

Assume initially that $f_{AB} = 0.018$. This equation becomes

$$60 \text{ ft} = \frac{\left(20 \frac{\text{lbf}}{\text{in}^2}\right)\left(144 \frac{\text{in}^2}{\text{ft}^2}\right)}{62.4 \frac{\text{lbf}}{\text{ft}^3}} + (0.018)\left(\frac{8000 \text{ ft}}{0.6667 \text{ ft}}\right) \times \left[\frac{Q^2}{(2)\left(32.2 \frac{\text{ft}}{\text{sec}^2}\right)\left(\pi\left[\frac{(0.6667 \text{ ft})^2}{4}\right]\right)^2}\right]$$

Solving for Q yields

$$Q = 0.71 \text{ ft}^3/\text{sec}$$

The velocity is

$$\text{v}_{AB} = \frac{Q}{\pi\left(\frac{D_{AB}^2}{4}\right)} = \frac{0.71 \frac{\text{ft}^3}{\text{sec}}}{\pi\left[\frac{(0.6667 \text{ ft})^2}{4}\right]} = 2.0 \text{ ft/sec}$$

The Reynolds number is

$$(N_{Re})_{AB} = \frac{\text{v}_{AB} D_{AB}}{\nu} = \frac{\left(2.0 \frac{\text{ft}}{\text{sec}}\right)(0.6667 \text{ ft})}{1.41 \times 10^{-5} \frac{\text{ft}^2}{\text{sec}}} = 9.5 \times 10^4$$

With the value ϵ/D_{AB} calculated in (a), the Moody diagram indicates $f_{AB} = 0.018$, which is the value used. The flow is

$$Q = \boxed{0.71 \text{ ft}^3/\text{sec } (319 \text{ gal/min})}$$

2. (a) The energy equation between the two reservoirs is

$$h_A = 350 \text{ ft} + f\left(\frac{L}{D}\right)\left(\frac{\text{v}^2}{2g}\right) + 2K\left(\frac{\text{v}^2}{2g}\right)$$

The local loss coefficient for one fully open gate valve is $K = 0.2$ (see Roberson, J.A., J.J. Cassidy, and M.H. Chaudry. *Hydraulic Engineering*, Houghton-Mifflin, 1989. p. 256). The velocity is

$$\text{v} = \frac{Q}{\pi\left(\frac{D^2}{4}\right)} = \frac{9.5 \frac{\text{ft}^3}{\text{sec}}}{\pi\left[\frac{(1 \text{ ft})^2}{4}\right]} = 12.1 \text{ ft/sec}$$

$$h_A = 350 \text{ ft} + (0.02)\left(\frac{2200 \text{ ft}}{1 \text{ ft}}\right) \times \left[\frac{\left(12.1 \frac{\text{ft}}{\text{sec}}\right)^2}{(2)\left(32.2 \frac{\text{ft}}{\text{sec}^2}\right)}\right] + (2)(0.2)\left[\frac{\left(12.1 \frac{\text{ft}}{\text{sec}}\right)^2}{(2)\left(32.2 \frac{\text{ft}}{\text{sec}^2}\right)}\right] = 450.9 \text{ ft}$$

The motor brake horsepower required by the pump is

$$\text{bhp} = \frac{\gamma h_A Q}{550\eta} = \frac{\left(62.4 \frac{\text{lbf}}{\text{ft}^3}\right)(450.9 \text{ ft})\left(9.5 \frac{\text{ft}^3}{\text{sec}}\right)}{\left(550 \frac{\text{ft-lbf}}{\text{hp-sec}}\right)(0.83)} = 586 \text{ hp}$$

The power requirement can be met with a set of three pumps driven by three 200-hp motors.

(b) Assume that the power is 600 hp, then

$$600 \text{ hp} = \frac{\left(62.4 \frac{\text{lbf}}{\text{ft}^3}\right) h_A Q}{(550)(0.83)}$$

$$h_A = \frac{4389.4}{Q}$$

Since the additional pipe is identical to the existing one, the flow through it will be half of the total, and the losses through each branch are the same. The energy equation for the new system is

$$h_A = 350 \text{ ft} + (0.02)\left(\frac{1100 \text{ ft}}{1 \text{ ft}}\right) \times \left[\frac{Q^2}{(2)\left(32.2 \frac{\text{ft}}{\text{sec}^2}\right)\left(\pi\left[\frac{(1 \text{ ft})^2}{4}\right]\right)^2}\right] + (0.02)\left(\frac{1100 \text{ ft}}{1 \text{ ft}}\right)\left[\frac{\left(\frac{Q}{2}\right)^2}{(2)\left(32.2 \frac{\text{ft}}{\text{sec}^2}\right)\left(\pi\left[\frac{(1 \text{ ft})^2}{4}\right]\right)^2}\right] + (2)(0.2)\left[\frac{Q^2}{(2)\left(32.2 \frac{\text{ft}}{\text{sec}^2}\right)\left(\pi\left[\frac{(1 \text{ ft})^2}{4}\right]\right)^2}\right] = 350 + 0.702Q^2$$

Substituting the expression for the power produces

$$\frac{4389.4}{Q} = 350 + 0.702Q^2$$

Solving for the flow by trial and error (or by other means) yields

$$Q = \boxed{10.3 \text{ ft}^3/\text{sec (4636 gal/min)}}$$

OPEN CHANNEL FLOW

1. Subdivide the channel into three sections as shown.

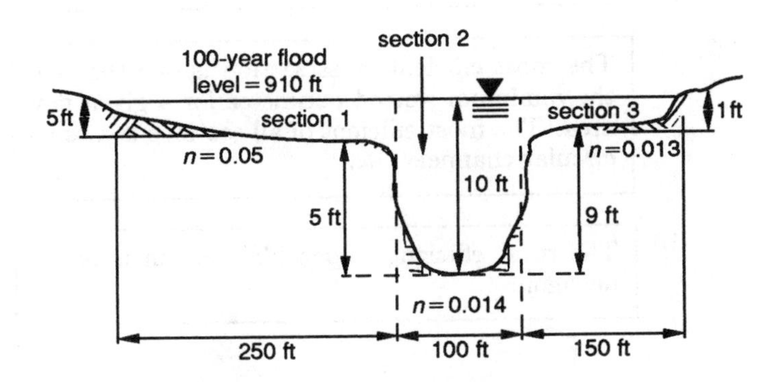

not to scale

(a) The total flow is assumed to be the sum of the individual flows through the main channel and the flood plains. The slope is the same for the three sections. Uniform flow has been established due to the channel length.

For section 1,

$$A_1 = wd \approx (250\text{ ft})(5\text{ ft}) = 1250\text{ ft}^2$$

Since this section can be considered to be a wide channel,

$$r_{H,1} \approx 5\text{ ft}$$

Manning's formula yields

$$\begin{aligned} Q_1 &= \left(\frac{1.49}{n_1}\right) A_1 (r_{H,1})^{2/3}\sqrt{S} \\ &= \left(\frac{1.49}{0.05}\right)(1250\text{ ft}^2)(5\text{ ft})^{2/3}\sqrt{0.0004} \\ &= 2178\text{ ft}^3/\text{sec} \end{aligned}$$

For section 2,

$$A_2 = wd \approx (100\text{ ft})(10\text{ ft}) = 1000\text{ ft}^2$$

From the figure, the wetted perimeter is

$$p_2 = 5\text{ ft} + 100\text{ ft} + 9\text{ ft} = 114\text{ ft}$$

The hydraulic radius is

$$\begin{aligned} r_{H,2} &= \frac{A_2}{p_2} = \frac{1000\text{ ft}^2}{114\text{ ft}} \\ &= 8.8\text{ ft} \end{aligned}$$

Manning's formula for this section is

$$\begin{aligned} Q_2 &= \left(\frac{1.49}{n_2}\right) A_2 (r_{H,2})^{2/3}\sqrt{S} \\ &= \left(\frac{1.49}{0.014}\right)(1000\text{ ft}^2)(8.8\text{ ft})^{2/3}\sqrt{0.0004} \\ &= 9073\text{ ft}^3/\text{sec} \end{aligned}$$

For section 3,

$$A_3 = wd \approx (150\text{ ft})(1\text{ ft}) = 150\text{ ft}^2$$

This section can also be considered wide, so

$$r_{H,3} \simeq 1\text{ ft}$$

The flow is given by Manning's formula.

$$\begin{aligned} Q_3 &= \left(\frac{1.49}{n_3}\right) A_3 (r_{H,3})^{2/3}\sqrt{S} \\ &= \left(\frac{1.49}{0.013}\right)(150\text{ ft}^2)(1\text{ ft})^{2/3}\sqrt{0.0004} \\ &= 344\text{ ft}^3/\text{sec} \end{aligned}$$

The total flow is

$$\begin{aligned} Q &= Q_1 + Q_2 + Q_3 \\ &= 2178\,\frac{\text{ft}^3}{\text{sec}} + 9073\,\frac{\text{ft}^3}{\text{sec}} + 344\,\frac{\text{ft}^3}{\text{sec}} \\ &= \boxed{11{,}595\text{ ft}^3/\text{sec}} \end{aligned}$$

(b) The average velocity is

$$\begin{aligned} \text{v} &= \frac{Q}{A} = \frac{11{,}595\,\frac{\text{ft}^3}{\text{sec}}}{1250\text{ ft}^2 + 1000\text{ ft}^2 + 150\text{ ft}^2} \\ &= \boxed{4.8\text{ ft/sec}} \end{aligned}$$

2.

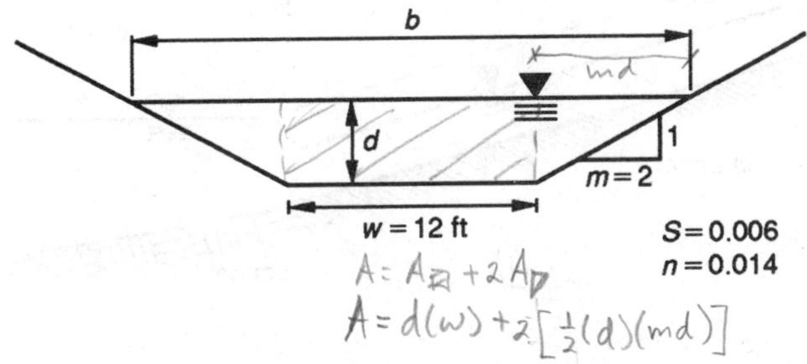

A = A + 2 A
A = d(w) + 2[½(d)(md)]
A = dw + d²m

Use the following formulas for geometric properties of a trapezoidal channel. (m is the horizontal to vertical side slope.)

$$\begin{aligned} \text{area:} \quad & A = d(w + md) \\ \text{wetted perimeter:} \quad & p = w + 2d\sqrt{1+m^2} \\ \text{top width:} \quad & b = w + 2md \\ \text{hydraulic radius:} \quad & r_H = \frac{A}{p} \end{aligned}$$

(a) Manning's formula holds for uniform flow—that is, when d equals the normal depth, d_o.

$$Q = \left(\frac{1.49}{n}\right) A (r_H)^{2/3} \sqrt{S}$$

$$m = \frac{2}{1} = 2$$

Replacing the known values yields

$$2300 \,\frac{\text{ft}^3}{\text{sec}} = \left(\frac{1.49}{0.014}\right) d_o(12\text{ ft} + 2d_o) \times \left[\frac{d_o(12\text{ ft} + 2d_o)}{12\text{ ft} + 2\sqrt{5}d_o}\right]^{2/3} \sqrt{0.006}$$

Solving by trial and error for d_o yields

$$d_o = \boxed{5.4\text{ ft}}$$

(b) The condition for critical flow is

$$\frac{Q^2}{g} = \frac{A^3}{b}$$

Replacing the known quantities yields

$$\frac{\left(2300 \,\frac{\text{ft}^3}{\text{sec}}\right)^2}{32.2 \,\frac{\text{ft}}{\text{sec}^2}} = \frac{[d_c(12\text{ ft} + 2d_c)]^3}{12\text{ ft} + (2)(2)d_c}$$

d_c is the critical depth. Solving for d_c by trial and error yields

$$d_c = \boxed{7.2\text{ ft}}$$

(c) The flow is supercritical since $d_c > d_o$.

(d) The flow profile at the junction is shown in the figure.

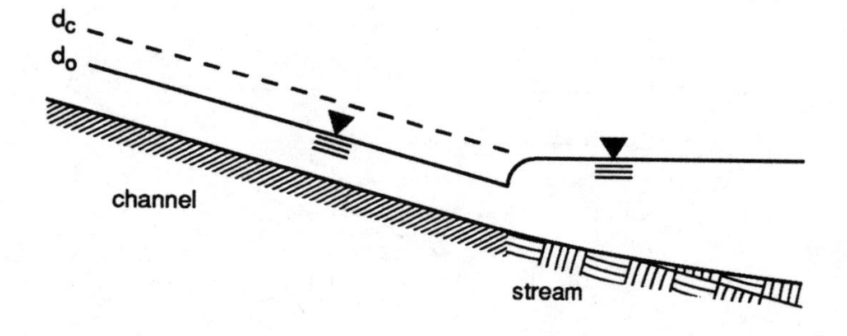

Since the flow is supercritical, the control section must be somewhere upstream, so the stream does not control the flow. A hydraulic jump takes place below the junction.

3. (a) The most efficient cross section is one that has the minimum wetted perimeter for a given flow area. The most efficient of all sections is a semicircular channel.

(b) The most efficient trapezoidal section is half a hexagon.

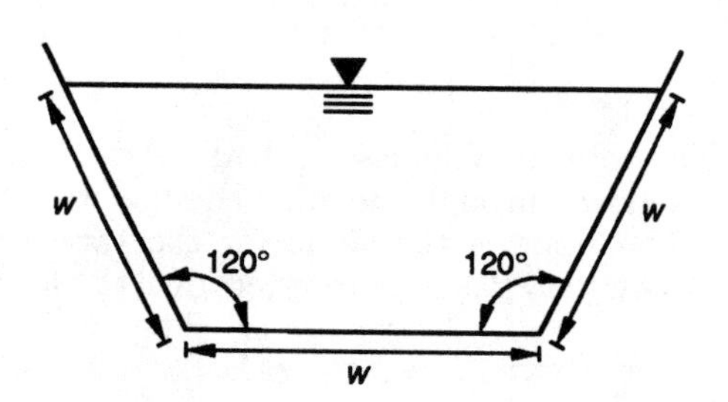

(c) A channel section is controlled upstream if the discharge is given by the flow depth at the upper end of the section, regardless of downstream features.

(d) A channel section is controlled downstream if the discharge is given by the flow depth at the lower end of the section.

4. Assume that the pipe operates as an open channel while flowing full.

(a) For a circular pipe with diameter $D = 22$ in,

$$D = \frac{22\text{ in}}{12\,\frac{\text{in}}{\text{ft}}} = 1.8333\text{ ft}$$

$$A = \tfrac{1}{4}\pi D^2 = \tfrac{1}{4}\pi(1.8333\text{ ft})^2 = 2.64\text{ ft}^2$$

$$r_H = \tfrac{1}{4}D = \left(\tfrac{1}{4}\right)(1.8333\text{ ft}) = 0.46\text{ ft}$$

If the pipe is long enough to allow development of uniform flow, Manning's formula is applicable.

$$\begin{aligned} Q &= \left(\frac{1.49}{n}\right) A(r_H)^{2/3}\sqrt{S} \\ &= \left(\frac{1.49}{0.01}\right)(2.64\ \text{ft}^2)(0.46\ \text{ft})^{2/3}\sqrt{0.02} \\ &= \boxed{33\ \text{ft}^3/\text{sec}} \end{aligned}$$

(b) To meet the 35 ft^3/sec demand, an additional PVC pipe should be installed to carry 2 ft^3/sec. The diameter of a circular pipe flowing full is

$$\begin{aligned} D &= (1.33)\left(\frac{nQ}{\sqrt{S}}\right)^{3/8} \\ &= (1.33)\left[\frac{(0.01)\left(2\ \frac{\text{ft}^3}{\text{sec}}\right)}{\sqrt{0.02}}\right]^{3/8} \\ &= \boxed{0.64\ \text{ft}\ (7.7\ \text{in})} \end{aligned}$$

The excess water can be carried by a pipe with a nominal diameter.

$$D = \boxed{8\ \text{in}}$$

(c)

$$\begin{aligned} D &= (1.33)\left(\frac{nQ}{\sqrt{S}}\right)^{3/8} \\ &= (1.33)\left[\frac{(0.01)\left(2\ \frac{\text{ft}^3}{\text{sec}}\right)}{\sqrt{0.025}}\right]^{3/8} \\ &= \boxed{0.61\ \text{ft}\ (7.3\ \text{in})} \end{aligned}$$

The new route will also require a nominal pipe size.

$$D = \boxed{8\ \text{in}}$$

(d) If 4257 ft is the upstream invert elevation, the downstream invert will be at elevation

$$4257\ \text{ft} - (0.025)(1000\ \text{ft}) = \boxed{4232\ \text{ft}}$$

5. The capacity of the section can be calculated assuming uniform flow.

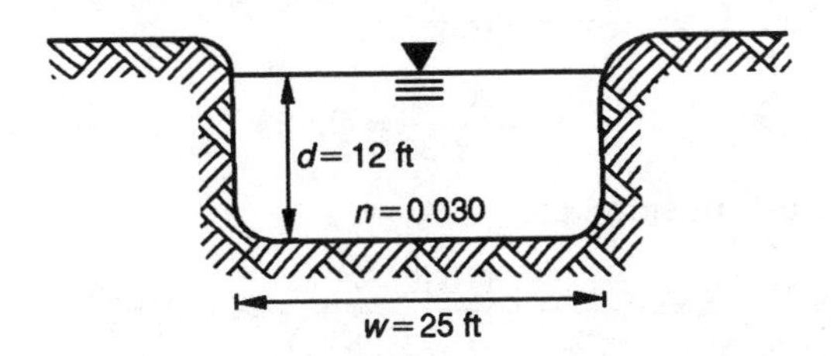

(a) The bottom slope is

$$S = \frac{470\ \text{ft}}{(8.4\ \text{mi})\left(5280\ \frac{\text{ft}}{\text{mi}}\right)} = 0.0106$$

The geometric properties are

$$\begin{aligned} A &= wd = (25\ \text{ft})(12\ \text{ft}) = 300\ \text{ft}^2 \\ p &= w + 2d \\ &= 25\ \text{ft} + (2)(12\ \text{ft}) = 49\ \text{ft} \\ r_H &= \frac{A}{p} = \frac{300\ \text{ft}^2}{49\ \text{ft}} \\ &= 6.122\ \text{ft} \end{aligned}$$

Manning's formula gives the capacity of the natural channel.

$$\begin{aligned} Q &= \left(\frac{1.49}{n}\right) A(r_H)^{2/3}\sqrt{S} \\ &= \left(\frac{1.49}{0.03}\right)(300\ \text{ft}^2)(6.122\ \text{ft})^{2/3}\sqrt{0.0106} \\ &= 5134\ \text{ft}^3/\text{sec} \end{aligned}$$

The nature of the flow regime is determined from the critical depth. For a rectangular channel,

$$\begin{aligned} d_c &= \left(\frac{Q^2}{gw^2}\right)^{1/3} \\ &= \left[\frac{\left(5134\ \frac{\text{ft}^3}{\text{sec}}\right)^2}{\left(32.2\ \frac{\text{ft}}{\text{sec}^2}\right)(25\ \text{ft})^2}\right]^{1/3} \\ &= 10.9\ \text{ft} \end{aligned}$$

Since $d > d_c$, the flow is subcritical. The artificial channel will have a smaller roughness coefficient because the effects of meandering and brush growth have been eliminated. Since the new length is 4.7 mi, the ratio of meander length to straight length is 8.4 mi/4.7 mi = 1.8, which indicates a severe degree of meandering. The roughness coefficient for the new channel (see Chow, V.T. *Open Channel Hydraulics*, McGraw-Hill, 1959, pp. 108–109) is approximately

8.4mi = .03

$$n = \frac{0.03}{1.3} = 0.023$$

The bottom slope is

$$S = \frac{470 \text{ ft}}{(4.7 \text{ mi})\left(5280 \frac{\text{ft}}{\text{mi}}\right)} = 0.0189$$

The following figure gives data for the artificial channel.

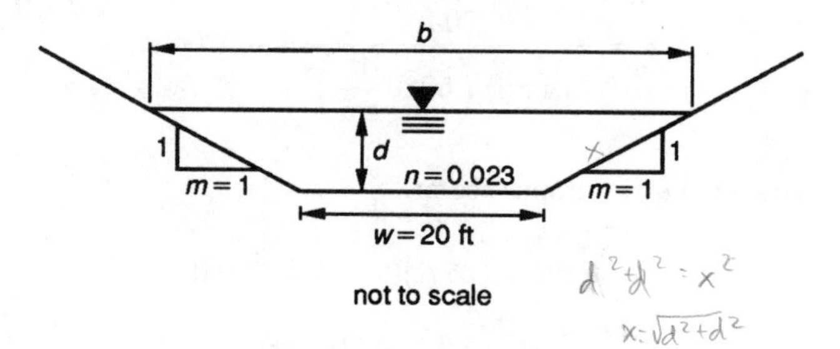

$d^2+d^2=x^2$
$x=\sqrt{d^2+d^2}$

The geometric properties of the section are

$$A = d(w+md)$$
$$p = w + 2d\sqrt{1+m^2}$$
$$b = w + 2md$$
$$r_H = \frac{A}{p}$$
$$m = \frac{1}{1} = 1$$

In order to determine if the channel will be able to carry the maximum flow of 5134 ft^3/sec, the hydraulic profile must be defined first. The normal depth, d_o, is calculated using Manning's formula.

$$Q = \left(\frac{1.49}{n}\right) A (r_H)^{2/3}\sqrt{S}$$

$$5134 \frac{\text{ft}^3}{\text{sec}} = \left(\frac{1.49}{0.023}\right) d_o(20 \text{ ft} + d_o) \times \left[\frac{d_o(20 \text{ ft} + (1)d_o)}{20 \text{ ft} + 2\sqrt{2}d_o)}\right]^{2/3} \sqrt{0.0189}$$

Solving by trial and error for d_o yields

$$d_o = 7.3 \text{ ft}$$

The critical depth, d_c, is calculated from

$$\frac{Q^2}{g} = \frac{A^3}{b}$$

$$\frac{\left(5134 \frac{\text{ft}^3}{\text{sec}}\right)^2}{32.2 \frac{\text{ft}}{\text{sec}^2}} = \frac{[d_c(20 \text{ ft} + d_c)]^3}{20 \text{ ft} + (2)(1)d_c}$$

Solving by trial and error for d_c yields

$$d_c = 10.6 \text{ ft}$$

The artificial channel has a steep slope. The likely flow profile is shown in the figure.

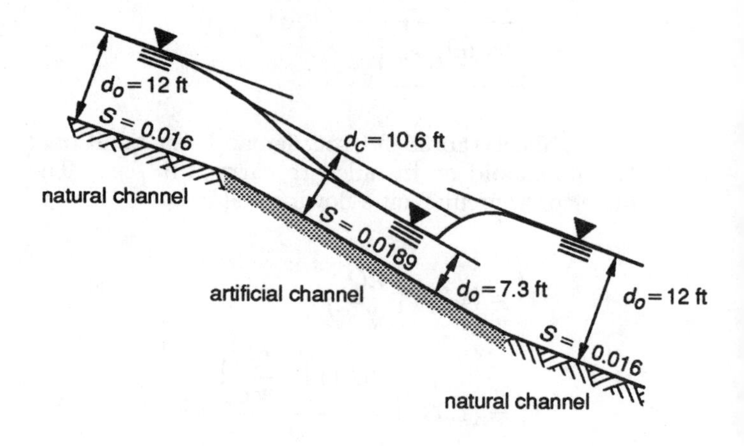

Since the flow depths in the artificial channel are within the range of natural depths, the new channel will be able to carry the maximum flow.

(b) The erosion potential of the flow can be asessed from the mean velocity.

For the natural channel,

$$\text{v} = \frac{Q}{A} = \frac{5134 \frac{\text{ft}^3}{\text{sec}}}{300 \text{ ft}^2} = 17.1 \text{ ft/sec}$$

In the artificial channel, the maximum velocity occurs when the flow becomes uniform. At this point,

$$A = d_o(w + md_o)$$
$$= (7.3 \text{ ft})[20 \text{ ft} + (1)(7.3 \text{ ft})]$$
$$= 199.3 \text{ ft}^2$$

$$\text{v} = \frac{Q}{A} = \frac{5134 \frac{\text{ft}^3}{\text{sec}}}{199.3 \text{ ft}^2} = 25.8 \text{ ft/sec}$$

Erosion problems are probably more severe in the artificial channel. However, both channels are well above maximum permissible velocities for the bed material (see Table 5.6 in the *Civil Engineering Reference Manual*).

6. (a) In order to guarantee the minimum freeboard, the configuration shown must hold for the minimum flow.

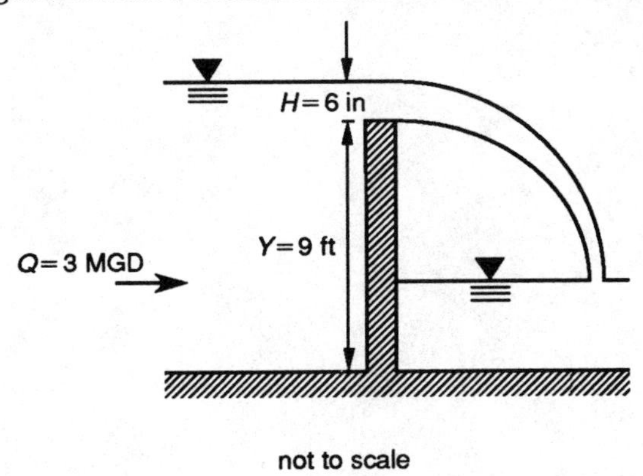

The flow area is

$$A = (22\text{ ft})(9.5\text{ ft}) = 209\text{ ft}^2$$

For a flow of 3 MGD = 4.6 ft^3/sec, the approach velocity is

$$v_1 = \frac{Q}{A} = \frac{4.6\,\frac{\text{ft}^3}{\text{sec}}}{209\text{ ft}^2} = 0.022\text{ ft/sec}$$

Since the velocity head is negligible, the following formula can be used.

$$Q = \frac{2}{3}C_1(b - 0.2H)\sqrt{2g}H^{3/2}$$

The coefficient C_1 is given by

$$\begin{aligned} C_1 &= \left(0.6035 + 0.0813\frac{H}{Y} + \frac{0.000295}{Y}\right) \\ &\quad\times\left(1+\frac{0.00361}{H}\right)^{3/2} \\ &= \left[0.6035 + (0.0813)\left(\frac{0.5\text{ ft}}{9\text{ ft}}\right) + \frac{0.000295}{9\text{ ft}}\right] \\ &\quad\times\left(1+\frac{0.00361}{0.5\text{ ft}}\right)^{3/2} \\ &= 0.615 \end{aligned}$$

Substituting this and the other numerical values in the discharge equation yields

$$4.6\,\frac{\text{ft}^3}{\text{sec}} = \left(\frac{2}{3}\right)(0.615)[b - (0.2)(0.5\text{ ft})] \times\sqrt{(2)\left(32.2\,\frac{\text{ft}}{\text{sec}^2}\right)}(0.5\text{ ft})^{3/2}$$

Solving for b results in

$$b = \boxed{4.1\text{ ft}}$$

(b) For the maximum flow at 12 MGD = 18.6 ft^3/sec, the approach velocity is

$$v_1 = \frac{18.6\,\frac{\text{ft}^3}{\text{sec}}}{209\text{ ft}^2} = 0.09\text{ ft/sec}$$

This is also negligible, so the following equation is valid and must be solved iteratively for the unknown H.

$$Q = \frac{2}{3}C_1(b - 0.2H)\sqrt{2g}H^{3/2}$$

As a first estimate, assume H = 16 in (1.33 ft). The coefficient C_1 is

$$\begin{aligned} C_1 &= \left(0.6035 + 0.0813\frac{H}{Y} + \frac{0.000295}{Y}\right) \\ &\quad\times\left(1+\frac{0.00361}{H}\right)^{3/2} \\ &= \left[0.6035 + (0.0813)\left(\frac{1.33\text{ ft}}{9\text{ ft}}\right) + \frac{0.000295}{9\text{ ft}}\right] \\ &\quad\times\left(1+\frac{0.00361}{1.33\text{ ft}}\right)^{3/2} \\ &= 0.618 \end{aligned}$$

A second estimate for the freeboard can be obtained by solving for H in the discharge equation. (The calculation is not very sensitive to the correction for contraction. Therefore, the previous estimate of H is used for computational convenience.)

$$18.6\,\frac{\text{ft}^3}{\text{sec}} = \left(\frac{2}{3}\right)(0.618)[4.1\text{ ft} - (0.2)(1.33\text{ ft})] \times\sqrt{(2)\left(32.2\,\frac{\text{ft}}{\text{sec}^2}\right)}H^{3/2}$$

This yields

$$H = 1.29\text{ ft}$$

With this new value, C_1 remains unchanged. An improved estimate of H comes from solving

$$18.6\,\frac{\text{ft}^3}{\text{sec}} = \left(\frac{2}{3}\right)(0.618)[4.1\text{ ft} - (0.2)(1.29\text{ ft})] \times\sqrt{(2)\left(32.2\,\frac{\text{ft}}{\text{sec}^2}\right)}H^{3/2}$$

This yields the same value of

$$H = \boxed{1.29\text{ ft (15.5 in)}}$$

(c) The usual rule is to measure the approach depth at a distance from the crest greater than $4H$ (see Daugherty, R.L., J.B. Franzini, and E.J. Finnemore. *Fluid Mechanics with Engineering Applications*, 8th ed., McGraw-Hill, 1985, p. 479).

The gauge station should be placed at least 6 ft upstream from the weir.

HYDROLOGY

1. (a) For a small drawdown compared with the initial thickness of the aquifer, the equation governing unsteady flow toward a fully penetrating well in an unconfined aquifer (see Bras, R. *Hydrology*, Addison-Wesley, 1990, p. 319) is

$$Y - y = \left(\frac{Q}{4\pi K_p Y}\right) W(u)$$

$W(u)$ is the well function and

$$u = \frac{r^2 S_y}{4K_p Y t}$$

Q is the flow rate, S_y is the yield, K_p is the hydraulic conductivity, t is the time, and the other variables are depicted in the figure shown.

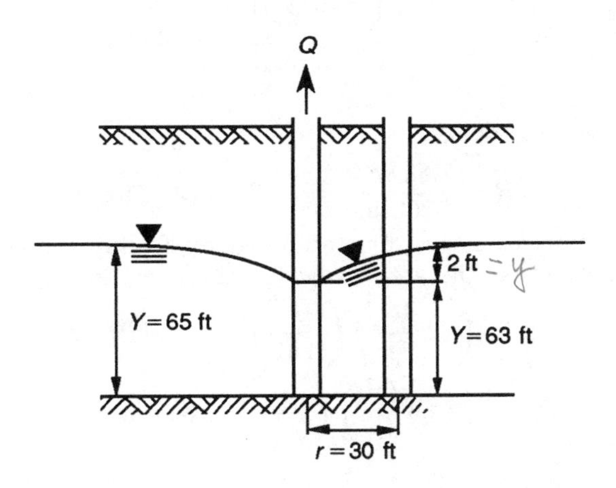

The pumping schedule is

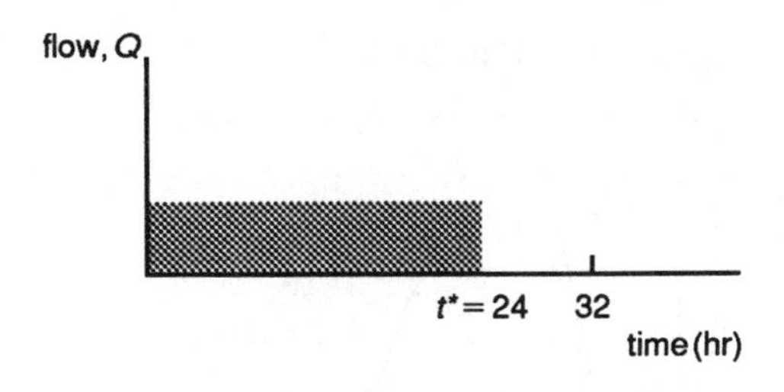

The principle of superposition indicates that the position of the water table is given by

$$Y - y = \left(\frac{Q}{4\pi K_p Y}\right)[W(u_1) - W(u_2)]$$

$$u_1 = \frac{r^2 S_y}{4K_p Y t}$$

$$u_2 = \frac{r^2 S_y}{4K_p Y (t - t^*)}$$

For $u < 0.1$, the well function is approximately

$$W(u) \simeq 0.5772 - \ln(u)$$

The equation becomes

$$Y - y = \left(\frac{Q}{4\pi K_p Y}\right) \ln\left(\frac{t}{t - t^*}\right)$$

The hydraulic conductivity is

$$K_p = \frac{120{,}000\ \frac{\text{gal}}{\text{ft-day}}}{65\ \text{ft}} = 1846\ \frac{\text{gal}}{\text{ft}^2\text{-day}}$$

For coarse sand, assume the yield is 85% of the porosity.

$$S_y = (0.85)(0.30) = 0.26$$

porosity

$$u_1 = \frac{(30\ \text{ft})^2(0.26)}{(4)\left(10.3\ \frac{\text{ft}}{\text{hr}}\right)(65\ \text{ft})(32\ \text{hr})} = 0.0027$$

$$u_2 = \frac{(30\ \text{ft})^2(0.26)}{(4)\left(10.3\ \frac{\text{ft}}{\text{hr}}\right)(65\ \text{ft})(32\ \text{hr} - 24\ \text{hr})} = 0.011$$

Since both u_1 and u_2 are less than 0.1, the simplified equation can be used.

$$65\ \text{ft} - 63\ \text{ft} = \left[\frac{Q}{(4\pi)\left(1846\ \frac{\text{gal}}{\text{ft}^2\text{-day}}\right)(65\ \text{ft})}\right] \times \ln\left(\frac{32\ \text{hr}}{32\ \text{hr} - 24\ \text{hr}}\right)$$

Solving for the flow rate,

$$Q = \boxed{2.175 \times 10^6\ \text{gal/day}\ (1511\ \text{gal/min})}$$

(b) For steady flow from a fully penetrating well into an unconfined aquifer, the following expression is valid at the production well (see Dawson, K. and J.D. Istok, *Aquifer Testing*, 1991, Lewis, p. 33).

$$\frac{Q}{d_o} = \frac{T}{2000}$$

Q is in gal/min, d_o is in ft, and T is in gal/ft-day.

$$d_o = \frac{(2000)\left(800\ \frac{\text{gal}}{\text{min}}\right)}{120{,}000\ \frac{\text{gal}}{\text{ft-day}}} = \boxed{13.3\ \text{ft}}$$

2. (a) Substituting the known data in the hydrograph equation yields

$$Q = (0.0132)(200\ \text{ac})^{0.9}(3000\ \text{ft})^{0.3}(0.35)^{1.1}(0.25)^{0.08}$$
$$= 4.8\ \text{ft}^3/\text{sec}$$
$$T_d = (0.44)(200\ \text{ac})^{1.1}(3000\ \text{ft})^{0.04}(0.35)^{1.2}(0.25)^{0.17}$$
$$= 46.2\ \text{hr}$$
$$T_p = (0.0024)(200\ \text{ac})^{1.7}(3000\ \text{ft})^{0.04}(0.35)^{0.7}(0.25)^{0.14}$$
$$= 10.7\ \text{hr}$$
$$T_{50} = (0.0367)(200\ \text{ac})^{1.3}(3000\ \text{ft})^{0.01}(0.35)^{0.7}(0.25)^{0.12}$$
$$= 15.8\ \text{hr}$$
$$T_{75} = (0.0671)(200\ \text{ac})^{1.2}(3000\ \text{ft})^{0.05}(0.35)^{0.7}(0.25)^{0.11}$$
$$= 23.8\ \text{hr}$$

The hydrograph variables are shown in the figure.

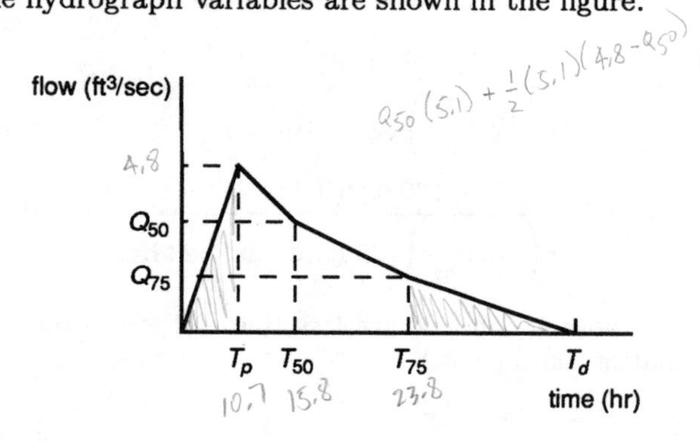

The total runoff, R, is equal to the area below the hydrograph.

$$R = \left(\frac{1}{2}\right)(10.7\ \text{hr})\left(4.8\ \frac{\text{ft}^3}{\text{sec}}\right)$$
$$+ \left(\frac{1}{2}\right)\left(4.8\ \frac{\text{ft}^3}{\text{sec}} + Q_{50}\right)(15.8\ \text{hr} - 10.7\ \text{hr})$$
$$+ \left(\frac{1}{2}\right)(Q_{50} + Q_{75})(23.8\ \text{hr} - 15.8\ \text{hr})$$
$$+ \frac{1}{2}Q_{75}(46.2\ \text{hr} - 23.8\ \text{hr})$$
$$= 37.9 + 6.55Q_{50} + 15.2Q_{75}$$

According to the definitions given, the area up to T_{50} must be 50% of the total runoff, and the area up to T_{75} must be 75% of the total runoff.

$$0.5R = \left(\frac{1}{2}\right)(10.7\ \text{hr})\left(4.8\ \frac{\text{ft}^3}{\text{sec}}\right)$$
$$+ \left(\frac{1}{2}\right)\left(4.8\ \frac{\text{ft}^3}{\text{sec}} + Q_{50}\right)(15.8\ \text{hr} - 10.7\ \text{hr})$$
$$0.75R = 0.5R + \left(\frac{1}{2}\right)(Q_{50} + Q_{75})(23.8\ \text{hr} - 15.8\ \text{hr})$$

Substituting R into these last two equations results in a system of two equations.

$$1.45Q_{50} + 15.2Q_{75} = 37.94\ \text{ft}^3\text{-hr/sec}$$
$$9.445Q_{50} + 0.8Q_{75} = 37.9\ \text{ft}^3\text{-hr/sec}$$

Solving for Q_{50} and Q_{75} yields

$$Q_{50} = 3.83\ \text{ft}^3/\text{sec}$$
$$Q_{75} = 2.13\ \text{ft}^3/\text{sec}$$

Substituting these values in the expression for the runoff volume,

$$R = 37.9\ \frac{\text{ft}^3\text{-hr}}{\text{sec}} + 6.55Q_{50} + 15.2Q_{75}$$
$$= 37.9\ \frac{\text{ft}^3\text{-hr}}{\text{sec}} + (6.55)\left(3.83\ \frac{\text{ft}^3}{\text{sec}}\right) + (15.2)\left(2.13\ \frac{\text{ft}^3}{\text{sec}}\right)$$
$$= 95.36\ \frac{\text{ft}^3\text{-hr}}{\text{sec}}$$
$$R = \frac{\left(95.36\ \frac{\text{ft}^3\text{-hr}}{\text{sec}}\right)\left(3600\ \frac{\text{sec}}{\text{hr}}\right)}{43{,}560\ \frac{\text{ft}^2}{\text{ac}}} = 7.88\ \text{ac-ft}$$

In terms of depth,

$$R = \frac{7.88\ \text{ac-ft}}{200\ \text{ac}} = 0.039\ \text{ft}\ (0.47\ \text{in})$$

The unit hydrograph is obtained by scaling the ordinates so that the area under the curve is 1 in.

$$Q = \left(4.8\ \frac{\text{ft}^3}{\text{sec}}\right)\left(\frac{1}{0.47}\right) = 10.2\ \text{ft}^3/\text{sec}$$
$$Q_{50} = \left(3.83\ \frac{\text{ft}^3}{\text{sec}}\right)\left(\frac{1}{0.47}\right) = 8.15\ \text{ft}^3/\text{sec}$$
$$Q_{75} = \left(2.13\ \frac{\text{ft}^3}{\text{sec}}\right)\left(\frac{1}{0.47}\right) = 4.53\ \text{ft}^3/\text{sec}$$

The resulting unit hydrograph is shown in the figure.

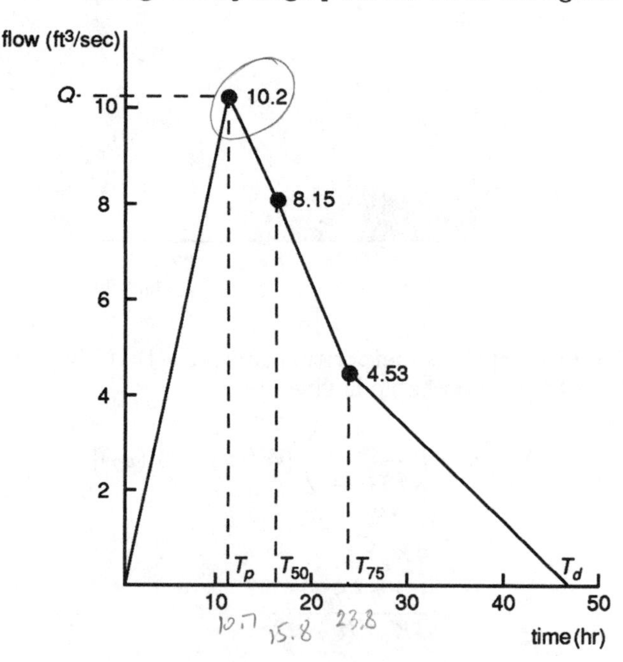

(b) The hydrograph for this storm is

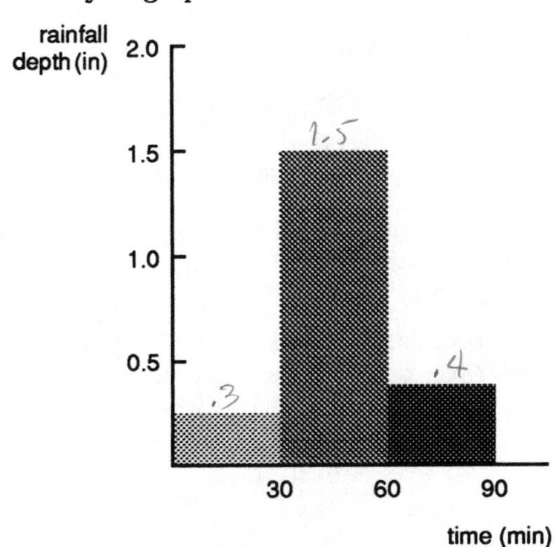

The duration of the storm is relatively short compared with the base of the unit hydrograph. Also, most of the precipitation occurs between 30 min and 60 min. This storm can be approximated by

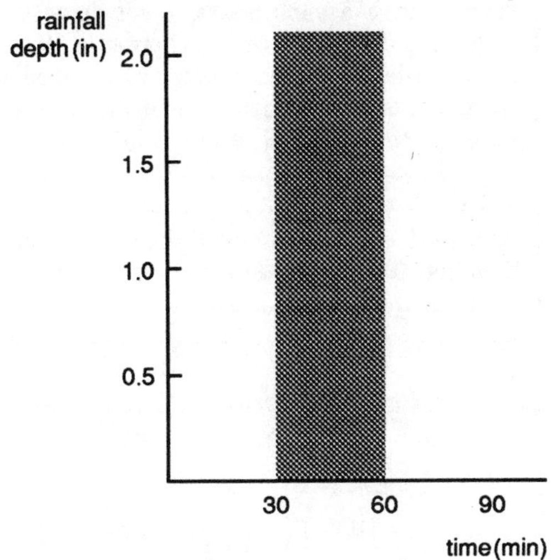

The total depth is equal to

$$0.3\text{ in} + 1.5\text{ in} + 0.4\text{ in} = 2.2\text{ in}$$

With this assumption, the peak will be

$$Q = (2.2)\left(10.2\ \frac{\text{ft}^3}{\text{sec}}\right) = \boxed{22.4\ \text{ft}^3/\text{sec}}$$

The characteristic times will be those of the unit hydrograph displaced by 30 min (0.5 hr).

$$T_d = 46.2\text{ hr} + 0.5\text{ hr} = 46.7\text{ hr}$$
$$T_p = 10.7\text{ hr} + 0.5\text{ hr} = 11.2\text{ hr}$$
$$T_{50} = 15.8\text{ hr} + 0.5\text{ hr} = 16.3\text{ hr}$$
$$T_{75} = 23.8\text{ hr} + 0.5\text{ hr} = 24.3\text{ hr}$$

3. The time of concentration is taken as the travel time along a diagonal toward the inlet. This distance is equal to $(200\text{ ft})\sqrt{2} = 283$ ft.

(a) For area A, the nomograph yields $t_c = 9$ min.

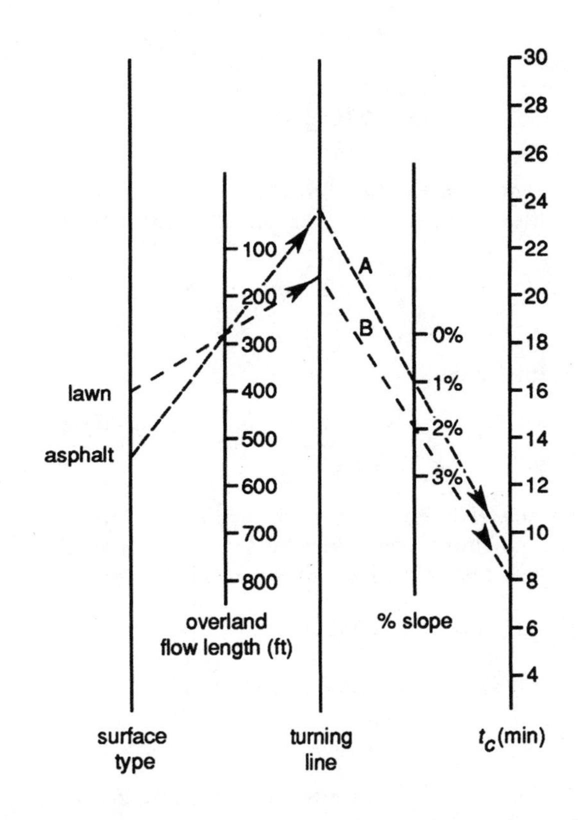

For a storm duration of 9 min, the 10-year IDF curve indicates an intensity of 2.3 in/hr.

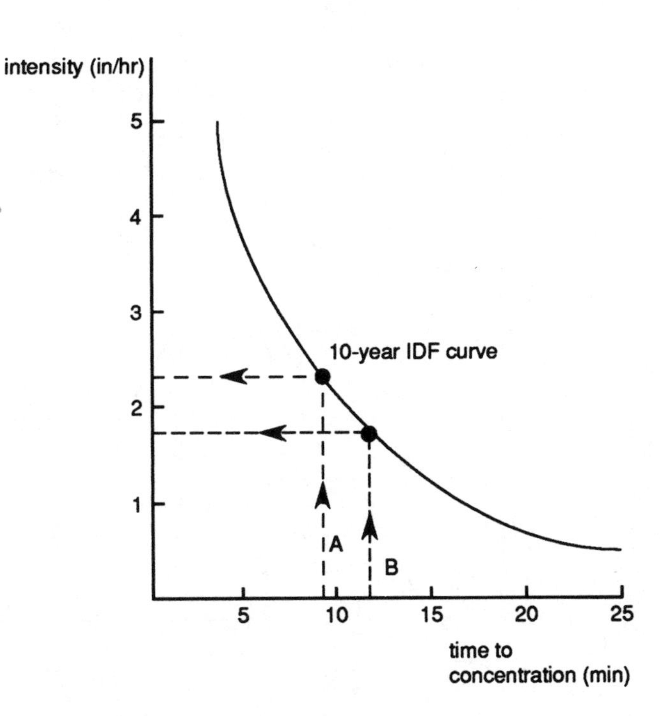

The surface areas of A and B are 3.7 ac. The runoff coefficient for asphalt area A is $C = 0.8$. The peak flow in pipe 1 is

$$\begin{aligned} Q_p &= CIA_d \\ &= (0.8)\left(2.3\ \frac{\text{in}}{\text{hr}}\right)(3.7\ \text{ac}) \\ &= \boxed{6.8\ \text{ft}^3/\text{sec}} \end{aligned}$$

(b) For area B, the nomograph yields a time of concentration (inlet time) of 8 min. The travel time in pipe 1 is

$$\frac{400\ \text{ft}}{2\ \frac{\text{ft}}{\text{sec}}} = 200\ \text{sec}\ (3.3\ \text{min})$$

The time of concentration for area B is

$$\begin{aligned} t_{\text{C,B}} &= \max\ \{9\ \text{min} + 3.3\ \text{min}, 8\ \text{min}\} \\ &= 12.3\ \text{min} \end{aligned}$$

From the IDF curve, the intensity is 1.7 in/hr. The estimated runoff coefficient for lawn area B is 0.07. The peak flow into pipe 2 produced by this storm is

$$\begin{aligned} Q_p &= C_{\text{A}}IA_{d,\text{A}} + C_{\text{B}}IA_{d,\text{B}} \\ &= (0.8)\left(1.7\ \frac{\text{in}}{\text{hr}}\right)(3.7\ \text{ac}) + (0.07)\left(1.7\ \frac{\text{in}}{\text{hr}}\right)(3.7\ \text{ac}) \\ &= 5.5\ \text{ft}^3/\text{sec} \end{aligned}$$

Since this flow is less than the peak flow in pipe 1, the design storm for area A must be used for pipe 2 as well—that is, a storm of intensity 2.3 in/hr and duration of 9 min. The figure shows the hydrographs for this storm into pipe 2.

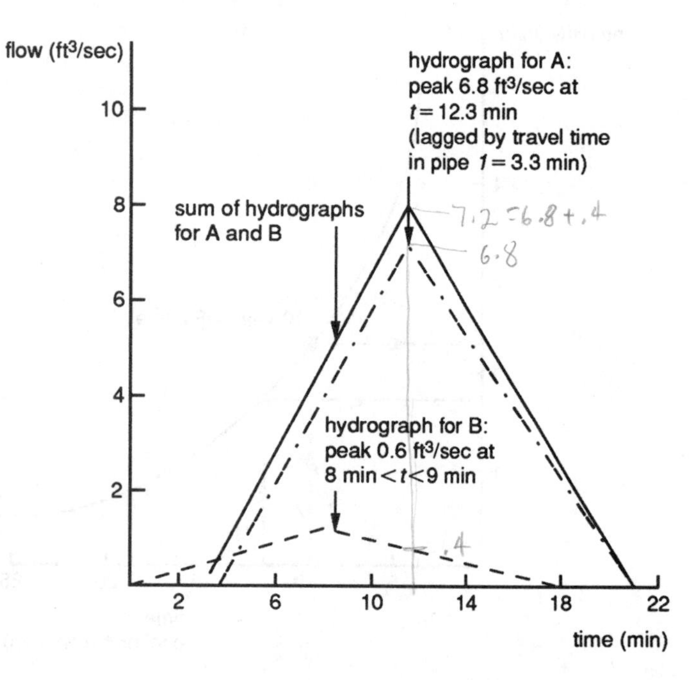

The peak flow caused by this storm in area B is

$$\begin{aligned} Q_{p,\text{B}} &= C_{\text{B}}IA_{d,\text{B}} \\ &= (0.07)\left(2.3\ \frac{\text{in}}{\text{hr}}\right)(3.7\ \text{ac}) \\ &= 0.6\ \text{ft}^3/\text{sec} \end{aligned}$$

This remains constant in the hydrograph between the time of concentration for area B (8 min) and the end of the storm (9 min).

The hydrograph for area A is lagged 3.3 min corresponding to the travel time in pipe 1. Adding the two hydrographs in the plot produces the flow into pipe 2. The peak is

$$Q_p = \boxed{7.2\ \text{ft}^3/\text{sec}}$$

4. (a) The return period for culvert design depends mainly on land and road use in the surrounding area. Federal or state guidelines for highway design recommend the return period. For instance, in Massachusetts, a sensitive area would require a 50-year storm, whereas a 10-year storm is suitable for a local road in a rural area (see Massachusetts Department of Public Works, *Highway Design Manual*, 1989 Chap. 10).

(b) A runoff coefficient for light industrial areas is between 0.5 to 0.8. For this problem, an average value of 0.65 will be used.

(c) The runoff is given by the rational formula.

$$\begin{aligned} Q_p &= CIA_d \\ &= (0.65)\left(2\ \frac{\text{in}}{\text{hr}}\right)(175\ \text{ac}) \\ &= \boxed{228\ \text{ft}^3/\text{sec}} \end{aligned}$$

(d) For a corrugated metal pipe,

$$n = 0.022$$

The hydraulic radius for a 54-in culvert flowing full is

$$\begin{aligned} r_H &= \frac{D}{4} = \frac{54\ \text{in}}{4} \\ &= 13.5\ \text{in}\ (1.13\ \text{ft}) \end{aligned}$$

The discharge through one culvert flowing full is

$$\begin{aligned} Q_{\text{full}} &= \frac{1.49}{n} A r_H^{2/3} \sqrt{S} \\ &= \left(\frac{1.49}{0.022}\right)\left(\pi\left[\frac{(4.5\ \text{ft})^2}{4}\right]\right)(1.13\ \text{ft})^{2/3}\sqrt{0.0075} \\ &= 101.2\ \text{ft}^3/\text{sec} \end{aligned}$$

The maximum flow that can be carried by a circular pipe operating as an open channel is

$$Q_{\text{max}} = 1.08 Q_{\text{full}}$$
$$= (1.08)\left(101.2\ \frac{\text{ft}^3}{\text{sec}}\right)$$
$$= 109.3\ \text{ft}^3/\text{sec}$$

Since each culvert must carry $(228\ \text{ft}^3/\text{sec})/2 = 114\ \text{ft}^3/\text{sec}$, the pipe must be under pressure to meet this flow. This could occur as shown.

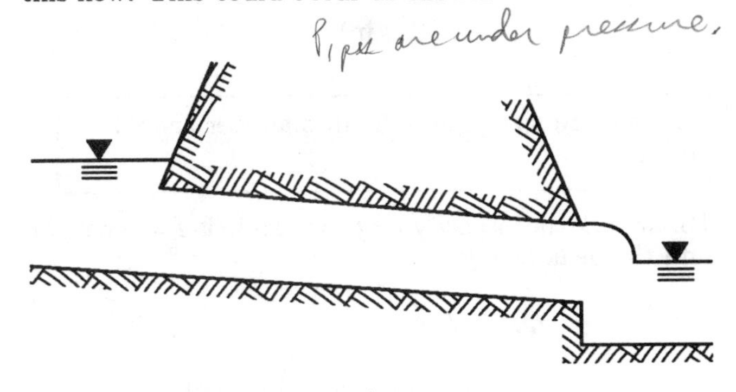

In this case, the culverts are controlled at the outlet.

5. Each storm duration has nine events as shown in the following tables.

10-min storm

rank	i (in/hr)
1	6.80
2	6.00
3	5.50
4	5.10
5	4.90
6	4.10
7	3.50
8	3.00
9	2.07

20-min storm

rank	i (in/hr)
1	4.90
2	4.40
3	4.07
4	3.85
5	3.55
6	3.05
7	2.60
8	2.15
9	1.60

30-min storm

rank	i (in/hr)
1	4.00
2	3.60
3	3.30
4	3.10
5	2.90
6	2.40
7	2.10
8	1.90
9	1.20

40-min storm

rank	i (in/hr)
1	3.40
2	3.10
3	2.90
4	2.60
5	2.30
6	2.00
7	1.75
8	1.50
9	1.05

60-min storm

rank	i (in/hr)
1	2.60
2	2.30
3	2.15
4	2.00
5	1.90
6	1.50
7	1.25
8	1.10
9	0.80

The probability of exceedance is calculated as

$$p = \frac{r}{N+1}$$

r is the rank when the storms are sorted from the highest intensity to the lowest, and N is the total number of events (nine in this case).

The return period is $T = 1/p$. For the 10-year IDF, $T = 10$ years, or $p = 0.1$. The curve is made up of all the events with rank

$$r = (0.1)(9+1) = 1$$

These storms are

duration (min)	i (in/hr)
10	6.80
20	4.90
30	4.00
40	3.40
60	2.60

The plot of the 10-year IDF curve is shown in the figure.

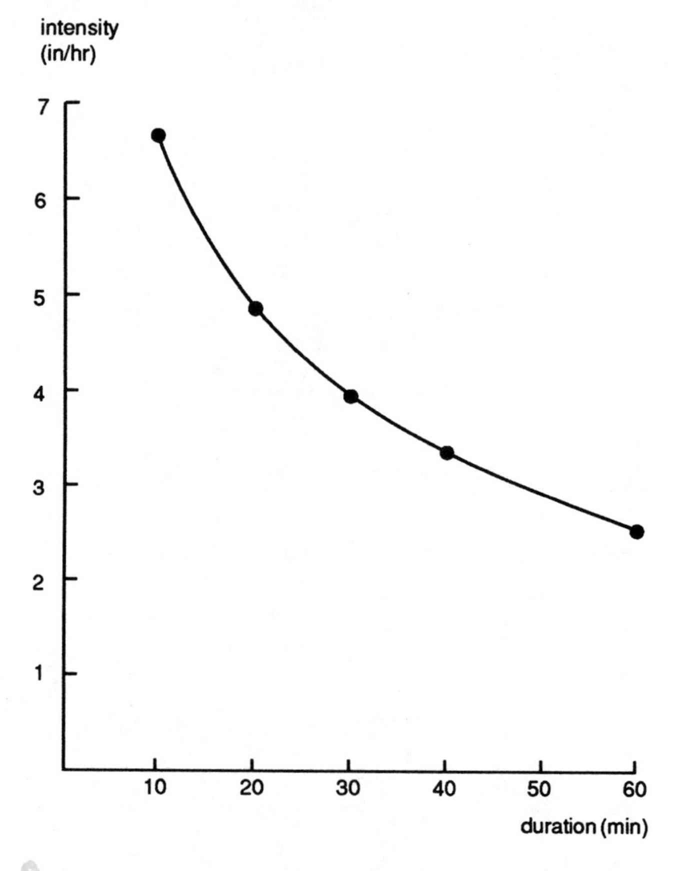

6. For area A, the 10-year IDF curve yields $I = 7.0$ in/hr for $t_c = 5$ min. The rational formula produces

$$\begin{aligned} Q_p &= CIA_d \\ &= (0.9)\left(7\,\frac{\text{in}}{\text{hr}}\right)(4\text{ ac}) \\ &= 25.2\ \text{ft}^3/\text{sec} \end{aligned}$$

Pipe 2 must carry 25.2 ft^3/sec while flowing full. Using the geometric properties of the full section,

$$r_H = \frac{D}{4}$$

$$A = \pi\left(\frac{D^2}{4}\right)$$

Manning's formula is used to size the pipe.

$$Q = Av = (A)\left(\frac{1.49}{n}\right)(r_H)^{2/3}\sqrt{S}$$

Replacing the geometric properties, this equation becomes

$$Q = \left(\frac{1.49}{n}\right)\pi\left(\frac{D^2}{4}\right)\left(\frac{D}{4}\right)^{2/3}\sqrt{S}$$

Solving for the diameter of the pipe,

$$D = (1.33)\left(\frac{nQ}{\sqrt{S}}\right)^{3/8}$$

Replacing the known values yields

$$\begin{aligned} D &= (1.33)\left[\frac{(0.012)\left(25.2\,\frac{\text{ft}^3}{\text{sec}}\right)}{\sqrt{0.035}}\right]^{3/8} \\ &= 1.59\text{ ft (19.1 in)} \end{aligned}$$

> A standard size pipe of 21-in diameter should be used.

For area B, the intensity for $t_c = 18$ min is $I = 3.8$ in/hr, and the peak flow is

$$\begin{aligned} Q_p &= CIA_d \\ &= (0.25)\left(3.8\,\frac{\text{in}}{\text{hr}}\right)(10\text{ ac}) \\ &= 9.5\ \text{ft}^3/\text{sec} \end{aligned}$$

The diameter is found from Manning's equation for a full pipe.

$$\begin{aligned} D &= (1.33)\left(\frac{nQ}{\sqrt{S}}\right)^{3/8} \\ &= (1.33)\left[\frac{(0.012)\left(9.5\,\frac{\text{ft}^3}{\text{sec}}\right)}{\sqrt{0.035}}\right]^{3/8} \\ &= 1.1\text{ ft (13 in)} \end{aligned}$$

> A standard size pipe of 15-in diameter should be used for pipe 1.

The flow through pipe 3 is given by the most severe of the following two storms.

Storm 1: duration = 5 min, $I = 7$ in/hr
Storm 2: duration = 18 min, $I = 3.8$ in/hr

The travel times in pipes 1 and 2 have been neglected since the pipes are short.

Storm 1:

After 5 min, all of area A responds to the storm, but only a portion of area B contributes to the flow. Assuming constant overland flow velocities, area B can be linearly divided into isochrones.

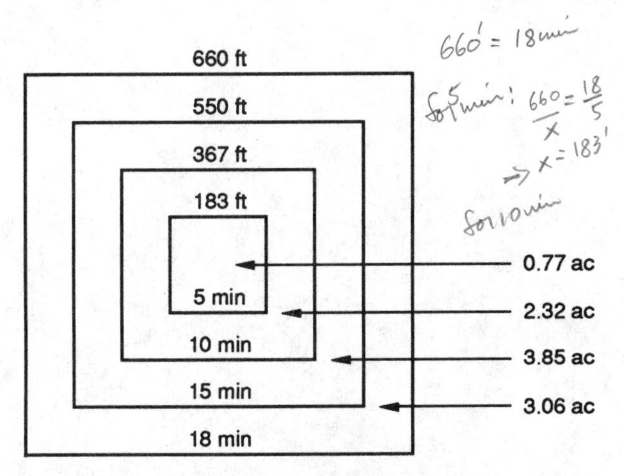

A square area of 10 ac has a dimension of 660 ft on the side. The isochrones are also square, and their dimensions are interpolated from the total concentration time. For instance, for the 5-min isochrone, the length of the side is

$$(660\ \text{ft})\left(\frac{5\ \text{min}}{18\ \text{min}}\right) = 183\ \text{ft}$$

At the end of the storm, only the 183-ft region will contribute to the flow. Consequently, the peak discharge during this storm will be

$$\begin{aligned} Q_p &= (0.25)\left(7\ \frac{\text{in}}{\text{hr}}\right)(0.77\ \text{ac}) + (0.9)\left(7\ \frac{\text{in}}{\text{hr}}\right)(4\ \text{ac}) \\ &= 26.5\ \text{ft}^3/\text{sec} \end{aligned}$$

Storm 2:

After 18 min, both areas A and B contribute fully to the flow in pipe 3. The peak discharge is

$$\begin{aligned} Q_p &= (0.25)\left(3.8\ \frac{\text{in}}{\text{hr}}\right)(10\ \text{ac}) + (0.9)\left(3.8\ \frac{\text{in}}{\text{hr}}\right)(4\ \text{ac}) \\ &= 23.2\ \text{ft}^3/\text{sec} \end{aligned}$$

The design flow is the greatest of the two peaks—that is, 26.5 ft^3/sec.

The diameter of pipe 3 is given by Manning's equation for a full pipe.

$$\begin{aligned} D &= (1.33)\left(\frac{nQ}{\sqrt{S}}\right)^{3/8} \\ &= (1.33)\left[\frac{(0.012)\left(26.5\ \frac{\text{ft}^3}{\text{sec}}\right)}{\sqrt{0.035}}\right]^{3/8} \\ &= 1.6\ \text{ft}\ (19.2\ \text{in}) \end{aligned}$$

A standard size pipe of 21-in diameter should be used for pipe 3.

7. From the IDF curve, the 10-year storm for the grassy area has an intensity of 5.8 in/hr. The peak flow is

$$\begin{aligned} (Q_p)_{\text{grass}} &= CIA_d \\ &= (0.3)\left(5.8\ \frac{\text{in}}{\text{hr}}\right)(20\ \text{ac}) \\ &= 34.8\ \text{ft}^3/\text{sec} \end{aligned}$$

This is the maximum flow handled by the creek. If the area is paved, the time of concentration decreases and the IDF curve yields an intensity of $I = 6.9$ in/hr. The new peak flow is

$$\begin{aligned} (Q_p)_{\text{paved}} &= CIA_d \\ &= (0.9)\left(6.9\ \frac{\text{in}}{\text{hr}}\right)(20\ \text{ac}) \\ &= 124.2\ \text{ft}^3/\text{sec} \end{aligned}$$

The approximate hydrograph is

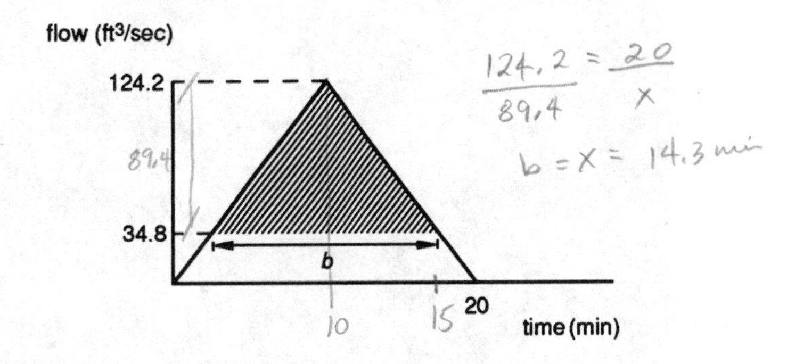

Since flows above 34.8 ft^3/sec are unacceptable, the detention basin must hold the volume in the shaded area. The base, b, of the shaded triangle is interpolated as

$$b = (20\ \text{min})\left(\frac{34.8\ \frac{\text{ft}^3}{\text{sec}}}{124.2\ \frac{\text{ft}^3}{\text{sec}}}\right) = 5.6\ \text{min}$$

The volume of the basin is

$$\begin{aligned} V &= \frac{\left(\frac{1}{2}\right)\left(124.2\ \frac{\text{ft}^3}{\text{sec}} - 34.8\ \frac{\text{ft}^3}{\text{sec}}\right)(5.6\ \text{min})\left(60\ \frac{\text{sec}}{\text{min}}\right)}{43{,}560\ \frac{\text{ft}^3}{\text{ac-ft}}} \\ &= \boxed{0.35\ \text{ac-ft}} \end{aligned}$$

WATER SUPPLY ENGINEERING

1. (a) The total dynamic head (TDH) can be found from the pump curve in the illustration using the flow given.

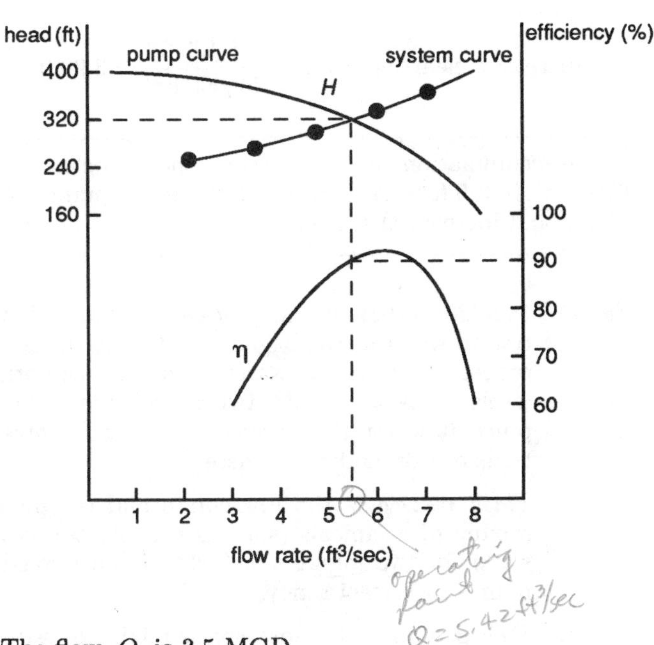

The flow, Q, is 3.5 MGD.

$$Q = (3.5\ \text{MGD})\left(1.55\ \frac{\frac{\text{ft}^3}{\text{sec}}}{\text{MGD}}\right) = 5.42\ \text{ft}^3/\text{sec}$$

TDH at 5.42 ft³/sec is $\boxed{320\ \text{ft}}$

(b) Static lift is the difference between the water elevation at the outlet (175 ft) and at the inlet (150 ft).

$$\text{static lift} = 175\ \text{ft} - 150\ \text{ft} = 25\ \text{ft}$$

$$\text{static exit pressure} = \left(80\ \frac{\text{lbf}}{\text{in}^2}\right)\left(2.31\ \frac{\text{ft}}{\frac{\text{lbf}}{\text{in}^2}}\right) = 184.8\ \text{ft}$$

$$\begin{aligned}\text{static head} &= \text{lift} + \text{exit pressure}\\ &= 25\ \text{ft} + 184.8\ \text{ft}\\ &= 209.9\ \text{ft}\end{aligned}$$

For this problem, minor losses are small and can be neglected.

$$\text{TDH} = \text{static head} + \text{friction head} + \text{minor losses}$$
$$320\ \text{ft} = 209.8\ \text{ft} + \text{friction head} + 0$$

The friction head, h_f, is the difference between the TDH and the static head.

$$h_f = 320\ \text{ft} - 209.8\ \text{ft} = 110.2\ \text{ft}$$

Use the Hazen-Williams equation to determine the pipe size, and calculate the system curve.

$$h_f = \frac{3.022\text{v}^{1.85}L}{C^{1.85}D^{1.165}} \qquad \text{[Eq. 1]}$$

Assume this system uses plain cast iron pipe. The roughness coefficient, C, is 100.

The pipe length, L, can be determined from the figure in the problem statement.

$$\begin{aligned}L &= 50\ \text{ft} + 200\ \text{ft} + 140\ \text{ft} + 20\ \text{ft} + 25\ \text{ft}\\ &\quad + 30\ \text{ft} + 15\ \text{ft} + 5\ \text{ft}\\ &= 485\ \text{ft}\end{aligned}$$

The velocity, v, is calculated from the continuity equation.

$$Q = A\text{v}$$

The area, A, in the continuity equation can be expressed as

$$A = \frac{\pi D^2}{4}$$

Substitute the equation for area in the continuity equation and solve for v.

$$\text{v} = \frac{4Q}{\pi D^2} \qquad \text{[Eq. 2]}$$

Substitute values for h_f, C, L, v, and Q in Eq. 1.

$$110.2\ \text{ft} = \frac{(3.022)\left[\frac{(4)\left(5.42\ \frac{\text{ft}^3}{\text{sec}}\right)}{\pi D^2}\right]^{1.85}(485\ \text{ft})}{(100)^{1.85}D^{1.165}}$$

$$= \frac{(0.292)\left(\frac{6.9}{D^2}\right)^{1.85}}{D^{1.165}} = \frac{(0.292)(6.9)^{1.85}}{D^{3.7}D^{1.165}}$$

Rearrange and simplify.

$$D^{4.865} = \frac{(0.292)(6.9)^{1.85}}{110.2\ \text{ft}} = 0.944$$

$$D = 0.616\ \text{ft}\ (7.39\ \text{in})$$

$$\begin{aligned}A &= \frac{\pi(0.616\ \text{ft})^2}{4}\\ &= 0.298\ \text{ft}^2\end{aligned}$$

Next, compute the system curve, flow (Q) versus head (TDH). Use values of Q from 3 ft^3/sec to 7 ft^3/sec and $D = 0.616$ ft to solve for v with Eq. 2. Use values of v, $L = 485$ ft, $C = 100$, and $D = 0.616$ ft to calculate values of h_f with Eq. 1. Finally, TDH = h_f + static head (209.8 ft). These calculations are shown in tabular form.

Q (ft^3/sec)	v (ft/sec)	h_f (ft)	static head (ft)	TDH (ft)
3	10.1	37.15	209.8	247.0
4	13.5	63.5	209.8	273.3
5	16.8	95.2	209.8	305.0
6	20.2	133.9	209.8	343.7
7	23.6	178.9	209.8	388.4

This system curve, Q versus TDH, is shown in the previous illustration.

(c) At the operating point (5.42 ft^3/sec) on the efficiency curve in the illustration, the pump efficiency, η_{pump}, is 90%.

(d) The required motor horsepower, P, can be found from the equation

$$P = \frac{h_A \dot{m}}{550\eta_{pump}} \times \frac{g}{g_c}$$

The heat added, h_A, equals the TDH found in (a).

$$h_A = 320 \text{ ft}$$

The water mass flow rate is

$$\begin{aligned}\dot{m} &= \text{v}\rho \\ &= \left(5.42\,\frac{\text{ft}^3}{\text{sec}}\right)\left(62.4\,\frac{\text{lbm}}{\text{ft}^3}\right) \\ &= 338.2 \text{ lbm/sec}\end{aligned}$$

$$P = \left[\frac{(320 \text{ ft})\left(338.2\,\frac{\text{lbm}}{\text{sec}}\right)}{\left(550\,\frac{\frac{\text{ft-lbf}}{\text{sec}}}{\text{hp}}\right)(0.90)}\right]\left(\frac{32.2\,\frac{\text{ft}}{\text{sec}^2}}{32.2\,\frac{\text{ft-lbm}}{\text{lbf-sec}^2}}\right)$$

$$= \boxed{218.6 \text{ hp} \quad [\text{use 225 hp motor}]}$$

Notice that motors are rated by their output power, and the motor efficiency is not needed.

(e) A reasonable assumption is that the inflow to the sand filter equals the outflow, Q. The volume, V, is the product of the flow, the time, and the overload rate.

$$\begin{aligned}V &= \left(5.42\,\frac{\text{ft}^3}{\text{sec}}\right)\left(3600\,\frac{\text{sec}}{\text{hr}}\right)(8 \text{ hr})(1.25) \\ &= 195{,}120 \text{ ft}^3\end{aligned}$$

$$\text{clearwell area} = (200 \text{ ft})(250 \text{ ft}) = 50{,}000 \text{ ft}^2$$

$$\text{increased clearwell level} = \frac{195{,}120 \text{ ft}^3}{50{,}000 \text{ ft}^2} = 3.90 \text{ ft}$$

The accumulation in the clearwell should be no more than 3.9 ft because it will decrease the pump head and increase the flow.

2. (a) 1) *Prechlorination* is the process of adding chlorine to water at the beginning of the treatment process to disinfect water with high coliform levels, remove certain tastes and odors, and pretreat water for removal of high concentrations of iron and manganese.

2) *Flash mixing* is the immediate and complete mixing of chemicals (such as coagulants) with water for two minutes or less to achieve maximum process efficiency.

3) *Flocculation* is the slow, gentle mixing of water with a coagulant to remove colloidal material. This allows the destabilized colloidal particles to agglomerate and form flocs. The mixing time is usually in the range of 20–60 min.

4) *Sedimentation* is usually 2–8 hr of low (or no) flow when flocs and other discrete particles settle out by gravity in a clarifier.

5) *Filtration* is the process of passing clarified water through a bed of granular media such as sand, anthracite, or garnet to remove particles that have not been removed by sedimentation.

6) *Storage* is provided by reservoirs or by water tanks/towers to account for daily and seasonal variations in water demand as well as for extra water demand for firefighting purposes. This excess storage capacity allows for the design capacity of the water treatment plant to be less than the peak water demand.

7) *Chlorination* is the final disinfection of water with chlorine before distribution and consumption. The chlorine dose is determined by local requirements to provide either a low free-chlorine residual or chloramines only at the points of use.

(b) Chlorinated organics would probably be present in the treated water because chlorine is added first—before any organic compounds are removed by flocculation, sedimentation, or filtration. Chlorinated organics are difficult to remove by these treatment processes.

(c) *Trihalomethanes* (THMs) are organic chemicals produced during water treatment when organic compounds (precursors) in the water react with disinfectants such as chlorine, bromine, and iodine. These organic compounds occur naturally from decaying plants.

(d) THMs can be reduced by

- using a raw water source with less organic THM precursors.
- disinfecting with chlorine, bromine, or iodine only at the end of the treatment process after most THM precursors have been removed.
- removing precursors before chlorination—for example, by treatment with granular activated carbon (GAC).
- using ozone, chlorine dioxide, or potassium permanganate for disinfection.
- adding ammonia to form chloramines, which do not form as many THMs as free chlorine.

(e) Trihalomethane treatments add more expense to the treatment process. The cost of alternate disinfectants is also greater. Alternate disinfectants should be evaluated for disinfecting power, residual stability, and toxicity.

3. (a) Power, P, is represented by

$$P = D\text{v} \quad \text{[Eq. 1]}$$

Drag, D, is calculated using

$$D = \frac{C_D A \rho \text{v}^2}{2g_c} \quad \text{[Eq. 2]}$$

$$\begin{aligned}
C_D &= \text{drag coefficient} = 1.8 \text{ for flat plates} \\
\rho &= \text{density of water} = 62.4 \text{ lbm/ft}^3 \\
g_c &= \text{gravitational constant} = 32.2 \text{ ft-lbm/sec}^2\text{-lbf} \\
A &= \text{total paddle area} \\
&= (2.81 \text{ ft}^2/\text{paddle})(4 \text{ paddles}) = 11.25 \text{ ft}^2 \\
\text{v} &= \text{mixing velocity} = 0.75 \text{ (tip speed)}
\end{aligned}$$

$$\begin{aligned}
\text{tip speed} &= (\text{rpm})(\text{perimeter circumference of paddle}) \\
&= (\text{rpm})(\pi D) \\
&= \left(\frac{100 \dfrac{\text{rev}}{\text{min}}}{60 \dfrac{\text{sec}}{\text{min}}}\right) \pi\,(3.75 \text{ ft}) \\
&= 19.63 \text{ ft/sec}
\end{aligned}$$

$$\begin{aligned}
\text{v} &= (0.75)\left(19.63 \frac{\text{ft}}{\text{sec}}\right) \\
&= 14.73 \text{ ft/sec}
\end{aligned}$$

Substitute Eq. 2 for drag into Eq. 1 for power.

$$\begin{aligned}
P &= \frac{C_D A \rho \text{v}^2}{2g_c} \\
&= \frac{(1.8)(11.25 \text{ ft}^2)\left(62.4 \dfrac{\text{lbm}}{\text{ft}^3}\right)\left(14.37 \dfrac{\text{ft}}{\text{sec}}\right)^2}{(2)\left(32.2 \dfrac{\text{ft-lbm}}{\text{sec}^2\text{-lbf}}\right)} \\
&= 62{,}709 \text{ ft-lbf/sec} \\
&= \frac{62{,}709 \dfrac{\text{ft-lbf}}{\text{sec}}}{550 \dfrac{\text{ft-lbf}}{\text{sec-hp}}} \\
&= \boxed{114.0 \text{ hp}}
\end{aligned}$$

(b) The mean velocity gradient, G, can be found from the equation

$$\begin{aligned}
G &= \sqrt{\frac{P}{\mu V_{\text{tank}}}} \\
P &= 62{,}709 \text{ ft-lbf/sec} \\
\mu &= 2.55 \times 10^{-5} \text{ lbf-sec/ft}^2 \text{ at } 55^\circ\text{F} \\
V_{\text{tank}} &= 185.7 \text{ ft}^3 \\
G &= \sqrt{\frac{62{,}709 \dfrac{\text{ft-lbf}}{\text{sec}}}{\left(2.55 \times 10^{-5} \dfrac{\text{lbf-sec}}{\text{ft}^2}\right)(185.7 \text{ ft}^3)}} \\
&= \boxed{3639/\text{sec}}
\end{aligned}$$

This value of G is high.

(c)

$$\begin{aligned}
Q &= (2 \text{ MGD})\left(\frac{1.55 \dfrac{\text{ft}^3}{\text{sec}}}{\text{MGD}}\right) \\
&= 3.1 \text{ ft}^3/\text{sec} \\
t_d &= \frac{V}{Q} = \frac{187.5 \text{ ft}^3}{3.1 \dfrac{\text{ft}^3}{\text{sec}}} \\
&= 60.5 \text{ sec} \quad \text{[high end]} \\
Gt_d &= \left(\frac{3639}{\text{sec}}\right)(60.5 \text{ sec}) \\
&= 2.2 \times 10^5
\end{aligned}$$

The Gt_d value is high. The typical range of Gt_d is $10^4 - 10^5$.

4. (a) When designing a treatment sequence for this water supply, the following items must be taken into account. The turbidity of 300 NTU is very high and should be reduced to 5 NTU or less. Assume that the water does not require treatment to remove hardness or strong tastes or odors. Iron (Fe) and managanese (Mn) are present in the water and require treatment to be removed. The bacteria count in the raw water is relatively low.

The treatment sequence is

1) intake
2) aeration to oxidize Fe and Mn
3) presedimentation to settle out large solids
4) coagulant addition and rapid mix
5) coagulation/flocculation
6) clarification to settle out flocs
7) filtration with dual media filters to remove non-settleable particles
8) disinfection with chlorine
9) fluoridation

(b) Treatment process profile:

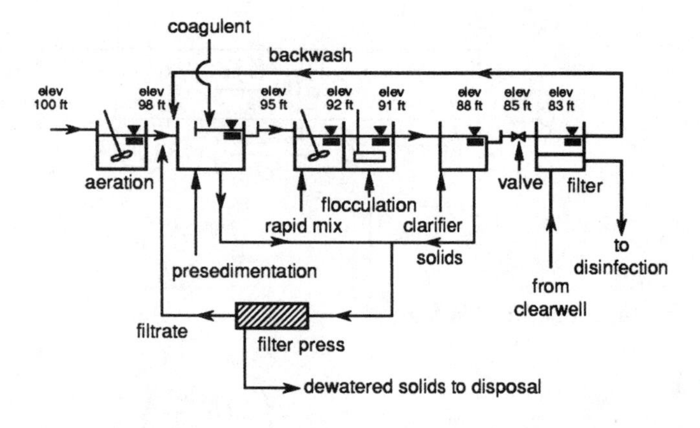

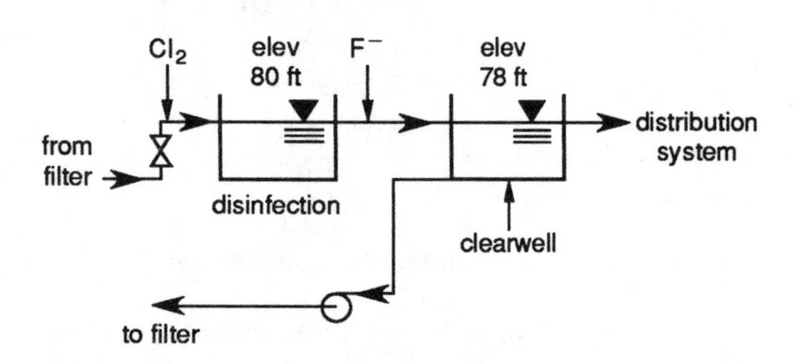

Note: All elevations are approximate to provide gravity flow between the units.

(c) When sizing the settling basins, the following assumptions should be made. The particles are fine sand and silt with specific gravities ranging from 1.2–2.65 and diameters ranging from 0.01–0.1 mm. For spherical particles with these characteristics, the settling velocity v_s, is typically 2×10^{-3} ft/sec. In this problem, the particles are not assumed to be spherical, so v_s will be reduced by a factor of 0.7.

$$v_s = (0.7)\left(2 \times 10^{-3}\ \frac{\text{ft}}{\text{sec}}\right) = 1.4 \times 10^{-3}\ \text{ft/sec}$$

Select the overflow rate v^* to be less than v_s.

$$v^* = 9.3 \times 10^{-4}\ \frac{\text{ft}}{\text{sec}} = 600\ \text{gal/day-ft}^2$$

Use two basins, each with half of the flow, Q.

$$\frac{Q}{2} = 1 \times 10^6\ \text{gal/day}$$

By the continuity equation,

$$A_{\text{surface}} = \frac{Q}{v^*} = \frac{1 \times 10^6\ \frac{\text{gal}}{\text{day}}}{600\ \frac{\text{gal}}{\text{day-ft}^2}} = 1667\ \text{ft}^2$$

Use a rectangular basin with a length-to-width ratio of 5:1.

$$A_{\text{surface}} = (5W)(W) = 1667\ \text{ft}^2$$

$$\boxed{\begin{aligned} W &= 18.25\ \text{ft} \\ L &= 91.33\ \text{ft} \\ \text{water depth} &= 10\ \text{ft} \end{aligned}}$$

$$\begin{aligned} \text{tank volume} &= (18.25\ \text{ft})(91.33\ \text{ft})(10\ \text{ft}) \\ &= (16{,}668\ \text{ft}^3)\left(7.48\ \frac{\text{gal}}{\text{ft}^3}\right) \\ &= \boxed{124{,}675\ \text{gal}} \end{aligned}$$

$$\begin{aligned} \text{deletion time} = t &= \frac{V}{Q} \\ &= \frac{124{,}675\ \text{gal}}{1 \times 10^6\ \frac{\text{gal}}{\text{day}}} \\ &= (0.125\ \text{day})\left(\frac{24\ \text{hr}}{\text{day}}\right) \\ &= 3.0\ \text{hr} \quad [\text{ok}] \end{aligned}$$

3-4hr ok

5. (a)

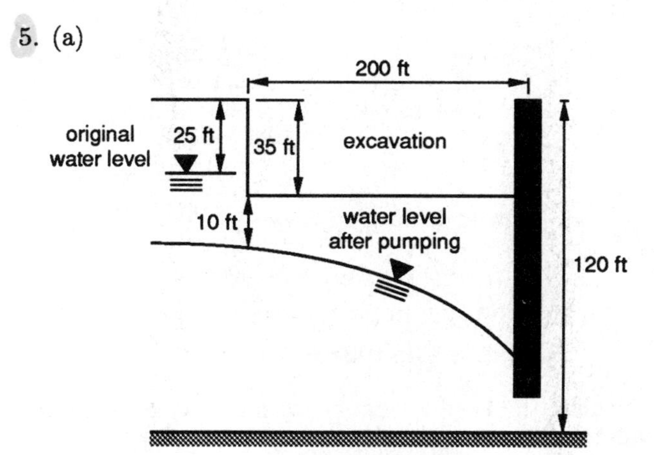

not to scale

$$\text{well flow, } Q = \frac{\pi K_p (y_1^2 - y_2^2)}{86{,}400 \ln\left(\frac{r_1}{r_2}\right)}$$

K_p = permeability = 190 ft/day
y_1 = original thickness of saturated layer
= 120 ft − 25 ft = 95 ft
y_2 = thickness of aquifer at edge of excavation
= 120 ft − 35 ft − 10 ft = 75 ft
r_1 = radius of influence of the well, assume 800 ft
r_2 = radius of excavation = 200 ft

$$Q = \frac{\pi\left(190\ \frac{\text{ft}}{\text{day}}\right)\left[(95\ \text{ft})^2 - (75\ \text{ft})^2\right]}{86{,}400 \ln\left(\frac{800\ \text{ft}}{200\ \text{ft}}\right)} = 16.9\ \frac{\text{ft}^3}{\text{sec}}$$

$$= \boxed{7604\ \text{gal/min}}$$

16.9 × 7.48 gal/ft³ × 60 sec/min = 7600 gal/min

(b) 7604 gal/min is too much water to draw from wells in the center of this excavation site. More than one well would be required. A 20-foot drawdown 200 ft from a well requires pumping a great deal of water. Wells in the center of a 400-ft by 400-ft excavation do not appear to be the best way to lower the water table 20 feet at this site.

(c) Other methods that could be used to provide a dry working site include

1) Wells spaced around the perimeter of the site, as well as one or two wells in the middle, would also dewater the site.

2) A bentonite slurry or concrete cutoff wall could be built around the site to a depth greater than the desired dewatered depth (such as 55 ft) to slow the flow of groundwater into the site. The excavation could then be dewatered by wells or trenches.

3) Dewatering trenches could be used around the site.

6. (a) The assumed average water demands are

residential:	100 gpcd
commercial:	80 gpcd
public:	20 gpcd
waste:	20 gpcd
total demand =	220 gpcd

Population records from the past 20–40 years for this town, as well as for surrounding towns, should be analyzed to help estimate future population growth.

Use the night population of 35,000, and assume a constant population growth of 20% per decade. Design a water treatment plant for a 30-year life. The multiplier for the design population is $1+0.2 = 1.2$. The exponent for the design population is 30 years/10 years = 3.0.

$$\text{design population} = (35{,}000)(1.2)^3 = 60{,}480 \quad [\text{use } 60{,}500]$$

$$\text{average design flow} = \frac{(220\ \text{gpcd})(60{,}500)}{1{,}000{,}000\ \frac{\text{gal}}{\text{MG}}} = \boxed{13.31\ \text{MGD}}$$

(b) The maximum hourly flow can be calculated using the average flow and a peak flow multiplier of 3.0.

$$\left(13.31 \times 10^6\ \frac{\text{gal}}{\text{day}}\right)(3.0)\left(\frac{\text{day}}{24\ \text{hr}}\right) = \boxed{1.66 \times 10^6\ \text{gal/hr}}$$

The maximum daily flow is 1.8 times the average flow.

$$\left(13.31 \times 10^6\ \frac{\text{gal}}{\text{day}}\right)(1.8) = 23.96 \times 10^6\ \text{gal/day}$$

(c) Firefighting water requirements:

$$Q = (1020\sqrt{P}) = (1 - 0.01\sqrt{P})\ \text{gal/min}$$

p = population in thousands of people

$$p = \frac{60{,}500}{1000} = 60.5$$

$$Q = (1020\sqrt{60.5})(1 - 0.01\sqrt{60.5}) = \left(7317\ \frac{\text{gal}}{\text{min}}\right)\left(\frac{60\ \text{min}}{\text{hr}}\right)\left(\frac{24\ \text{hr}}{\text{day}}\right) = \boxed{10.54 \times 10^6\ \text{gal/day}}$$

(d) The minimum water pressure requirement is 50 lbf/in². This can be maintained by pumping water to storage tanks that are high enough to provide adequate pressure, such as

$$\left(50\ \frac{\text{lbf}}{\text{in}^2}\right)\left(2.31\ \frac{\text{ft}}{\frac{\text{lbf}}{\text{in}^2}}\right) = 115.5\ \text{ft} + \text{head losses}$$

The maximum water pressure requirement is 80 lbf/in^2 = 184.7 ft. This can be maintained (not exceeded) by proper design of the storage tanks and distribution system, and by using pressure-regulating or altitude valves where appropriate.

(e) The storage flow to equalize the pumping rate is 0.25 times the maximum daily flow.

$$\text{storage flow} = (0.25)\left(23.96 \times 10^6 \ \frac{\text{gal}}{\text{day}}\right)(1 \text{ day}) = 6.0 \times 10^6 \text{ gal}$$

$$\text{fire-flow storage} = (7 \text{ hr})\left(7317 \ \frac{\text{gal}}{\text{min}}\right)\left(\frac{60 \text{ min}}{\text{hr}}\right) = 3.07 \times 10^6 \text{ gal}$$

$$\text{emergency storage} = (3 \text{ days})\left(13.31 \times 10^6 \ \frac{\text{gal}}{\text{day}}\right) = 39.93 \times 10^6 \text{ gal}$$

$$\text{total storage} = 6.0 \times 10^6 \text{ gal} + 3.07 \times 10^6 \text{ gal} + 39.93 \times 10^6 \text{ gal} = \boxed{49 \times 10^6 \text{ gal (49 MG)}}$$

(f) System layout:

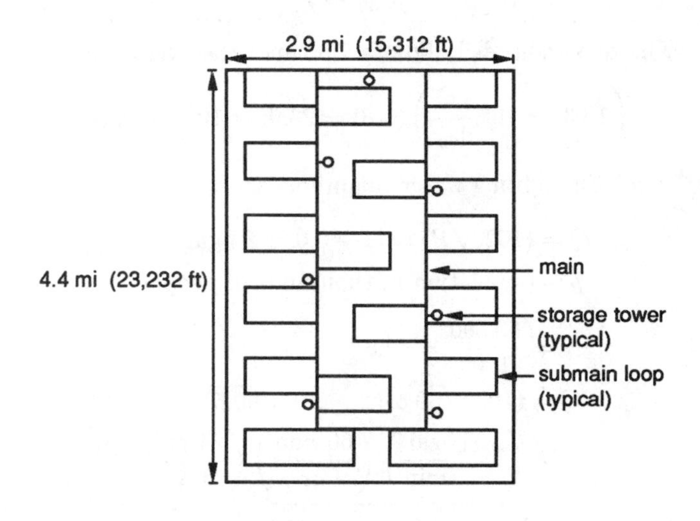

(g) Using seven tanks,

$$\text{storage in each tank} = \frac{49 \times 10^6 \text{ gal}}{7 \text{ tanks}} = 7 \times 10^6 \text{ gal/tank}$$

$$\text{maximum flow} = \left(1.66 \times 10^6 \ \frac{\text{gal}}{\text{hr}}\right)\left(\frac{\text{hr}}{60 \text{ min}}\right) = 27{,}730 \text{ gal/min}$$

$$\text{flow in 16-in pipe} = \frac{27{,}730 \ \frac{\text{gal}}{\text{min}}}{14} = 1980 \text{ gal/min}$$

$$\text{flow in 12-in pipe} = \frac{1980 \ \frac{\text{gal}}{\text{min}}}{3} = 660 \text{ gal/min}$$

$$\text{flow in 6-in pipe} = 200 \text{ gal/min}$$

$$\text{flow in } \tfrac{3}{4}\text{-in tube} = 8 \text{ gal/min}$$

Calculate the head losses, h_f, with the Hazen-Williams equation.

$$h_f = \frac{10.44 L Q^{1.85}}{C^{1.85} d^{4.8655}}$$

pipe diameter (in)	length (ft)	flow, Q (gal/min)	C	h_f (ft)
16	4000	1980	100	14.5
12	4000	660	100	7.7
6	1000	200	100	6.2
$\frac{3}{4}$	80	8	130	19.5

total friction losses = 47.9 ft

Minimum pressure at house = 50 lbf/in^2.

$$\text{static head} = \left(50 \ \frac{\text{lbf}}{\text{in}^2}\right)\left(2.31 \ \frac{\text{ft}}{\text{lbf-in}^2}\right) = 115.5 \text{ ft}$$

Add static head and friction losses. The working tank elevation is

$$115.5 \text{ ft} + 47.9 \text{ ft} = \boxed{163.4 \text{ ft}}$$

(h) The water tanks should be filled at periods of low demand such as early morning (2:00–5:00 a.m.) and afternoon (2:00–4:00 p.m.). In addition to being filled twice a day, the tanks may need to have water pumped in during peak demand periods, such as morning (6:00–9:00 a.m.) and evening (4:30–8:30 p.m.).

(i) The minimum water pressure requirement is

$$\left(50 \ \frac{\text{lbf}}{\text{in}^2}\right)\left(2.31 \ \frac{\text{ft}}{\text{lbf-in}^2}\right) = 115.5 \text{ ft}$$

The working static elevation in a tank 80 ft above the ground is the static head. Subtract the friction losses to get the total available head: 80 ft < 115.5 ft.

The water pressure requirements will not be met.

7. (a) Hardness is contributed by Ca^{++}, Fe^{++}, and Mg^{++}. Convert mg/ℓ of substance to mg/ℓ as $CaCO_3$.

$$\text{multiplier} = \frac{\text{equivalent weight } CaCO_3}{\text{equivalent weight substance}} = \frac{50}{\frac{\text{MW}}{\text{change}}}$$

For Ca^{++}, MW = 40.

$$\text{multiplier} = \frac{50}{\frac{40}{2}} = 2.5$$

compound	mg/ℓ	multiplier	mg/ℓ as $CaCO_3$
Ca^{++}	80	2.5	200.0
Fe^{++}	3	1.79	5.4
Mg^{++}	30	4.1	123.0

$$\begin{aligned}\text{total hardness} &= 200.0 + 5.4 + 123.0\\ &= \boxed{328.4 \text{ mg/ℓ as } CaCO_3}\end{aligned}$$

(b) Alkalinity is mainly caused by HCO_3^-. Acidity is mainly caused by $H_2CO_3^*$ (i.e., the sum of H_2CO_3 and CO_2). Assume the water is alkaline—check in (e).

> From (e), pH = 7.5. The water is alkaline.

(c) alkalinity (eq/ℓ) = $[HCO_3^-] + 2[CO_3^{--}] + [OH^-] - [H^+]$

From (e),

$$\begin{aligned}[HCO_3^-] &= 5.51 \times 10^{-3} \text{moles/ℓ}\\ [H^+] &= 3.1 \times 10^{-8} \text{moles/ℓ}\\ [H^+][OH^-] &= (1 \times 10^{-14})[OH^-]\\ &= \frac{1 \times 10^{-14}}{3.1 \times 10^{-8}}\\ &= 3.23 \times 10^{-7} \text{ moles/ℓ}\\ HCO_3^- &\rightleftharpoons CO_3^{--} + H^+\\ pK_2 &= 10.3\\ \frac{[H^+][CO_3^{--}]}{[HCO_3^-]} &= (10)^{-10.3}\end{aligned}$$

$$\begin{aligned}[CO_3^{--}] &= \frac{(10^{-10.3})\left(5.51 \times 10^{-3} \frac{\text{moles}}{ℓ}\right)}{3.1 \times 10^{-8} \frac{\text{moles}}{ℓ}}\\ &= 8.91 \times 10^{-6} \text{ moles/ℓ}\\ \text{alkalinity} &= \left(5.51 \times 10^{-3} \frac{\text{moles}}{ℓ}\right)\\ &\quad + (2)\left(8.91 \times 10^{-6} \frac{\text{moles}}{ℓ}\right)\\ &\quad + \left(3.23 \times 10^{-7} \frac{\text{moles}}{ℓ}\right)\\ &\quad - \left(3.1 \times 10^{-8} \frac{\text{moles}}{ℓ}\right)\\ &= 5.528 \times 10^{-3} \text{ eq/ℓ}\\ &= \left(5.528 \times 10^{-3} \frac{\text{eq}}{ℓ}\right)\\ &\quad \times \left(50 \times 10^{3} \frac{\text{mg as } CaCO_3}{\text{eq}}\right)\\ &= \boxed{276.4 \text{ mg/ℓ as } CaCO_3}\end{aligned}$$

(c) acidity (eq/ℓ) = $2[H_2CO_3^*] + [HCO_3^-] + [H^+] - [OH^-]$

From (e),

$$[H_2CO_3^*] = 3.41 \times 10^{-4} \frac{\text{mole}}{ℓ}$$

$$\begin{aligned}\text{acidity} &= (2)\left(3.41 \times 10^{-4} \frac{\text{moles}}{ℓ}\right)\\ &\quad + \left(5.51 \times 10^{-3} \frac{\text{moles}}{ℓ}\right)\\ &\quad + \left(3.1 \times 10^{-8} \frac{\text{moles}}{ℓ}\right)\\ &\quad + \left(3.23 \times 10^{-7} \frac{\text{moles}}{ℓ}\right)\\ &= 6.19 \times 10^{-3} \text{eq/ℓ}\\ &= \left(6.19 \times 10^{-3} \frac{\text{eq}}{ℓ}\right)\\ &\quad \left(50 \times 10^{3} \frac{\text{mg as } CaCO_3}{\text{eq}}\right)\\ &= \boxed{309.6 \text{ mg/ℓ as } CaCO_3}\end{aligned}$$

(d) This water would probably be the color of rust or rusty tea. The iron would contribute a rust color. Any organics (such as TDS) may also contribute a tea color.

(e)

$$[CO_2] = \frac{15 \times 10^{-3} \frac{g}{\ell}}{44 \frac{g}{mole}} = 3.41 \times 10^{-4} \text{ moles}/\ell$$

$$[HCO_3^-] = \frac{336 \times 10^{-3} \frac{g}{\ell}}{61 \frac{g}{mole}} = 5.51 \times 10^{-3} \text{ moles}/\ell$$

Assume

$$[H_2CO_3^*] = [CO_{2\ aq}] = 3.41 \times 10^{-4} \text{ moles}/\ell$$

$$H_2CO_3^* \rightleftharpoons HCO_3^- + H^+$$

$$pK_1 = 6.3$$

$$\frac{[H^+][HCO_3^-]}{[H_2CO_3^*]} = 10^{-6.3}$$

$$[H^+] = \frac{(10^{-6.3})[H_2CO_3^*]}{[HCO_3^-]}$$

$$= \frac{(10^{-6.3})\left(3.41 \times 10^{-4} \frac{moles}{\ell}\right)}{5.51 \times 10^{-3} \frac{moles}{\ell}}$$

$$= 3.10 \times 10^{-8} \text{ moles}/\ell$$

$$pH = -\log[H^*]$$

$$= -\log\left(3.10 \times 10^{-8} \frac{moles}{\ell}\right)$$

$$= \boxed{7.5}$$

(f) This water should be softened to reduce the hardness to 60 mg/ℓ or less (which is considered soft water).

(g)

$$\text{total hardness} = 328.4 \text{ mg}/\ell \text{ as } CaCO_3$$

$$\text{alkalinity} = 276.4 \text{ mg}/\ell \text{ as } CaCO_3$$

$$CO_2 = \left(15 \frac{mg}{\ell}\right)\left(\frac{50}{\frac{44}{2}}\right) = 34.1 \text{ mg}/\ell \text{ as } CaCO_3$$

In the first stage, add hydrated lime (93% pure $Ca(OH)_2$) to remove CO_2, alkalinity $Ca(HCO_3)_2$, and $Mg(HCO_3)_2$).

Hydrated lime must be converted to $CaCO_3$ equivalents.

$$\text{formula weight, } Ca(OH)_2 = 74$$

$$\text{valence} = 2$$

$$\text{equivalence, } Ca(OH)_2 = \frac{50}{\frac{74}{2}} = 1.35$$

The amount of lime required to remove CO_2 and HCO_3^- is

$$\frac{34.1 \frac{mg}{\ell} + 276.4 \frac{mg}{\ell}}{(1.35)(0.93)} = 247.3 \text{ mg}/\ell \quad [Ca(OH)_2]$$

Add another 50 mg/ℓ of lime to raise the pH above 10.8 to precipitate $Mg(OH)_2$.

$$\text{extra lime} = \frac{50 \frac{mg}{\ell}}{0.93} = 53.8 \text{ mg}/\ell$$

$$\text{total lime } (Ca(OH)_2) = 247.3 \frac{mg}{\ell} + 53.8 \frac{mg}{\ell} = \boxed{301.1 \text{ mg}/\ell}$$

In the second stage, add soda ash (98% Na_2CO_3) to remove $CaSO_4$ and $MgSO_4$.

$$\text{noncarbonate hardness} = \text{total hardness} - \text{alkalinity}$$

$$= 328.4 \frac{mg}{\ell} - 276.4 \frac{mg}{\ell}$$

$$= 52.0 \text{ mg}/\ell \text{ as } CaCO_3$$

The hardness needs to be reduced to 80 mg/ℓ as $CaCO_3$. Remove

$$52.0 \frac{mg}{\ell} - 80 \frac{mg}{\ell} = -28 \text{ mg}/\ell \text{ as } CaCO_3$$

$$CaCO_3 \text{ equivalence of } Na_2CO_3 = \frac{50}{\frac{106}{2}} = 0.94$$

No soda ash is required because hardness is less than 80 mg/ℓ after lime treatment.

WASTEWATER ENGINEERING

1. The influent wastewater flow is

$$(4.5\ \text{MGD})\left(950\,\frac{\text{lbm BOD}}{\text{MG}}\right) = 4275\ \text{lbm BOD/day}$$

85% of the BOD is removed in this process.

$$(0.85)\left(4275\,\frac{\text{lbm BOD}}{\text{day}}\right) = 3634\ \text{lbm BOD/day}$$

The O_2 required is

$$\begin{array}{r} 1.15\ \text{O}_2\text{/lbm BOD removed} \\ +0.03\ \text{O}_2\text{/lbm BOD removed} \\ \hline 1.18\ \text{O}_2\text{/lbm BOD removed} \end{array}$$

$$\left(1.18\,\frac{\text{lbm O}_2}{\text{lbm BOD}}\right)\left(3634\,\frac{\text{lbm BOD}}{\text{day}}\right) = 4288\ \text{lbm O}_2\text{/day}$$

Air is 23.2% O_2 by weight and has a density of 0.075 lbm/ft^3. The oxygen transfer efficiency of the diffusers in wastewater is 15%. The airflow required is

$$\frac{4288\,\frac{\text{lbm O}_2}{\text{day}}}{\left(0.075\,\frac{\text{lbm}}{\text{ft}^3}\right)(0.232)(0.15)} = 1.643\times 10^6\ \text{ft}^3\text{/day}$$

The air blowers have 100% excess capacity, so use a factor of safety of 2.0.

$$\frac{\left(1.643\times 10^6\,\frac{\text{ft}^3}{\text{day}}\right)(2.0)}{1440\,\frac{\text{min}}{\text{day}}} = 2282\ \text{ft}^3\text{/min}$$

The blower power can be calculated from the following equation.

$$P = \frac{mRT_1}{550\eta e}\left[\left(\frac{p_2}{p_1}\right)^{0.283} - 1\right]$$

P = power (hp)
m = mass flow rate of air (lbm/sec)
$= \left(2282\,\frac{\text{ft}^3}{\text{min}}\right)\left(0.075\,\frac{\text{lbm}}{\text{ft}^3}\right)\left(\frac{\text{min}}{60\ \text{sec}}\right)$
$= 2.85$ lbm/sec
R = gas constant = 53.3 ft-lbf/lbm air-°R
T = temperature °R = °F + 460 = 70°F + 460 = 530°R
η = 0.283 for air
e = efficiency of blowers
p_1 = inlet pressure = standard atmospheric pressure = 14.7 lbf/in^2
p_2 = outlet pressure

$$P = \left[\frac{\left(2.85\,\frac{\text{lbm}}{\text{sec}}\right)\left(53.3\,\frac{\text{ft-lbf}}{\text{lbm-°R}}\right)(530\text{°R})}{\left(550\,\frac{\text{ft-lbf}}{\text{sec-hp}}\right)(0.283)(0.80)}\right] \times\left[\left(\frac{22.7\,\frac{\text{lbf}}{\text{in}^2}}{14.7\,\frac{\text{lbf}}{\text{in}^2}}\right)^{0.283} - 1\right]$$

$= 84.6$ hp [use 85 hp]

The power cost for 24 hr of constant aeration is

$$(85\ \text{hp})\left(0.746\,\frac{\text{kW}}{\text{hp}}\right)\left(\frac{24\ \text{hr}}{\text{day}}\right)(1\ \text{day})\left(\frac{\$0.08}{\text{kW-hr}}\right) = \boxed{\$121.75/\text{day}}$$

conversion factor

2. (a) The heat required for raw sludge is calculated from the following equation.

$$q_s = Vc_p(T_2 - T_1)$$

V = 30,000 gal/day of sludge
c_p = 1.0 BTU/lbm-°F (water)
T_1 = 61°F
T_2 = 91°F

$$q_s = \left(\frac{30{,}000\,\frac{\text{gal}}{\text{day}}}{24\,\frac{\text{hr}}{\text{day}}}\right)\left(8.34\,\frac{\text{lbm}}{\text{gal}}\right)\left(1.0\,\frac{\text{BTU}}{\text{lbm-°F}}\right)\times(91\text{°F} - 61\text{°F})$$

$= 312{,}750$ BTU/hr

The digester shape is

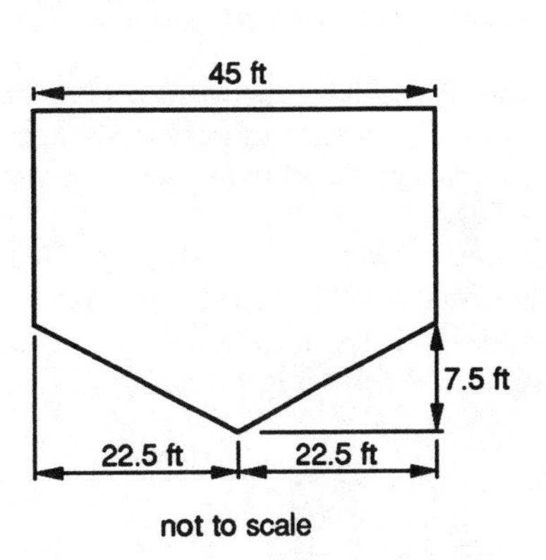

The heat required to make up for heat losses from the walls, floor, and roof is

walls:
$$q_w = (q_{\text{digester}})(\text{no. of digesters})$$
$$= \left(110{,}000\ \frac{\text{BTU}}{\text{hr-digester}}\right)(3 \text{ digesters})$$
$$= 330{,}000 \text{ BTU/hr}$$

floors:
$$A = \pi r\sqrt{r^2 + b^2}$$
$$= \pi(22.5 \text{ ft})\sqrt{(22.5 \text{ ft})^2 + (7.5 \text{ ft})^2}$$
$$= 1676 \text{ ft}^2/\text{digester}$$

roof:
$$A = \pi r^2$$
$$= \pi(22.5 \text{ ft})^2 = 1590 \text{ ft}^2/\text{digester}$$

$$q = AU(T_1 - T_2)$$

floors: $U = 0.15$ BTU/ft^2-°F-hr
roof: $U = 0.16$ BTU/ft^2-°F-hr

$$T_2 = 91°\text{F}$$
$$T_1 = 58°\text{F}$$

$$q_f = \left(1676\ \frac{\text{ft}^2}{\text{digester}}\right)(3 \text{ digesters})\left(0.15\ \frac{\text{BTU}}{\text{ft}^2\text{-°F-hr}}\right) \times (91°\text{F} - 58°\text{F})$$
$$= 24{,}889 \text{ BTU/hr}$$

$$q_r = \left(1590\ \frac{\text{ft}^2}{\text{digester}}\right)(3 \text{ digesters})\left(0.16\ \frac{\text{BTU}}{\text{ft}^2\text{-°F-hr}}\right) \times (91°\text{F} - 58°\text{F})$$
$$= 25{,}186 \text{ BTU/hr}$$

$$q_t = q_w + q_f + q_r$$
$$= 330{,}000\ \frac{\text{BTU}}{\text{hr}} + 24{,}889\ \frac{\text{BTU}}{\text{hr}} + 25{,}186\ \frac{\text{BTU}}{\text{hr}}$$
$$= \boxed{380{,}075 \text{ BTU/hr}}$$

(b) The volume of methane produced from the sludge is calculated by

$$V_{CH_4} = \left(5.62\ \frac{\text{ft}^3}{\text{lbm}}\right)(EQS_0 - 1.42P_x)$$

(Reference: *Wastewater Engineering*, p. 818, eq. 12-7.)

V_{CH_4} = volume of methane produced (ft^3/day)
5.61 = theoretical conversion factor for amount of methane produced from conversion of 1 lbm of BOD_L
E = efficiency of waste utilization = 70%
Q = sludge flow rate = 30,000 gal/day
S_0 = ultimate BOD_L (lbm/gal)
P_x = net mass of cell tissue produced (lbm/day)

$$BOD_L = \frac{BOD_5}{1 - e^{-Rt}}$$

$R = 0.2$/day and $t = 5$ days.

$$BOD_L = \frac{12{,}000\ \frac{\text{lbm}}{\text{day}}}{1 - e^{-(0.2/\text{day})(5 \text{ days})}}$$
$$= 18{,}984 \text{ lbm/day}$$

$$S_0 = \frac{BOD_L}{Q} = \frac{18{,}984\ \frac{\text{lbm}}{\text{day}}}{30{,}000\ \frac{\text{gal}}{\text{day}}}$$
$$= 0.63 \text{ lbm/gal}$$

$$P_x = \frac{YQES_0}{1 + R_d\theta_c}$$

yield coefficient, $Y = 0.06$ lbm/lbm
endogenous coefficient, $R_d = 0.03$/day
mean cell residence time, $\theta_c = 10$ days

$$P_x = \frac{\left(0.06\ \frac{\text{lbm}}{\text{lbm}}\right)\left(30{,}000\ \frac{\text{gal}}{\text{day}}\right)(0.70)\left(0.63\ \frac{\text{lbm}}{\text{gal}}\right)}{1 + \left(\frac{0.03}{\text{day}}\right)(10 \text{ days})}$$
$$= 611 \text{ lbm/day}$$

$$V_{CH_4} = \left(5.62\ \frac{\text{ft}^3}{\text{lbm}}\right)\left[(0.70)\left(30{,}000\ \frac{\text{gal}}{\text{day}}\right)\left(0.63\ \frac{\text{lbm}}{\text{gal}}\right) - (1.42)\left(611\ \frac{\text{lbm}}{\text{gal}}\right)\right]$$
$$= 69{,}477 \text{ ft}^3/\text{day}$$

The heating value of methane is approximately 960 BTU/ft^3.

$$\text{heat produced} = \left(960\ \frac{\text{BTU}}{\text{ft}^3}\right)\left(69{,}477\ \frac{\text{ft}^3}{\text{day}}\right)$$
$$= 6.67 \times 10^7 \text{ BTU/day}$$

$$\left(6.67 \times 10^7\ \frac{\text{BTU}}{\text{day}}\right)\left(\frac{\text{day}}{24 \text{ hr}}\right) = 2.78 \times 10^6 \text{ BTU/hr}$$
$$2.78 \times 10^6 \text{ BTU/hr} > 380{,}075 \text{ BTU/hr}$$

Since the heating value of the methane produced is greater than the heat required to raise the sludge temperature, the methane produced is sufficient to provide the energy required.

3. (a) flow, $Q = (\text{population})(\text{flow per capita})$
$$= (300{,}000 \text{ capita})\left(250\ \frac{\text{gal}}{\text{capita-day}}\right)$$
$$= 75 \times 10^6 \text{ gal/day}$$

For the existing 24-in (inside diameter) unlined concrete sewer pipe,

$$\text{length} = 4000 \text{ ft}$$

$$\text{vertical drop} = 100 \text{ ft}$$

$$\text{slope} = \frac{100 \text{ ft}}{4000 \text{ ft}} = 0.025 \text{ ft/ft}$$

$$\text{diameter} = (24 \text{ in})\left(\frac{\text{ft}}{12 \text{ in}}\right) = 2 \text{ ft}$$

For an unlined concrete pipe, assume $n = 0.013$.

Calculate the capacity of the 2-ft-diameter pipe flowing full.

$$A = \frac{\pi D^2}{4} = \frac{\pi(2 \text{ ft}^2)}{4}$$
$$= 3.14 \text{ ft}^2$$

$$r_H = \frac{A}{p_w} = \frac{\pi\left(\frac{D^2}{4}\right)}{\pi D}$$
$$= \frac{D}{4} = \frac{2 \text{ ft}}{4} = 0.5 \text{ ft}$$

Use the Chezy-Manning equation.

$$Q = \left(\frac{1.49}{n}\right) A (r_H)^{2/3}\sqrt{S}$$
$$= \left(\frac{1.49}{0.013}\right)(3.14 \text{ ft}^2)(0.5 \text{ ft})^{2/3}\sqrt{0.025}$$
$$= 35.85 \text{ ft}^3/\text{sec}$$

Compare this to the flow rate.

$$Q = \left(75 \times 10^6 \frac{\text{gal}}{\text{day}}\right)\left(\frac{1.55 \frac{\text{ft}^3}{\text{sec}}}{10^6 \frac{\text{gal}}{\text{day}}}\right)$$
$$= 116 \text{ ft}^3/\text{sec}$$

A second pipe is required to handle the flow.

$$116 \frac{\text{ft}^3}{\text{sec}} - 35.8 \frac{\text{ft}^3}{\text{sec}} = 80.2 \text{ ft}^3/\text{sec}$$

Assume that the second pipe is flowing full with the same slope (0.025 ft/ft) and is also an unlined concrete pipe ($n = 0.013$).

$$r_H = \frac{D}{4}$$
$$A = \frac{\pi D^2}{4}$$

Use the Chezy-Manning equation.

$$80.2 \frac{\text{ft}^3}{\text{sec}} = \left(\frac{1.49}{0.013}\right)\left(\frac{\pi D^2}{4}\right)\left(\frac{D}{4}\right)^{2/3}\sqrt{0.025}$$
$$= (14.23 D^2)\left(\frac{D}{4}\right)^{2/3}$$
$$= 5.65 D^{8/3}$$
$$D = \left(\frac{80.2}{5.65}\right)^{3/8} = 2.70 \text{ ft (32.5 in)}$$

The 32.5-in pipe size is not standard, and sewers are not normally designed to flow full.

Use a 36-in diameter pipe.

(b) Two 24-in-diameter unlined concrete pipes with a slope of 0.025 ft/ft will flow full and will not handle a peak flow of 75×10^5 gal/day (116 ft^3/sec).

4.

wastewater source	units	volume (gal/unit-day)	flow (gal/day)
offices (sanitary)	1000 employees	13	13,000
hospital (medical)	250 beds	165	41,250
plating plant (chemical)	–	–	5000
dairy animal	–	–	15,000

(a)

wastewater source	expected BOD
sanitary	250 mg/ℓ
hospital	350 mg/ℓ
plating plant	0 mg/ℓ
dairy animal	1200 mg/ℓ

(b) In addition to organic loading (measured as BOD), the following wastewater characteristics should be considered because these conditions may upset biological treatment processes.

- the presence of heavy metals
- the presence of hexavalent chromium
- the presence of cyanide
- a high or low pH (less than 6 or greater than 9)

The following compounds should also be checked for, although their presence is less likely: phenol, excessive oil and grease, and high levels of hydrogen sulfide.

All these compounds must be removed or neutralized in pretreatment before biological treatment to reduce the organic waste load (BOD).

To help design biological pretreatment, analyze the following constituents.

- dissolved oxygen in incoming waste, COD
- nutrients such as nitrogen and phosphorus (deficient or rich)
- chlorides and sodium, total dissolved solids (TDS), suspended solids, and the settleability of the waste

(c) The pretreatment process is as follows.

1) (a) Use a basin (with aeration if required) to equalize the flow and waste strength.
   (b) In the same basin, allow solids to settle out (during quiescent times). A skimmer and/or absorbent boom will be used to remove floating material (especially oil and grease).
2) Use chemical treatment and precipitation to remove heavy metals such as Cr^{+6}, CN^-, and phenol.
3) Adjust the pH to 6.0–9.0 (if required).
4) Add nutrients (N or P) or carbon (if required).
5) Use a trickling filter or aerated lagoon to provide a moderate amount of biological treatment—i.e., to reduce BOD and NH_3–N or organic nitrogen to levels acceptable to the municipal wastewater treatment facility.
6) Use final clarification and turbulent discharge to the municipal sewer system to reoxygenate wastewater.

5. Peak summer wastewater flows:

wastewater source	unit volume	units[a] (gal/unit-day)	flow (gal/day)
summer population	10,000	30	300,000
hospital (beds)	100	165	16,500
laundromat (washers)	20	550	11,000
dining room (meals)	4500[b]	7	31,500
gas station (employees)	4	12	48
		total	359,048

[a](Reference: *Wastewater Engineering*, pp. 28–29, tables 2-10, 2-11, and 2-12.)

[b]The dining room serves:
(3 meals/day)(3 servings)(500 seats) = 4500 meals/day

Winter wastewater flows:

wastewater source	unit volume	units (gal/unit-day)	flow (gal/day)
winter population	300	30	9000
hospital (beds)	10	165	1650
laundromat (washers)	4	550	2200
dining room (meals)	900[a]	7	6300
gas station (employees)	2	12	24
		total	19,174

[a]The dining room serves:
(3 meals/day)(300 seats) = 900 meals/day

Adjust k for winter and summer temperatures.

$$k_T = k_{20°C}\theta^{T-20}$$

$$\text{winter: } k_{10°C} = \left(\frac{0.25}{\text{day}}\right)(1.06)^{10°-20°} = 0.14/\text{day}$$

$$\text{summer: } k_{32°C} = \left(\frac{0.25}{\text{day}}\right)(1.06)^{32°-20°} = 0.5/\text{day}$$

The detention time in winter is

$$t = \frac{kt}{k_{10}} = \frac{5}{\dfrac{0.14}{\text{day}}} = 35.7 \text{ days}$$

The detention time in summer is

$$t = \frac{5}{\dfrac{0.5}{\text{day}}} = 10 \text{ days}$$

Determine the pond surface area requirements in both summer and winter.

$$A_{\text{summer}} = \frac{\left(Q\,\dfrac{\text{gal}}{\text{day}}\right)(t \text{ day})}{(2 \text{ ft})\left(7.48\,\dfrac{\text{gal}}{\text{ft}^3}\right)\left(43{,}560\,\dfrac{\text{ft}^3}{\text{ac}}\right)}$$

$$= \frac{\left(359{,}048\,\dfrac{\text{gal}}{\text{day}}\right)(10 \text{ days})}{(2 \text{ ft})\left(7.48\,\dfrac{\text{gal}}{\text{ft}^3}\right)\left(43{,}560\,\dfrac{\text{ft}^3}{\text{ac}}\right)}$$

$$= 5.5 \text{ ac}$$

$$A_{\text{winter}} = \frac{\left(19{,}174\,\dfrac{\text{gal}}{\text{day}}\right)(35.7 \text{ days})}{(2 \text{ ft})\left(7.48\,\dfrac{\text{gal}}{\text{ft}^3}\right)\left(43{,}560\,\dfrac{\text{ft}^3}{\text{ac}}\right)}$$

$$= 1.05 \text{ ac}$$

Check the BOD loading.

$$L_{\text{BOD, summer}} = \frac{\left(250\,\dfrac{\text{mg}}{\ell}\right)(0.36 \text{ MGD})\left(8.34\,\dfrac{\text{lbm-}\ell}{\text{mg-MG}}\right)}{5.5 \text{ ac}}$$

$$= 136 \text{ lbm/ac-day}$$

$$L_{\text{BOD, winter}} = \frac{\left(250\,\dfrac{\text{mg}}{\ell}\right)\left(19{,}174\,\dfrac{\text{gal}}{\text{day}}\right)\left(8.34\,\dfrac{\text{lbm-}\ell}{\text{mg-MG}}\right)}{\left(1\times 10^6\,\dfrac{\text{gal}}{\text{MG}}\right)(1.1 \text{ ac})}$$

$$= 36.3 \text{ lbm/ac-day}$$

The summer conditions control the design because the surface-area and BOD-loading requirements are greater than in winter.

For this facility, use an aerobic stabilization pond with a maximum working depth of 2 ft that has five cells, each 1.1 ac in surface area. The total area is 5.5 ac. Only one cell will be required during the winter months. Additional cells can be brought on-line as the wastewater flows increase in the spring and early summer. All the cells are used during the peak summer season.

6. *Landfill:* (a) The principle advantage of landfills is that they can handle all types of waste. (b) The disadvantages include the requirement for large land areas, the slow decomposition of wastes, and the possibility of leachate contaminating nearby groundwater and surface water. (c) The relative cost of landfills is average but increasing. (d) Landfills can serve any size population.

Open burning: (a) The main advantages of open burning are that it can be done on both small and large scales with relatively little cost, and that it reduces the volume of waste, leaving only ash. (b) The disadvantages are that it only works with combustible waste such as paper and wood products. The combustion of the waste may not be complete. Combustion of large amounts of waste may be difficult to control. No energy is recovered from the combustion of wastes, and the process produces air pollution. (c) The relative cost is low. (d) The population served is usually small.

Incineration: (a) The advantages of incineration include more complete destruction of the wastes and less ash production than in open burning. This process can destroy less-burnable wastes such as plastics and food. Energy from incineration can be captured and used to make steam and electricity. (b) The disadvantages are the high cost and large volumes of waste required. If the waste combustion is not self-sustaining, supplemental fuel may be required to maintain incineration. This process does not work for incombustible materials such as glass and steel. The ash may contain hazardous compounds. This process also produces air pollution. (c) The cost of incineration is high. (d) It best serves a large population.

Composting: (a) The advantages of composting are that it uses the process of natural degradation of wastes and that it can be performed on a small or large scale. (b) The main disadvantage is that this process is only applicable to easily degradable wastes. It is not applicable to plastic, glass, or metal. The addition of moisture and readily degradable material such as yard wastes may be needed to help this process. (c) The relative costs are low to medium. (d) Composting best serves a small to medium population.

Recycling: (a) The main advantages of recycling are that it reuses process materials and saves natural resources. It also reduces the amount of waste that must be disposed by other means. (b) The disadvantage is that it uses more labor than other methods. This method is viable only if there is a company willing to perform the recycling and if there is a market for the recycled materials. The trash must be separated into different waste streams. (c) The relative costs of recycling are medium to low and should decrease as the demand for recycled materials and the cost of other disposal methods increase. (d) The population served is medium to large depending on the market for recycled materials.

7. (a) Use the trash disposal in the eighth year as the average over the past 16 years.

Population in year 8:

$$\frac{500{,}000}{(1.03)^8} = 394{,}705 \text{ capita}$$

Solid waste disposed (per capita in year 8):

$$\frac{5.4\,\dfrac{\text{lbm}}{\text{day-capita}}}{(1.01)^8} = 4.99 \text{ lbm/day-capita}$$

Solid waste disposed over the past 16 years:

$$\begin{aligned}(394{,}705 \text{ capita})&\left(4.99\,\frac{\text{lbm}}{\text{day-capita}}\right)(16 \text{ yr})\\ &= 31{,}513{,}247 \text{ lbm-yr/day}\end{aligned}$$

Over the next four years, use the average trash disposal as the second year.

Population in 2 years:

$$(500{,}000)(1.03)^2 = 530{,}450 \text{ capita}$$

Solid waste disposed:

$$\begin{aligned}\left(5.4\,\frac{\text{lbm}}{\text{day-capita}}\right)&(1-0.03)(1.01)^2\\ &= 5.34 \text{ lbm/day-capita}\end{aligned}$$

Solid waste disposed over the next 4 years:

$$\begin{aligned}(530{,}450 \text{ capita})&\left(5.34\,\frac{\text{lbm}}{\text{day-capita}}\right)(4 \text{ yr})\\ &= 11{,}330{,}412 \text{ lbm-yr/day}\end{aligned}$$

The solid waste occupies 75% of the site. The depth of the waste is 40 ft.

The area filled by 16 years of trash is

$$A = \frac{\left(31{,}513{,}247\ \frac{\text{lbm-yr}}{\text{day}}\right)\left(\frac{365\ \text{days}}{\text{yr}}\right)}{\left(1000\ \frac{\text{lbm}}{\text{yd}^3}\right)\left(\frac{\text{yd}^3}{27\ \text{ft}^3}\right)(40\ \text{ft})\left(43{,}560\ \frac{\text{ft}^3}{\text{ac}}\right)(0.75)}$$
$$= 237.65\ \text{ac}$$

The area filled in the next four years is

$$A = \frac{\left(11{,}330{,}412\ \frac{\text{lbm-yr}}{\text{day}}\right)\left(\frac{365\ \text{days}}{\text{yr}}\right)}{\left(1000\ \frac{\text{lbm}}{\text{yd}^3}\right)\left(\frac{\text{yd}^3}{27\ \text{ft}^3}\right)(40\ \text{ft})\left(43{,}560\ \frac{\text{ft}^3}{\text{ac}}\right)(0.75)}$$
$$= 85.45\ \text{ac} \quad [135.5\ \text{ac with buffer zone}]$$

Add the areas and include the 250-ft buffer zone.

$$\sqrt{(237.65\ \text{ac} + 85.45\ \text{ac})\left(43{,}560\ \frac{\text{ft}^2}{\text{ac}}\right)} = 3752\ \text{ft}$$

$$\frac{(3752\ \text{ft} + 250\ \text{ft} + 250\ \text{ft})^2}{43{,}560\ \frac{\text{ft}^2}{\text{ac}}} = \boxed{415\ \text{ac}}$$

(b) Replacement site:

$$4\ \text{yr} + 5\ \text{yr capacity} = 9\ \text{yr}$$

Average trash disposal at 4.5 years:

$$\text{population in 4.5 years} = (500{,}000)(1.03)^{4.5} = 571{,}133\ \text{capita}$$

$$\text{solid waste disposed} = \left(5.4\ \frac{\text{lbm}}{\text{capita-day}}\right)(1 - 0.03) \times (1.01)^{4.5} = 5.48\ \text{lbm/capita-day}$$

Solid waste disposed over the next 9 years:

$$(571{,}133\ \text{capita})\left(5.48\ \frac{\text{lbm}}{\text{capita-day}}\right)(9\ \text{yr}) = 28{,}168{,}280\ \text{lbm-yr/day}$$

Area of the site:

$$A = \frac{\left(28{,}168{,}280\ \frac{\text{lbm-yr}}{\text{day}}\right)\left(\frac{365\ \text{days}}{\text{yr}}\right)}{\left(1000\ \frac{\text{lbm}}{\text{yd}^3}\right)\left(\frac{\text{yd}^3}{27\ \text{ft}^3}\right)(40\ \text{ft})(0.75)}$$
$$= 9{,}253{,}280\ \text{ft}^2$$
$$= \frac{9{,}253{,}280\ \text{ft}^2}{43{,}560\ \frac{\text{ft}^2}{\text{ac}}}$$
$$= 212.43\ \text{ac}$$

Include the 250-ft buffer zone.

$$\sqrt{9{,}253{,}280\ \text{ft}^2} = 3042\ \text{ft}$$

$$\frac{(3042\ \text{ft} + 250\ \text{ft} + 250\ \text{ft})^2}{43{,}560\ \frac{\text{ft}^2}{\text{ac}}} = \boxed{288\ \text{ac}}$$

8. (a) The spray volume is

$$\left(2.5\ \frac{\text{ac-in}}{\text{ac-week}}\right)\left(43{,}560\ \frac{\text{ft}^2}{\text{ac}}\right)\left(\frac{\text{ft}}{12\ \text{in}}\right) \times \left(7.48\ \frac{\text{gal}}{\text{ft}^3}\right)\left(\frac{\text{week}}{7\ \text{days}}\right) = 9697\ \text{gal/day-ac}$$

The total area required is

$$A_t = \frac{1 \times 10^6\ \frac{\text{gal}}{\text{day}}}{9697\ \frac{\text{gal}}{\text{day-ac}}} = 103.1\ \text{ac}$$

Only 25% of the nozzles are used at one time, so 25% of the total area is used at one time. The area used at any moment is

$$(103.1\ \text{ac})(0.25) = \boxed{25.8\ \text{ac}}$$

(b) The area covered by each nozzle is

$$A_n = \frac{\pi(350\ \text{ft})^2}{(4)\left(43{,}560\ \frac{\text{ft}^2}{\text{ac}}\right)} = 2.21\ \text{ac/nozzle}$$

The number of nozzles is

$$\frac{103.1\ \text{ac}}{2.21\ \frac{\text{ac}}{\text{nozzle}}} = \boxed{46.7\ \text{nozzles} \quad [\text{say } 48]}$$

(c) Each nozzle delivers 275 gal/min at 75 lbf/in^2. There are 48 nozzles in use, but only 25% are used at one time.

$$(48)(0.25) = 12\ \text{nozzles at once}$$

The pump flow, Q, is

$$\left(275\ \frac{\text{gal}}{\text{min}}\right)(12) = 3300\ \text{gal/min} = 4.75\ \text{MGD} > 1\ \text{MGD} \quad [\text{ok}]$$

The exit head at the nozzles is

$$\left(75\ \frac{\text{lbf}}{\text{in}^2}\right)\left(2.31\ \frac{\text{ft}}{\frac{\text{lbf}}{\text{in}^2}}\right) = 173.25\ \text{ft}$$

The head added by the pumps is

$$h_a = 173.25 \text{ ft} - 45 \text{ ft} + 30 \text{ ft} = 158.25 \text{ ft}$$

$$\text{water hp} = \frac{(h_a \text{ ft})\left(Q \frac{\text{gal}}{\text{min}}\right)(\text{SG of water})}{3956}$$

$$= \frac{(158.25 \text{ ft})\left(3300 \frac{\text{gal}}{\text{min}}\right)(1.0)}{3956}$$

$$= 132.0 \text{ hp}$$

$$\text{motor hp} = \frac{\text{water hp}}{\eta_{\text{pump}}}$$

$$= \frac{132}{0.82} = 175 \text{ hp}$$

Use one 175-hp pump motor or two 90-hp pump motors. Note that motors are rated by their output power and that motor efficiency is irrelevant.

(d) The following should be considered when determining the application rate of effluent to spray the fields.

- soil permeability
- soil crop cover (crop uptake of moisture and nutrients)
- average monthly/annual precipitation and evaporation
- organic and nutrient content of effluent to be sprayed
- metal/toxic content of effluent to be sprayed
- depth to groundwater table
- subsurface geology
- site topography and drainage
- distance to nearest surface water

The first five considerations are primarily used in determining the application rate. All the considerations are used in determining the overall site applicability.

9. $$A = \frac{(90 \text{ ft})(120 \text{ ft})}{43{,}560 \frac{\text{ft}^2}{\text{ac}}} = 0.25 \text{ ac}$$

$$\frac{(12 \text{ sites})(\text{block})}{\text{block}} = \frac{(540 \text{ ft})(240 \text{ ft})}{43{,}560 \frac{\text{ft}^2}{\text{ac}}}$$

$$= 3 \text{ ac}$$

$$\frac{480 \text{ sites}}{(12 \text{ sites})(\text{block})} = 40 \text{ blocks}$$

Assume wastewater flow is 100 gal/capita-day. Wastewater flow from the houses is

$$\left(4.1 \frac{\text{capita}}{\text{site}}\right)\left(100 \frac{\text{gal}}{\text{capita-day}}\right)(480 \text{ sites})$$

$$= 196{,}800 \text{ gal/day}$$

$$\text{infiltration + inflow} = \frac{200 \text{ gal}}{\text{in-mi-day}} + \frac{2000 \text{ gal}}{\text{in-mi-day}}$$

$$= 2200 \text{ gal/in-mi-day}$$

The length of the 6-in feeder from the houses is

$$L = 12\text{-ft street} + 50\text{-ft lot}$$

$$= 62 \text{ ft}$$

$$\frac{(62 \text{ ft})(480 \text{ sites})}{5280 \frac{\text{ft}}{\text{mi}}} = 5.64 \text{ mi}$$

Infiltration/inflow from the 6-in pipe is

$$\left(\frac{2200 \text{ gal}}{\text{in-mi-day}}\right)(6 \text{ in})(5.64 \text{ mi}) = 74{,}448 \text{ gal/day}$$

Assume the 8-in main is eight blocks long and there are five mains.

$$\frac{(8 \text{ in})(540 \text{ ft}) + (8 \text{ blocks})(24 \text{ ft})(5 \text{ mains})}{5280 \frac{\text{ft}}{\text{mi}}} = 4.27 \text{ mi}$$

Infiltration/inflow from the 8-in pipe is

$$\left(\frac{2200 \text{ gal}}{\text{in-mi-day}}\right)(8 \text{ in})(4.27 \text{ mi}) = 75{,}152 \text{ gal/day}$$

Assume that the 10-in main is five blocks long and 200 ft to the pump station.

$$\frac{(5 \text{ blocks})(240 \text{ ft}) + (5 \text{ blocks})(24 \text{ ft}) + 200 \text{ ft}}{5280 \frac{\text{ft}}{\text{mi}}} = 0.29 \text{ mi}$$

Infiltration/inflow from the 10-in main is

$$\left(\frac{2200 \text{ gal}}{\text{in-mi-day}}\right)(10 \text{ in})(0.29 \text{ mi}) = 6380 \text{ gal/day}$$

The total flow is

$$196{,}800 \frac{\text{gal}}{\text{day}} + 74{,}448 \frac{\text{gal}}{\text{day}} + 75{,}152 \frac{\text{gal}}{\text{day}} + 6380 \frac{\text{gal}}{\text{day}}$$

$$= \boxed{352{,}780 \text{ gal/day}}$$

10. (a) The effluent BOD_5 is the untreated influent soluble BOD_5, S, plus the BOD_5 of the effluent biological solids.

$$20 \frac{\text{mg}}{\ell} = S + \left(20 \frac{\text{mg}}{\ell}\right)(0.65)(1.42)(0.68)$$

$$= S + 12.6 \frac{\text{mg}}{\ell}$$

$$S = 20 \frac{\text{mg}}{\ell} - 12.6 \frac{\text{mg}}{\ell}$$

$$= 7.4 \text{ mg}/\ell$$

The biological efficiency, η, is

$$\eta_b = \left(\frac{\text{influent BOD}_5 - \text{effluent BOD}_5}{\text{influent BOD}_5}\right)(100\%)$$

$$\left(\frac{250\,\frac{\text{mg}}{\ell} - 7.4\,\frac{\text{mg}}{\ell}}{250\,\frac{\text{mg}}{\ell}}\right)(100\%) = \boxed{97\%}$$

(b) The overall efficiency, η, is

$$\eta = \left(\frac{250\,\frac{\text{mg}}{\ell} - 20\,\frac{\text{mg}}{\ell}}{250\,\frac{\text{mg}}{\ell}}\right)(100\%) = \boxed{92\%}$$

(c) Calculate the reactor volume.

$$V_{\text{reactor}} = \theta Q$$

$$X = \frac{\theta_c Y(S_0 - S)}{\theta(1 + R_d\theta_c)}$$

mass of cells

Substitute and solve for V_{reactor}.

$\frac{Q\theta_c(S_0-S)}{(1+R_d\theta_c)}$

$$V_{\text{reactor}} = \frac{(5\text{ MGD})(10\text{ days})\left(0.6\,\frac{\text{lbm VSS}}{\text{lbm BOD}_5}\right) \times \left(250\,\frac{\text{mg}}{\ell} - 7.4\,\frac{\text{mg}}{\ell}\right)}{\left(3500\,\frac{\text{mg}}{\ell}\right)\left[1 + \left(\frac{0.06}{\text{day}}\right)(10\text{ days})\right]}$$

$$= 1.3\text{ MG}$$

Sludge production can be calculated from the observed yield.

biodegradable yield coef.

$$Y_{\text{obs}} = \frac{Y}{1 + k_d\theta_c}$$

$$= \frac{0.6\,\frac{\text{lbm VSS}}{\text{lbm BOD}}}{1 + \left(\frac{0.06}{\text{day}}\right)(10\text{ days})}$$

$$= 0.375$$

The biomass production is

$$P_x(\text{MLVSS}) = Y_{\text{obs}}(S_0 - S)Q$$

$$= (0.375)\left(250\,\frac{\text{mg}}{\ell} - 7.4\,\frac{\text{mg}}{\ell}\right) \times (5\text{ MGD})\left(8.34\,\frac{\text{lbm-}\ell}{\text{mg-MGD}}\right)$$

$$= \boxed{3794\text{ lbm/day}}$$

(d) The increase in MLSS is

$$\frac{\text{MLVSS}}{0.8} = \frac{3794\,\frac{\text{lbm}}{\text{day}}}{0.8}$$

$$= \boxed{4743\text{ lbm/day}}$$

(e) The sludge to be wasted is the increase in MLSS − MLSS in the effluent.

$$4743\,\frac{\text{lbm}}{\text{day}} - \left[\left(20\,\frac{\text{mg}}{\ell}\right)(5\text{ MGD})\left(8.34\,\frac{\text{lbm-}\ell}{\text{mg-MGD}}\right)\right]$$

$$= \boxed{3909\text{ lbm/day}}$$

(f) If sludge is wasted from the reactor, the sludge wasting rate is Q_w.

$$\theta_c = \frac{V_{\text{reactor}}X}{Q_wX + Q_eX_e}$$

Rearrange this equation, solving for Q_w.

$$Q_wX + Q_eX_e = \frac{V_{\text{reactor}}X}{\theta_c}$$

$$Q_w = \left(\frac{V_{\text{reactor}}X}{\theta_c} - Q_eX_e\right)\left(\frac{1}{X}\right)$$

Assume that $Q_e = Q$.

$$(\text{VSS})X_e = (0.8)\left(20\,\frac{\text{mg}}{\ell}\right) = 16\text{ mg}/\ell$$

$$Q_w = \left[\frac{(1.3\text{ MG})\left(3500\,\frac{\text{mg}}{\ell}\right)}{10\text{ days}} - (5\text{ MGD})\left(16\,\frac{\text{mg}}{\ell}\right)\right] \times \left(\frac{1}{3500\,\frac{\text{mg}}{\ell}}\right)$$

$$= \boxed{0.107\text{ MGD}}$$

(g) Calculate the recirculation ratio, R, using a mass balance around the reactor.

$$\text{reactor MLVSS} = 3500\text{ mg}/\ell$$

$$\text{recycle MLVSS} = 8000\text{ mg}/\ell$$

$$\left(3500\,\frac{\text{mg}}{\ell}\right)(Q + Q_{\text{reactor}}) = \left(8000\,\frac{\text{mg}}{\ell}\right)(Q_{\text{reactor}})$$

$$\frac{Q + Q_{\text{reactor}}}{Q_{\text{reactor}}} = \frac{8000\,\frac{\text{mg}}{\ell}}{3500\,\frac{\text{mg}}{\ell}}$$

$$= 2.29\text{ mg}/\ell$$

$$\frac{Q}{Q_{\text{reactor}}} + 1 = 2.29$$

$$\frac{Q}{Q_{\text{reactor}}} = 2.29 - 1$$

$$= 1.29$$

$$R = \frac{Q_{\text{reactor}}}{Q}$$

$$= \frac{1}{1.29} = \boxed{0.78}$$

(h) The hydraulic retention time, θ, is

$$\theta = \frac{V}{Q} = \frac{1.3\ \text{Mgal}}{5\ \text{MGD}} = (0.26\ \text{days})\left(\frac{24\ \text{hr}}{\text{day}}\right) = \boxed{6.24\ \text{hr}}$$

(i) Check the specific substrate utilization rate, u.

$$u = \frac{S_0 - S}{\theta X} = \frac{250\ \frac{\text{mg}}{\ell} - 7.4\ \frac{\text{mg}}{\ell}}{(0.26\ \text{day})\left(3500\ \frac{\text{mg}}{\ell}\right)} = \boxed{\frac{0.27\ \text{mg BOD}_5\ \text{utilized}}{\text{mg MLVSS-day}}}$$

(j) Check the food-to-microorganism ratio, $R_{\text{F-M}}$.

$$R_{\text{F-M}} = \frac{S_0}{\theta X} = \frac{250\ \frac{\text{mg}}{\ell}}{(0.26\ \text{day})\left(3500\ \frac{\text{mg}}{\ell}\right)} = \boxed{\frac{0.275\ \text{mg BOD}}{\text{mg MLVSS-day}}}$$

(k) Check the volumetric loading rate, VLR.

$$\text{VLR} = \frac{S_0 Q}{V} = \frac{\left(250\ \frac{\text{mg}}{\ell}\right)(5\ \text{MGD})\left(8.34\ \frac{\text{lbm-}\ell}{\text{mg-MG}}\right)(1000)}{\frac{1.3 \times 10^6\ \text{gal}}{7.48\ \frac{\text{gal}}{\text{ft}^3}}} = \frac{60\ \frac{\text{lbm}}{\text{day-BOD}}}{1000\ \text{ft}^3}$$

Calculate the oxygen required.

$$m_{O_2} = \frac{Q(S_0 - S)}{f} - 1.42 P_x$$

$$f = \frac{\text{BOD}_5}{\text{BOD}_L} = 0.68$$

$$m_{O_2} = \frac{(5\ \text{MGD})\left(250\ \frac{\text{mg}}{\ell} - 7.4\ \frac{\text{mg}}{\ell}\right)\left(8.34\ \frac{\text{lbm-}\ell}{\text{mg-MG}}\right)}{0.68} - (1.42)\left(3794\ \frac{\text{lbm}}{\text{day}}\right) = 9490\ \text{lbm/day}$$

Air is 23.2% O_2 by weight. Since the transfer efficiency is 8%, the normal air flow is

$$\text{air flow} = \frac{9490\ \frac{\text{lbm}}{\text{day}}}{\left(0.075\ \frac{\text{lbm}}{\text{ft}^3}\right)(0.232)(0.08)\left(1440\ \frac{\text{min}}{\text{day}}\right)} = 4734\ \text{ft}^3/\text{min}$$

With the 100% excess capacity, the blower capacity is

$$(2)\left(4734\ \frac{\text{ft}^3}{\text{min}}\right) = \boxed{9468\ \text{ft}^3/\text{min}}$$

11. (a) Reuse as much flume/wash water as possible after sedimentation and chlorination to reduce the wastewater flow.

The sedimentation tank volume is

$$V = Qt = \left(2.2 \times 10^6\ \frac{\text{gal}}{\text{day}}\right)(2\ \text{hr})\left(\frac{\text{day}}{24\ \text{hr}}\right) = 1.83 \times 10^5\ \text{gal}$$

$$A = \frac{V}{D} = \frac{1.83 \times 10^5\ \text{gal}}{\left(7.48\ \frac{\text{gal}}{\text{ft}^3}\right)(10\ \text{ft})} = 2451\ \text{ft}^2$$

Use a length-to-width ratio of 4:1.

$$LW = (4W)W = 2451\ \text{ft}^2$$

$$L = 99\ \text{ft}$$

$$W = \frac{L}{4} = \frac{99\ \text{ft}}{4} = 24.75\ \text{ft}$$

The surface overflow rate is

$$\frac{2.2 \times 10^6\ \frac{\text{gal}}{\text{day}}}{2451\ \text{ft}^2} = 898\ \text{gal/day-ft}^2$$

The required weir length is

$$\frac{2.2 \times 10^6\ \frac{\text{gal}}{\text{day}}}{15{,}000\ \frac{\text{gal}}{\text{day-ft}}} = 147\ \text{ft}$$

The design summary of the sedimentation basin for the flume/wash water is

detention time, $t = 2$ hr
volume, $V = 1.83 \times 10^5$ gal
dimensions, $L \times W \times D = 99\ \text{ft} \times 24.75\ \text{ft} \times 10\ \text{ft}$
weir length $= 147$ ft
overflow rate $= 900$ gal/day-ft^2

(b) Design an aerobic lagoon to treat all the wastewater.

The composition of the combined wastes is

$$Q = 2.2 \times 10^6 \, \frac{\text{gal}}{\text{day}} + 6.6 \times 10^5 \, \frac{\text{gal}}{\text{day}} + 7.5 \times 10^4 \, \frac{\text{gal}}{\text{day}}$$
$$= 2.935 \times 10^6 \text{ gal/day}$$

$$\text{BOD} = \frac{\left(2.2 \times 10^6 \, \frac{\text{gal}}{\text{day}}\right)\left(200 \, \frac{\text{mg}}{\ell}\right) + \left(6.6 \times 10^5 \, \frac{\text{gal}}{\text{day}}\right) \times \left(1230 \, \frac{\text{mg}}{\ell}\right) + \left(7.5 \times 10^4 \, \frac{\text{gal}}{\text{day}}\right)\left(1420 \, \frac{\text{mg}}{\ell}\right)}{2.935 \times 10^6 \, \frac{\text{gal}}{\text{day}}}$$
$$= 463 \text{ mg}/\ell$$

The BOD organic loading is

$$(2.935 \text{ MGD})\left(463 \, \frac{\text{mg}}{\ell}\right)\left(8.34 \, \frac{\text{lbm-}\ell}{\text{mg MG}}\right) = 11{,}333 \text{ lbm/day}$$

Size the aerobic lagoon for a two-day detention time at 1.5-ft depth.

$$A = \frac{Qt}{D} = \frac{\left(2.935 \times 10^6 \, \frac{\text{gal}}{\text{day}}\right)(2 \text{ days})}{(1.5 \text{ ft})\left(7.48 \, \frac{\text{gal}}{\text{ft}^3}\right)\left(43{,}560 \, \frac{\text{ft}^2}{\text{ac}}\right)}$$
$$= 12.0 \text{ ac}$$

Check the organic loading.

$$\frac{11{,}333 \, \frac{\text{lbm}}{\text{day}}}{12.0 \text{ ac}} = 944.4 \text{ lbm/ac-day}$$

Check the lagoon size based on an organic loading of 86 lbm/ac-day and 1.5-ft depth.

$$A = \frac{11{,}333 \, \frac{\text{lbm}}{\text{day}}}{86 \, \frac{\text{lbm}}{\text{ac-day}}} = 131.8 \text{ ac}$$

$$V = (131.8 \text{ ac})\left(43{,}560 \, \frac{\text{ft}^2}{\text{ac}}\right)(1.5 \text{ ft})\left(7.48 \, \frac{\text{gal}}{\text{ft}^3}\right)$$
$$= 64.4 \times 10^6 \text{ gal}$$

$$t = \frac{V}{Q} = \frac{64.4 \times 10^6 \text{ gal}}{2935 \times 10^6 \, \frac{\text{gal}}{\text{day}}} = 21.9 \text{ days}$$

A lagoon that is 64.4×10^6 gal in volume with an area of 131.8 ac is too large to be practical or economical for this facility.

Design a lagoon for a composite of process water and lime drainage only.

$$Q = 6.6 \times 10^5 \, \frac{\text{gal}}{\text{day}} + 7.5 \times 10^4 \, \frac{\text{gal}}{\text{day}}$$
$$= 7.35 \times 10^5 \text{ gal/day}$$

$$\text{BOD} = \frac{\left(6.6 \times 10^5 \, \frac{\text{gal}}{\text{day}}\right)\left(1230 \, \frac{\text{mg}}{\ell}\right) + \left(7.5 \times 10^4 \, \frac{\text{gal}}{\text{day}}\right)\left(1420 \, \frac{\text{mg}}{\ell}\right)}{7.35 \times 10^5 \, \frac{\text{gal}}{\text{day}}}$$
$$= 1249 \text{ mg}/\ell$$

The BOD organic loading is

$$(0.735 \text{ MGD})\left(1249 \, \frac{\text{mg}}{\ell}\right)\left(8.34 \, \frac{\text{lbm-}\ell}{\text{mg MG}}\right) = 7658 \text{ lbm/day}$$

Design a lagoon for a two-day detention time at 1.5-ft depth.

$$A = \frac{\left(7.35 \times 10^5 \, \frac{\text{gal}}{\text{day}}\right)(2 \text{ days})}{(1.5 \text{ ft})\left(7.48 \, \frac{\text{gal}}{\text{ft}^3}\right)\left(43{,}560 \, \frac{\text{ft}^2}{\text{ac}}\right)}$$
$$= 3.0 \text{ ac}$$

$$\text{organic loading} = \frac{7658 \, \frac{\text{lbm}}{\text{day}}}{3 \text{ ac}}$$
$$= 2553 \text{ lbm/ac-day}$$

A lagoon that is only three acres in area has a very high organic loading and will not effectively treat the waste.

Use an aerobic lagoon, built in four cells, to treat all the waste. Each cell will be 3.0 ac in area to accommodate the flow of the process water and lime drainage only.

The design summary for an aerobic lagoon to treat all the wastewater is

$$\begin{aligned} \text{total area, } A &= 12 \text{ ac} \\ \text{number of cells and size} &= \text{four 3-ac cells} \\ \text{detention time, } t &= 2 \text{ days} \\ \text{working depth, } D &= 1.5 \text{ ft} \\ \text{organic loading} &= 944.4 \text{ lbm/ac-day} \end{aligned}$$

The treated effluent will be disposed by land application.

(c) The following alternatives may also be considered for treating the wastewater from this facility.

1) Since the beet sugar plant operates only three months of the year, the wastewater could be stored and treated slowly over the entire year. More storage capacity would be required. This option would reduce the organic loading on the aerobic lagoon.

2) Treat the composite flow of process water and lime drainage by a high-rate-activated sludge system.

12. (a) The treatment process is shown in the following schematic diagram.

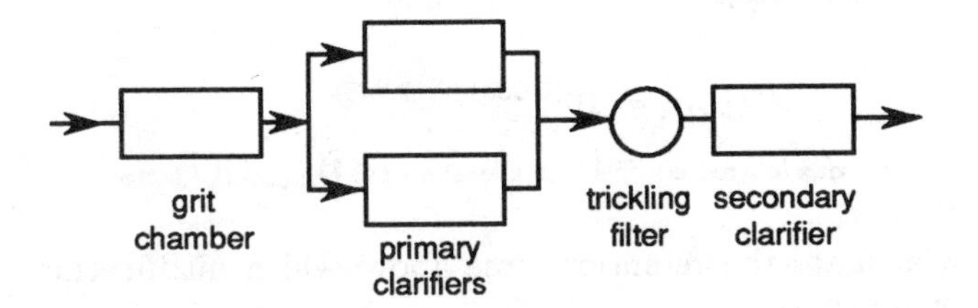

(b) According to the *Ten States' Standards*, this facility requires two primary clarifiers, each designed for full flow.

Assume a surface loading rate, SLR, of 1000 gal/day-ft^2.

$$A = \frac{Q}{\text{SLR}} = \frac{2 \times 10^6 \,\frac{\text{gal}}{\text{day}}}{1000 \,\frac{\text{gal}}{\text{day-ft}^2}} = 2000 \text{ ft}^2$$

Assume a detention time, t, of two hours.

$$V = tQ = (2\,\text{hr})\left(2 \times 10^6 \,\frac{\text{gal}}{\text{day}}\right)\left(\frac{\text{day}}{24\text{ hr}}\right)$$
$$= 22{,}282 \text{ ft}^3$$
$$D = \frac{V}{A} = \frac{22{,}282 \text{ ft}^3}{2000 \text{ ft}^2} = 11.1 \text{ ft}$$

For the clarifier, the length-to-width ratio is 4:1.

$$LW = (4W)W = 2000 \text{ ft}^2$$
$$L = 89.4 \text{ ft}$$
$$W = 22.4 \text{ ft}$$

Assume a weir loading rate of 20,000 gal-day/ft.

$$\text{weir length} = \frac{2 \times 10^6 \,\frac{\text{gal}}{\text{day}}}{20{,}000 \,\frac{\text{gal-day}}{\text{ft}}} = 100 \text{ ft}$$

The design summary of the clarifiers is

$$\text{dimensions, } L \times W \times D = 89.4 \text{ ft} \times 22.4 \text{ ft} \times 11.1 \text{ ft}$$
$$\text{detention time, } t = 2 \text{ hr}$$
$$\text{volume, } V = 22{,}282 \text{ ft}^3$$
$$\text{weir length} = 100 \text{ ft}$$

In the primary clarifiers, assume the BOD$_5$ reduction is 30% and the TSS reduction is 60%.

(c) For a trickling filter using plastic media, the NRC formulas are not applicable. (Reference: *Wastewater Engineering*, pp. 412–415.)

The influent BOD$_5$ to the trickling filter, S_i, is

$$\left(250 \,\frac{\text{mg}}{\ell}\right)(1 - 0.3) = 175 \text{ mg}/\ell$$

Assume the effluent BOD$_5$ from the trickling filter: $S_e \leq$ 30 mg/ℓ. The treatment efficiency is

$$\left(\frac{S_i - S_e}{S_i}\right)(100\%) = \left(\frac{175 \,\frac{\text{mg}}{\ell} - 30 \,\frac{\text{mg}}{\ell}}{175 \,\frac{\text{mg}}{\ell}}\right)(100\%) = 82.9\%$$

Assume the depth, D, of the trickling filter is 10 ft.

The area, A, of the trickling filter can be calculated by

$$\frac{S_e}{S_i} = e^{\left[-k_{20}D(Q/A)^{-n}\right]}$$

Rearranging,

$$A = Q\left[\frac{-\ln\left(\frac{S_e}{S_i}\right)}{k_{20}D}\right]^{1/n}$$

Assume $k_{20} = 0.085$ gal/min and $n = 0.5$.

$$Q = \frac{2 \times 10^6 \,\frac{\text{gal}}{\text{day}}}{1440 \,\frac{\text{min}}{\text{day}}} = 1389 \text{ gal/min}$$

$$A = \left(1389 \,\frac{\text{gal}}{\text{min}}\right)\left[\frac{-\ln\left(\frac{30}{175}\right)}{\left(0.085 \,\frac{\text{gal}}{\text{min}}\right)(10\text{ ft})}\right]^{1/0.5}$$
$$= 5979 \text{ ft}^2$$
$$D = \sqrt{\frac{4A}{\pi}} = \sqrt{\frac{(4)(5979 \text{ ft}^2)}{\pi}}$$
$$= 87.25 \text{ ft}$$

The trickling filter dimensions are 87.25 ft diameter × 10 ft deep.

$$V = (5979 \text{ ft}^2)(10 \text{ ft}) = 59{,}790 \text{ ft}^3$$

The organic loading is

$$\frac{(2\text{ MGD})\left(175 \,\frac{\text{mg}}{\ell}\right)\left(8.34 \,\frac{\text{lbm-}\ell}{\text{mg MG}}\right)}{59{,}790 \text{ ft}^3}$$
$$= \boxed{48.8 \text{ lbm/day-1000 ft}^3}$$

The hydraulic loading is

$$\frac{1389\ \frac{\text{gal}}{\text{min}}}{5979\ \text{ft}^2} = \boxed{0.23\ \text{gal/min-ft}^2}$$

These values fit the criteria for a high-rate trickling filter.

13. (a) The design procedure is as follows. Calculate the cell volume, V, from the flow and detention time. (Assume all three ponds have the same volume.)

$$V = \boxed{Qt}$$

Calculate the cell surface area, A.

$$A = \boxed{\frac{V}{D}}$$

(b)
$$T_{\text{winter}} = \boxed{\frac{(A_{\text{ft}^2})(f)(T_{\text{aw},°\text{F}}) + (Q_{\text{MGD}})(T_{\text{ww},°\text{F}})}{(A_{\text{ft}^2})(f) + Q_{\text{MGD}}}}$$

$$T_{\text{summer}} = \boxed{\frac{(A_{\text{ft}^2})(f)(T_{\text{as},°\text{F}}) + (Q_{\text{MGD}})(T_{\text{ws},°\text{F}})}{(A_{\text{ft}^2})(f) + Q_{\text{MGD}}}}$$

(c)
$$\theta = \text{temperature coefficient}$$

$$k_2 = \boxed{k_1 \theta^{T_2 - T_1}}$$

(d) Assume there is no algal growth in any of the three cells.

$$S_1 = \frac{1 + k_d \theta_c}{Y_T k \theta_c}$$

(e)
$$X_{1,\text{VSS}} = \boxed{\frac{(S_0 - S_1)Q}{S_1 k V}}$$

(f) Assume all the soluble BOD_u entering the cell is removed.

$$X_{2,\text{VSS}} = \boxed{\frac{X_1 + Y_T S_1}{\frac{k_d}{t} + 1}}$$

(g) Assume the soluble BOD_u entering the cell is zero.

$$X_3 = \frac{X_2}{\frac{k_d}{t} + 1}$$

(h)
$$\text{BOD}_{5,\text{eff}} = \boxed{0.54 X_3 (\text{VSS}_{\text{eff}}) + S_e}$$

(i) The correction factor for pH is

$$\text{pH factor} = \frac{1}{1 + (0.04)\left(10^{\text{pH}_{\text{opt}} - \text{pH}} - 1\right)}$$

The correction for temperature is

$$T_{\text{factor}} = 10^{(0.033)(T - 20°\text{C})}$$

$$(\mu_{\text{max}})_{\text{corrected}} = (\mu_{\text{max}})_{\text{assumed}}(\text{pH}_{\text{factor}})(T_{\text{factor}})$$

Calculate the retention time above which nitrification will occur.

$$\theta_c = \boxed{\frac{1}{\mu_{\text{max}}}}$$

(j)
$$\text{NOD} = (3.81Q)(N_{\text{available mg}/\ell})$$

$$\Delta O_2 = \boxed{\begin{array}{r}(8.34Q)(1 - 1.424T)(S_o - S_e) \\ +(8.34)(1.42)k_d V + \text{NOD}\end{array}}$$

(k)
$$\text{hp} = \boxed{\frac{\text{lbm } O_2 \text{ required}}{\left(1.8\ \frac{\text{lbm } O_2}{\text{hp-hr}}\right)(24\ \text{hr})}}$$

(l) Assume the mixing power is approximately 30 horsepower per million gallons.

$$P = \left(30\ \frac{\text{hp}}{\text{MG}}\right)(V_{\text{MG}})$$

Use the greater horsepower of steps (k) and (l) in each cell. Determine the number of aerators needed to achieve the total horsepower required.

(m) Use a length-to-width ratio of 3:1 minimum and determine the position of the aerators.

(n) The first cell controls the nutrient requirements.

$$\Delta X = 8.34 Q X_1$$

$$N_{\text{requirement}} = 0.122 \Delta X$$

$$P_{\text{requirement}} = 0.023 \Delta X$$

$$N_{\text{available}} = 8.34 Q N_{\text{influent, mg}/\ell}$$

$$P_{\text{available}} = 8.34 Q P_{\text{influent, mg}/\ell}$$

If the amount of N or P available is less than the amount required, add the difference of N or P.

14. (a) For the pipe,

$$A = \frac{\pi D^2}{4} = \frac{\pi\left(\frac{8\text{ in}}{12\,\frac{\text{in}}{\text{ft}}}\right)^2}{4}$$
$$= 0.349\text{ ft}^2$$
$$Q = vA = \left(0.6\,\frac{\text{ft}}{\text{sec}}\right)(0.349\text{ ft}^2)$$
$$= 0.21\text{ ft}^3/\text{sec}$$
$$Q = \left(0.21\,\frac{\text{ft}^3}{\text{sec}}\right)\left(\frac{7.48\text{ gal}}{\text{ft}^3}\right)\left(\frac{60\text{ sec}}{\text{min}}\right)$$
$$= \boxed{94\text{ gal/min}}$$

(b) The diluted characteristics can be calculated from

$$C = \frac{Q_1C_1 + Q_2C_2}{Q_1 + Q_2}$$
$$\text{BOD}_5 = \frac{\left(150\,\frac{\text{mg}}{\ell}\right)\left(0.21\,\frac{\text{ft}^3}{\text{sec}}\right) + \left(4\,\frac{\text{mg}}{\ell}\right)\left(2\,\frac{\text{ft}^3}{\text{sec}}\right)}{0.21\,\frac{\text{ft}^3}{\text{sec}} + 2\,\frac{\text{ft}^3}{\text{sec}}}$$
$$= \boxed{17.9\text{ mg}/\ell}$$

(c) The dissolved oxygen, DO, is

$$\frac{\left(1\,\frac{\text{mg}}{\ell}\right)\left(0.21\,\frac{\text{ft}^3}{\text{sec}}\right) + \left(10\,\frac{\text{mg}}{\ell}\right)\left(2\,\frac{\text{ft}^3}{\text{sec}}\right)}{0.21\,\frac{\text{ft}^3}{\text{sec}} + 2\,\frac{\text{ft}^3}{\text{sec}}}$$
$$= \boxed{9.1\text{ mg}/\ell}$$

(d) The summer temperature is

$$\frac{(22°\text{C})\left(0.21\,\frac{\text{ft}^3}{\text{sec}}\right) + (11.1°\text{C})\left(2\,\frac{\text{ft}^3}{\text{sec}}\right)}{0.21\,\frac{\text{ft}^3}{\text{sec}} + 2\,\frac{\text{ft}^3}{\text{sec}}} = \boxed{12.14°\text{C}}$$

(e) The winter temperature is

$$\frac{(22°\text{C})\left(0.21\,\frac{\text{ft}^3}{\text{sec}}\right) + (7.2°\text{C})\left(2\,\frac{\text{ft}^3}{\text{sec}}\right)}{0.21\,\frac{\text{ft}^3}{\text{sec}} + 2\,\frac{\text{ft}^3}{\text{sec}}} = \boxed{8.61°\text{C}}$$

(f) Calculate the dissolved oxygen concentration 20 mi downstream using the summer temperature and the following equation.

$$K_{\text{D,T}} = 1.047^{(T-20)}K_{D,20°\text{C}}$$
$$K_{D,12.1°} = \left[(1.047)^{12.1-20}\right]\left(\frac{0.1}{\text{day}}\right)$$
$$= 0.07/\text{day}$$
$$K_{R,T} = (1.016)^{T-20}K_{R,20°\text{C}}$$
$$K_{R,12.1°} = \left[(1.016)^{12.1-20}\right]\left(\frac{0.25}{\text{day}}\right)$$
$$= 0.22/\text{day}$$

$$t = \frac{\text{distance}}{\text{velocity}} = \frac{(20\text{ mi})\left(5280\,\frac{\text{ft}}{\text{mi}}\right)}{\left(0.2\,\frac{\text{ft}}{\text{sec}}\right)\left(\frac{3600\text{ sec}}{\text{hr}}\right)\left(\frac{24\text{ hr}}{\text{day}}\right)}$$
$$= 6.1\text{ days}$$
$$\text{BOD}_u = \frac{\text{BOD}_t}{1 - 10^{-K_Dt}}$$
$$= \frac{17.9\,\frac{\text{mg}}{\ell}}{1 - 10^{-(0.07/\text{day})(5\text{ days})}}$$
$$= 32.35\text{ mg}/\ell$$

Use the Streeter-Phelps equation to calculate the dissolved oxygen concentration 20 mi downstream.

$$D_t = \frac{K_D\text{BOD}_u}{K_R - K_D}\left(10^{-K_Dt} - 10^{-K_Rt}\right) + D_o\left(10^{-K_Rt}\right)$$

D_t is the oxygen deficit at time, t, and D_o is the oxygen deficit after initial mixing.

$$\text{DO}_{\text{sat}}\text{ at }12.1°\text{C} = 10.8\text{ mg}/\ell$$
$$D_o = 10.8\,\frac{\text{mg}}{\ell} - 9.1\,\frac{\text{mg}}{\ell}$$
$$= 1.7\text{ mg}/\ell$$
$$D_t = \left[\frac{\left(\frac{0.07}{\text{day}}\right)\left(32.35\,\frac{\text{mg}}{\ell}\right)}{0.22\,\frac{1}{\text{day}} - 0.07\,\frac{1}{\text{day}}}\right] \times \left[10^{-(0.07/\text{day})(6.1\text{ days})} - 10^{-(0.22/\text{day})(6.1\text{ days})}\right] + \left(1.7\,\frac{\text{mg}}{\ell}\right)\left[10^{-(0.22/\text{day})(6.1\text{ days})}\right]$$
$$= 5.04\text{ mg}/\ell$$

The DO at 20 mi is

$$\mathrm{DO_{sat}} - D_t = 10.8\,\frac{\text{mg}}{\ell} - 5.04\,\frac{\text{mg}}{\ell}$$
$$= \boxed{5.76 \text{ mg}/\ell}$$

15. (a) Assume the wastewater flow, Q, is 125 gal/capita-day. Add 10% for infiltration.

$$Q = \left(125\,\frac{\text{gal}}{\text{capita-day}}\right)(2000 \text{ capita})(1 + 0.10)$$
$$= 275{,}000 \text{ gal/day}$$

Design a pipe to flow half full at a rate of 275,000 gal/day. Use the Chezy-Manning equation.

$$Q = \left(\frac{1.49}{n}\right) A (r_H)^{2/3}\sqrt{S}$$

Assume $n = 0.013$.

$$S = \frac{0.25 \text{ in}}{12 \text{ in}} = 0.021 \text{ ft/ft}$$

The flow area, A, is

$$A_{\text{flow}} = \frac{\pi D^2}{(4)(2)} = \frac{\pi D^2}{8}$$

$$r_H = \frac{A}{P_w} = \frac{\frac{\pi D^2}{4}}{\frac{\pi D}{2}} = \frac{D}{4}$$

$$Q_1 = \left(275{,}000\,\frac{\text{gal}}{\text{day}}\right)\left(\frac{\text{ft}^3}{7.48 \text{ gal}}\right)\left(\frac{\text{day}}{24 \text{ hr}}\right)\left(\frac{\text{hr}}{3600 \text{ sec}}\right)$$
$$= 0.426 \text{ ft}^3/\text{sec}$$

$$0.426\,\frac{\text{ft}^3}{\text{sec}} = \left(\frac{1.49}{0.013}\right)\left(\frac{\pi D^2}{8}\right)\left(\frac{D}{4}\right)^{2/3}\sqrt{0.021}$$

$$(D^2)\left(\frac{D}{4}\right)^{2/3} = 0.0653$$

Simplifying,

$$D^{8/3} = 0.1646$$
$$D = 0.51 \text{ ft } (6.1 \text{ in})$$

Use an 8-in pipe (the recommended minimum size for a sewer main).

(b) Calculate the full pipe flow, Q_2, for an 8-in pipe at a slope $S = 0.021$ and $n = 0.013$.

$$A = \frac{\pi D^2}{4}$$
$$r_H = \frac{D}{4}$$

$$Q_2 = \left(\frac{1.49}{0.013}\right)\left[\frac{\pi\left(\frac{8 \text{ in}}{\frac{12 \text{ in}}{\text{ft}}}\right)^2}{4}\right]\left(\frac{\frac{8 \text{ in}}{\frac{12 \text{ in}}{\text{ft}}}}{4}\right)^{2/3}\sqrt{0.021}$$
$$= 1.76 \text{ ft}^3/\text{sec}$$

$$\frac{Q_1}{Q_2} = \frac{0.426\,\frac{\text{ft}^3}{\text{sec}}}{1.76\,\frac{\text{ft}^3}{\text{sec}}} = 0.242$$

From a graph or table of circular channel ratios,

$$\frac{d}{D} = 0.38$$

The depth of flow, d, is

$$(0.38)(8 \text{ in}) = \boxed{3.0 \text{ in}}$$

From a table of properties of partially-filled circular pipes, for $d/D = 0.38$, $A/D^2 = 0.2739$.

$$A = (0.2739)\left(\frac{8 \text{ in}}{\frac{12 \text{ in}}{\text{ft}}}\right)^2 = 0.122 \text{ ft}^2$$

$$v = \frac{Q}{A} = \frac{0.426\,\frac{\text{ft}^3}{\text{sec}}}{0.122 \text{ ft}^2} = \boxed{3.49 \text{ ft/sec}}$$

16. The following steps should be taken to neutralize and clean up the site.

1) Secure the site by a fence and/or other barrier to prevent unauthorized entry, possible injury, and damage to the confinement dike.
2) Implement the required health and safety plan, measures, and procedures. These include protective clothing and equipment, monitors, an exclusion zone, a decontamination area, etc.
3) Remove the liquid chemical from the diked area by pumping it into a tanker truck for transport off-site for treatment and disposal or reuse.
4) Visually assess the area of contamination. Perform soil borings or excavation of pits or trenches to characterize the soil, determine the groundwater elevation, locate confining layers, and define the depth and area of contamination. Sample the contaminated soil and water to determine the concentrations of the toxic chemical around the site. This will help define the area that must be treated.

5) Install groundwater-monitoring wells to determine if the groundwater becomes contaminated and to monitor its depth and movement. This step assumes that the groundwater is fairly close to the surface.

6) Cover the contaminated area and install hay bales or silt fences around it to minimize the spread of contamination by rainwater and the formation of leachate. Perform the same measures for any excavated piles of contaminated soil during the site cleanup process. This step is based on the assumption that the area of contamination is small.

7) Determine the best methods to treat the contaminated soil, and begin treatment. These methods may include on-site, in situ treatment, excavation and on-site treatment, or excavation and off-site treatment.

8) Begin treating any contaminated groundwater and surface water by on-site pumping, if necessary.

9) Once the contaminated soil is removed and/or treated, replace the treated soil on-site to fill any excavations (if permitted by the regulatory agencies), or fill the excavations with clean soil. Finish cleaning up the site and return it to its original condition.

10) Regulatory agencies (U.S. EPA as well as state and local) must be notified of the spill and the cleanup activities. Regulatory approval may be required before site cleanup can begin. If this spill is in a heavily populated area, a public relations/information campaign may be warranted to gain the support of the residents.

17. A lagoon is typically designed for at least several days' detention time of wastewater flow. Detention times of lagoons range from one to more than 100 days. This large volume of wastewater can readily absorb peak flow rates of several times the average flow rate, as long as the capacity of the lagoon is not exceeded. The wastewater in the lagoon can be discharged at lower flow rates to other treatment processes. A lagoon serves as a good equalization basin to dampen peak flow rates to lower average flow rates.

18. (a) The self-cleansing velocity is used in sewer design. It is the minimum velocity required to pick up and carry solid particles of a typical size with the flow. This velocity is usually between 2.0–2.5 ft/sec.

The scour velocity is used in the design of aerated grit chambers. It is the velocity required to scour organic particles off the grit. This velocity is usually between 1.5–2.0 ft/sec.

(b) For a spherical particle of a 0.5-in diameter, the minimum velocity should be the self-cleansing velocity, which can be calculated with the following equation.

$$v_{\text{minimum}} = C\sqrt{\frac{k(\gamma_s - \gamma)}{\gamma}d}$$

γ = specific weight of water = 62.4 lbf/ft^3
γ_s = specific weight of the particles = 100 lbf/ft^3
C = Chezy coefficient = 100
k = 0.04 (initiation of scour) to 0.8 (effective cleansing)
d = particle diameter

Use $k = 0.8$.

$$v_{\text{minimum}} = (100)\sqrt{\frac{(0.8)\left(100\ \frac{\text{lbf}}{\text{ft}^3} - 62.4\ \frac{\text{lbf}}{\text{ft}^3}\right)}{62.4\ \frac{\text{lbf}}{\text{ft}^3}}\left(\frac{0.5\text{ in}}{12\ \frac{\text{in}}{\text{ft}}}\right)}$$

$$= \boxed{2.9\text{ ft/sec}}$$

19. (a) The state regulatory agencies and the U.S. EPA control sanitary landfill design.

(b) The average density of shredded waste is approximately 450 lbm/yd^3 before compaction in the landfill.

(c) The average density of baled and compacted waste is 1600–1800 lbm/yd^3.

(d) Daily soil cover is typically 6 in deep.

(e) Intermediate soil cover is typically 2 ft deep.

(f) Final soil cover is typically 2 ft deep.

(g) Intermediate soil cover can be used for up to 30 days.

(h) The clay liner serves to separate the waste from the underlying soil and groundwater. It reduces the chance of landfill leachate seeping into the nearby groundwater and surface water.

(i) The limitation of clay liners is that they are somewhat permeable and their properties can change considerably over time due to interactions with materials in the landfill and surrounding forces. When clay dries out (becomes desiccated) it is much more permeable. Chemicals in the landfill may break down and permeate the clay. The clay liner may fracture or rupture due to the weight of the landfill, shifting of the underlying soil or landfill material, or other subsurface disturbances.

(j) NIMBY stands for "not in my backyard." This is the prevailing attitude of people regarding the siting of landfills (and other waste-treatment and disposal facilities). People agree that these facilities are necessary, but they do not want to live near them. The NIMBY attitude makes landfill siting difficult.

SOILS/SOIL MECHANICS

1. (a) Plot the zero air voids curve. Use the following formula for which the degree of saturation, s, equals one. The dry density is

$$\gamma_d = \frac{\gamma_w s}{w + \frac{s}{\text{SG}}}$$

γ_w = density of water (62.4 lbf/ft^3)

w = water content expressed as a decimal

SG = specific gravity of solids

Choose values for w and solve for γ_d.

w	γ_d (lbf/ft^3)
15%	118.3
20%	108.1
25%	99.5
30%	92.1

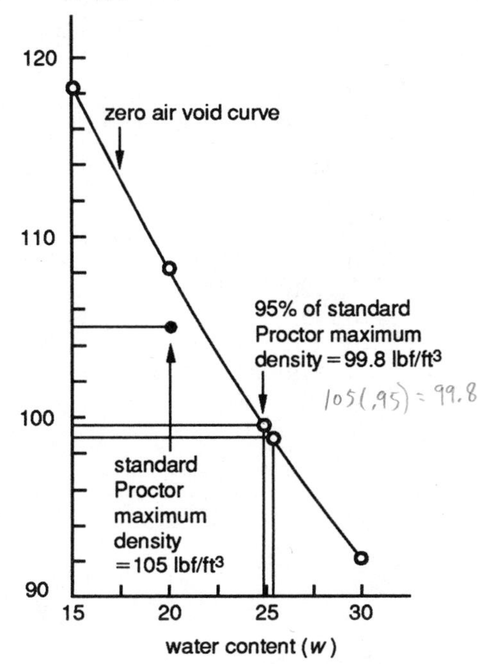

(b) 95% of the standard Proctor maximum density is

$$\gamma = (0.95)\left(105\ \frac{\text{lbf}}{\text{ft}^3}\right) = 99.8\ \text{lbf/ft}^3$$

Soil can be no more than fully saturated ($s = 1$). Find the intersection of the zero air voids curve with 95% of the standard Proctor maximum.

> From the plot, the maximum water content at 95% of the standard Proctor maximum is approximately 25%.

(c) Maximum density without pumping at 26% water content, w, is the point on the zero air voids curve at 26%. From the plot, $\gamma \approx 98$ lbf/ft^3.

The maximum density can also be found by

$$\gamma_d = \frac{62.4\ \frac{\text{lbf}}{\text{ft}^3}}{0.26 + \frac{1}{2.65}} = \boxed{97.9\ \text{lbf/ft}^3}$$

(d) The earthwork contractor has a valid point. At 26% moisture content, any density greater than 98 lbf/ft^3 will cause the soil to pump and rut.

(e) The contractor should dry the soil back to near-optimum water content, w (20%). Each lift should be spread and then disked until it dries to about 20% water content. Disking should be done prior to compaction.

Drying all the way back to 20% is recommended, but not required. Anything less than or equal to about 24% should be acceptable.

Finally, the contractor should be reminded that the nearer the soil is to optimum w, the easier it will be to reach 95% of the standard Proctor maximum. Fewer compactor passes are required at 20% water content than would be required at 24% or 25% water content.

2. (a) Use the AASHTO soil classification chart. Proceed from left to right in the chart. The correct group will be found by the process of elimination. The first group meeting all the test data is the group classification.

soil	AASHTO group	Unified Soils Classification
A	A-1-a	SW or SP (C_u undetermined)
B	A-2-6	SC (CL fines)
C	A-2-4	SW–SM (ML fines; $C_u \geq 4$)
D	A-6	SC (CL fines)
E	A-7-6	CH (CH fines)

(b) Use the Unified Soil Classification flow charts (found in ASTM test method D-2487-90—standard test method for classification of soils for engineering purposes). Choose the appropriate flow chart based on 50% or more passing the number 200 sieve, or 50% or more retained on the number 200 sieve. Enter the flow chart from the left and follow the path that meets all the test data. Assume all soils have 50% or more passing the number 4 sieve.

Note that C_u is not specified and cannot be determined for soil A. Classify fines using ASTM D-2487-90 or any standard plasticity chart.

$$C_u = \frac{D_{60}}{D_{10}}$$

(c) Refer to the AASHTO soil classification chart. Soils B, D, and E would not be suitable for roadway base material because they are rated A-2-6 or worse.

(d) Soils D and E are clayey soils rated A-6 or worse. They would not be suitable for a roadway embankment.

(e) Soils rated A-2-6 or worse (such as soils B, D, and E) would not make good subbase borrow soil.

(Reference: American Society for Testing and Materials. *Annual Book of ASTM Standards: Soil and Rock, Dimension Stone, Geosynthetics*, Volume 4. 08. Philadelphia, March, 1991.)

3. (a) The mass diagram is constructed by first modifying the fill values by the shrinkage factor given for the soil. Next, add all the cut values and subtract all the modified fill values, then tabulate the cumulative values for each station as shown in the following table. Plot the cumulative values as shown in the following figure.

(b) Balance points may be found by drawing a horizontal line at any location on the mass diagram. Cut

sta	cut (+)	fill (−)	fill (−) ×1.25	cumulative cut or fill	haul volume (sta yd)	total volume at balance points (sta yd)
0	–	0	–	0	–	–
1	0	1	1.25	−1.25	−0.625	–
1.5	–	–	–	0	−0.3125	−0.9375
2	2	0	0	0.75	0.1875	–
3	1	0	0	1.75	1.25	–
4	0	1.5	1.875	−0.125	0.8125	2.25
5	0.5	0	0	0.375	0.125	–
6	1	0	0	1.375	0.875	–
7	1.5	0	0	2.875	2.125	–
8	0	0.5	0.625	2.25	2.5625	–
9	0	1.5	1.875	0.375	1.3125	–
9.5	–	–	–	0	0.09375	7.09375
10	0	0.5	0.625	−0.25	−0.0625	–
11	0	1	1.25	−1.5	−0.875	–
11.7	–	–	–	0	−0.525	−1.4625
12	2	0	0	0.5	0.075	–
13	2.5	0	0	3	1.75	–
14	1	0	0	4	3.5	–
15	0	0	0	4	4	–
16	0	1	1.25	2.75	3.375	–
17	0	1.5	1.875	0.875	1.8125	–
17.4	–	–	–	0	0.175	14.6875
18	0	2	2.5	−1.625	−0.4875	–
19	0	2.5	3.125	−4.75	−3.1875	–
20	0	2	2.5	−7.25	−6	−9.675

(All cut, fill, and volumes are in thousands of cubic yards.)

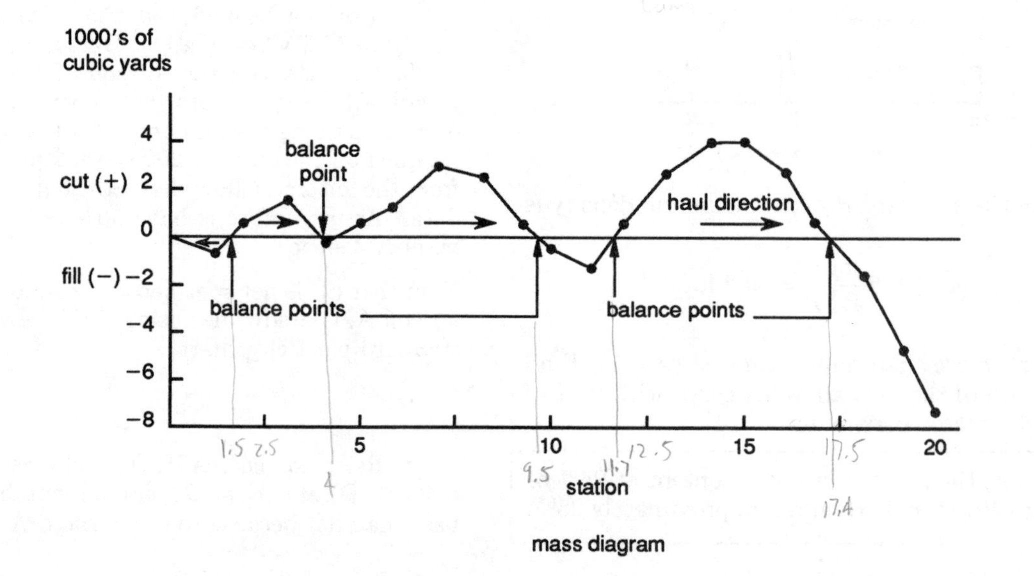

mass diagram

and fill are balanced between any two adjacent stationswhere the mass diagram is intersected by the horizontal. Consider the points where the mass diagram crosses the zero cut and fill horizontal. Balance points along this line are indicated by vertical arrows on the mass diagram. These points occur approximately at stations 1+50, 4+00, 9+50, 11+70, and 17+40.

Haul directions are indicated on the mass diagram. In general, a positive slope indicates cutting, and a negative slope indicates filling.

(c) Haul volumes may be calculated as the areas below the mass diagram between balance points. For example, the volume between stations 1+50 and 4+00 is found by

$$\begin{aligned}&\left[\text{sta }(2\text{–}1.5)\left(\frac{750\text{ yd}^3}{2}\right)\right]\\&\quad+\left(\text{sta }(3\text{–}2)\left[\frac{(1750\text{ yd}^3+750\text{ yd}^3)}{2}\right]\right)\\&\quad+\left(\text{sta }(4\text{–}3)\left[\frac{(-125\text{ yd}^3+1750\text{ yd}^3)}{2}\right]\right)\\&=188\text{ sta yd}^3+1250\text{ sta yd}^3+813\text{ sta yd}^3\\&=\boxed{2251\text{ sta yd}^3\text{ (or sta yd)}}\end{aligned}$$

Other haul volumes for balance points are shown in the table.

(d) The overhaul volume between station 1+00 and the first balance point is zero.

(e) A bank yard is defined as cubic yardage in place or in the ground.

(f) Measures to control slope stability and erosion include reducing slope angles, hydroseeding or vegetating slopes, and using silt fences.

4. (a) The factor of safety, FS, for the material alone is calculated as the ratio of soil friction angle, ϕ, to slope angle, β.

$$\text{FS}_\text{sand}=\frac{\phi}{\beta}=\frac{30^\circ}{45^\circ}=\boxed{0.67}$$

For homogeneous soft clay, the friction angle is taken as zero.

$$\text{FS}_\text{clay}=\frac{\phi}{\beta}=\frac{0}{45^\circ}=\boxed{0}$$

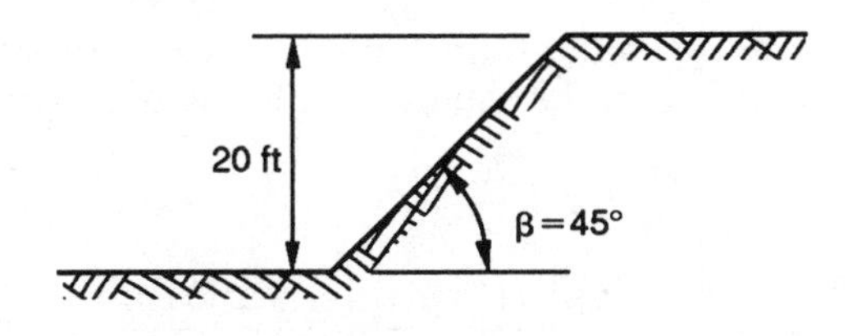

(b) The factor of safety against a soil block translation or rotation is computed by a computer program or taken from a slope stability chart. For the sand, there is no surcharge and no water inside or outside of the slope.

$$\begin{aligned}\text{FS}&=\frac{\tan\phi}{\tan\beta}\\&=\frac{\tan 30^\circ}{\tan 45^\circ}=\boxed{0.58}\end{aligned}$$

For the clay, use the Taylor slope stability chart ($\phi=0$). The factor of safety can be computed as

$$\text{FS}=\frac{N_o c}{\gamma H}$$

N_o is the stability number from the chart, c is the cohesion intercept taken as one-half the unconfined compressive strength for $\phi=0$, γ is the specific weight of the soil, and H is the slope height.

From the Taylor slope stability chart, an average value of N_o is approximately 6 for $\beta=45^\circ$. The factor of safety is

$$\begin{aligned}\text{FS}=\frac{N_o c}{\gamma H}&=\frac{(6)\left(\dfrac{1600\,\dfrac{\text{lbf}}{\text{ft}^2}}{2}\right)}{\left(110\,\dfrac{\text{lbf}}{\text{ft}^3}\right)(20\text{ ft})}\\&=\boxed{2.2}\end{aligned}$$

(c) Any factor of safety less than 1 is unacceptable because failure is indicated. Generally, a factor of safety of 1.25–1.5 is considered acceptable. The clay slope should be stable as shown in (b), but the sand slope will be unstable.

(d) The clay slope is stable with a factor of safety of 2.2. The sand slope should be revised to have a factor of safety of approximately 1.5. From the previous equation, a maximum stable slope angle may be determined by

$$\begin{aligned}\beta&=\tan^{-1}\left(\frac{\tan\phi}{\text{FS}}\right)\\&=\tan^{-1}\left(\frac{\tan 30^\circ}{1.5}\right)\\&=\boxed{21^\circ}\end{aligned}$$

5. (a) When the hard clay layer is deeper than $0.7B$ (B = excavation width), the factor of safety, FS, against the bottom heave is

$$\text{FS}=\frac{Q_u}{Q}=\frac{5.7cB_1}{\gamma HB_1-cH}$$

c = cohesion

$B_1=0.7B$

γ = soil specific weight

H = excavation height

$$\text{FS} = \frac{(5.7)\left(450\ \frac{\text{lbf}}{\text{ft}^2}\right)(0.7)(40\ \text{ft})}{\left(115\ \frac{\text{lbf}}{\text{ft}^3}\right)(18\ \text{ft})(0.7)(40\ \text{ft}) - \left(450\ \frac{\text{lbf}}{\text{ft}^2}\right)(18\ \text{ft})}$$

$$= \boxed{1.44}$$

(Reference: Das, Braja M. *Principles of Foundation Engineering*, 2nd ed., Boston: PWS-Kent Publishers, 1990.)

(b) If all other parameters remain the same but the excavation is carried deeper, the bottom will heave when the factor of safety equals 1.

$$\text{FS} = \left(\frac{1}{H}\right)\left(\frac{5.7c}{\gamma - \frac{c}{0.7B}}\right) = 1$$

Solve for H.

$$H = \frac{5.7c}{\gamma - \frac{c}{0.7B}}$$

$$= \frac{(5.7)\left(450\ \frac{\text{lbf}}{\text{ft}^2}\right)}{115\ \frac{\text{lbf}}{\text{ft}^3} - \frac{450\ \frac{\text{lbf}}{\text{ft}^2}}{(0.7)(40\ \text{ft})}}$$

$$= \boxed{25.9\ \text{ft}}$$

(c) Assume that the original conditions apply but that the hard clay is now 10 ft below the bottom of the excavation. For a depth to hard layer, D, of less than $0.7B$, use

$$\text{FS} = \left(\frac{1}{H}\right)\left(\frac{5.7c}{\gamma - \frac{c}{D}}\right)$$

$$= \left(\frac{1}{18\ \text{ft}}\right)\left[\frac{(5.7)\left(450\ \frac{\text{lbf}}{\text{ft}^2}\right)}{115\ \frac{\text{lbf}}{\text{ft}^3} - \frac{450\ \frac{\text{lbf}}{\text{ft}^2}}{(10\ \text{ft})}}\right]$$

$$= \boxed{2.03}$$

6. Compute the dead load per unit length, w, and the dead load pressure, $p_{\text{dead load}}$.

$$w = C_w \gamma B^2$$

$$C_w = \text{correction coefficient}$$

$$B = \text{trench width}$$

For a 20-in-diameter pipe, a trench of approximately 36 in would be required. The C_w is approximately 1, so

$$w = (1)\left(120\ \frac{\text{lbf}}{\text{ft}^3}\right)(3\ \text{ft})^2$$

$$= 1080\ \text{lbf/ft}$$

$$p_{\text{dead load}} = \frac{1080\ \frac{\text{lbf}}{\text{ft}}}{3\ \text{ft}} + 100\ \frac{\text{lbf}}{\text{ft}^2}$$

$$= 460\ \text{lbf/ft}^2$$

The live load, $p_{\text{live load}}$, for H-20 loading is determined from NAVFAC DM 7-1, Fig. 19. For 4 ft of cover, the live load on the pipe is approximately 350 lbf/ft^2. Note that this value includes impact effects.

$$p_{\text{live load}} = 350\ \text{lbf/ft}^2$$

The allowable load, $p_{\text{allowable}}$, that the pipe should be able to withstand without excessive cracking is

$$p_{\text{allowable}} = (p_{\text{dead load}} + p_{\text{live load}})\left(\frac{\text{FS}}{L_f}\right)$$

L_f is a load factor depending on bedding conditions.

For bedding class B, $L_f = 1.9$; for bedding class D, $L_f = 1.1$. For class B bedding,

$$p_{\text{allowable}} = \left(460\ \frac{\text{lbf}}{\text{ft}^2} + 350\ \frac{\text{lbf}}{\text{ft}^2}\right)\left(\frac{1.25}{1.9}\right)$$

$$= \boxed{533\ \text{lbf/ft}^2 < 1000\ \text{lbf/ft}^2 \quad \text{[safe]}}$$

For class D bedding,

$$p_{\text{allowable}} = \left(460\ \frac{\text{lbf}}{\text{ft}^2} + 350\ \frac{\text{lbf}}{\text{ft}^2}\right)\left(\frac{1.25}{1.1}\right)$$

$$= \boxed{920\ \text{lbf/ft}^2 < 1000\ \text{lbf/ft}^2 \quad \text{[safe]}}$$

However, if a factor of safety of 1.5 is desired, the pipe would not be safe for class D bedding.

(Reference: U.S. Department of the Navy. *Soil Mechanics, Design Manual 7.1*. Alexandria, VA: Naval Facilities Engineering Command, May, 1982, figure 18.)

7. (a) The hydraulic conductivity of the two soils may be computed from the equation for the falling-head test.

$$k = \left(\frac{A'}{At}\right)\ln\left(\frac{h_i}{h_f}\right)$$

$$k_{\text{soil A}} = \left[\frac{(5\ \text{cm}^2)(15\ \text{cm})}{(40\ \text{cm}^2)(300\ \text{cm})}\right]\ln\left(\frac{100\ \text{cm}}{75\ \text{cm}}\right)$$

$$= 0.0018\ \text{cm/min}$$

Converting the units,

$$k_{\text{soil A}} = \left(0.0018\ \frac{\text{cm}}{\text{min}}\right)\left(\frac{525{,}600\ \frac{\text{min}}{\text{yr}}}{30.5\ \frac{\text{cm}}{\text{ft}}}\right)$$

$$= \boxed{31.0\ \text{ft/yr}}$$

$$k_{\text{soil B}} = \left[\frac{(2\ \text{cm}^2)(15\ \text{cm})}{(40\ \text{cm}^2)(6000\ \text{cm})}\right]\ln\left(\frac{200\ \text{cm}}{190\ \text{cm}}\right)$$
$$= 6.4 \times 10^{-6}\ \text{cm/min}$$

$$k_{\text{soil B}} = \left(6.4 \times 10^{-6}\ \frac{\text{cm}}{\text{min}}\right)\left(\frac{525{,}600\ \frac{\text{min}}{\text{yr}}}{30.5\ \frac{\text{cm}}{\text{ft}}}\right)$$

$$= \boxed{0.11\ \text{ft/yr}}$$

(b) Most regulations require a soil liner to have an inplace hydraulic conductivity of 1×10^{-7} cm/sec (or 0.1 ft/yr). Soil B would be acceptable as a liner.

FOUNDATIONS

1. (a) Compute the pile load.

$$\begin{aligned}\text{pile load} &= (\text{area})(\text{live load} + \text{dead load})\\ &= (12\text{ ft})(12\text{ ft})\left[1100\,\frac{\text{lbf}}{\text{ft}^2} + (2\text{ ft})\left(150\,\frac{\text{lbf}}{\text{ft}^3}\right)\right]\\ &= 201{,}600\text{ lbf}\end{aligned}$$

Pile capacity is derived from side (skin) resistance and tip (end) resistance. In this problem, negative skin friction or downdrag will occur as the clay consolidates. Additional load will be applied to the pile. Downdrag will occur between the top of the clay layer and the neutral axis (generally taken as the center of the layer, although a more rigorous solution may be employed). The average lateral force applied by the clay is

$$\begin{aligned}f_{\text{ave}} &= \overline{\sigma}_{\text{v}} k \tan\delta\\ \overline{\sigma}_{\text{v}} &= \text{effective vertical stress}\\ k &= \text{earth pressure coefficient}\\ &= 1 - \sin\phi\\ \delta &= \text{soil-pile friction angle}\\ &\approx 0.5\phi \text{ to } 0.7\phi\end{aligned}$$

The average vertical stress acting over the top half of the layer is computed at a depth of 6 ft below the top of the clay.

$$\begin{aligned}f_{\text{ave}} &= \overline{\sigma}_{\text{v}} k \tan\delta\\ &= \left[(10\text{ ft})\left(110\,\frac{\text{lbf}}{\text{ft}^3}\right) + (6\text{ ft})\left(95\,\frac{\text{lbf}}{\text{ft}^3}\right)\right]\\ &\quad\times (1 - \sin 30^\circ)[\tan(0.6)(30^\circ)]\\ &= 271\text{ lbf/ft}^2\end{aligned}$$

The downdrag force may be computed as

$$\begin{aligned}Q_n &= (f_{\text{ave}})(\text{pile surface area})\\ &= \left(271\,\frac{\text{lbf}}{\text{ft}^2}\right)(12\text{ ft})\pi\left(\frac{20\text{ in}}{12\,\frac{\text{in}}{\text{ft}}}\right)\\ &= 17{,}027\text{ lbf}\end{aligned}$$

The load applied to the pile is

$$\begin{aligned}\text{load applied} &= 201{,}600\text{ lbf} + 17{,}027\text{ lbf}\\ &= 218{,}627\text{ lbf}\end{aligned}$$

The tip capacity of the pile may be computed as

$$\begin{aligned}Q_p &= q_p A_p = A_p q' N_q\\ A_p &= \text{pile tip cross-sectional area}\\ q' &= \text{effective vertical stress at pile tip}\\ N_q &= \text{bearing capacity factor}\end{aligned}$$

The critical embedment depth in the dense sand may be determined.

$$\begin{aligned}\left(\frac{L_b}{D}\right)_{\text{critical}} &\approx 10\\ L_b &= \text{embedment length}\\ D &= \text{pile diameter}\end{aligned}$$

The maximum tip capacity is obtained at

$$\begin{aligned}L_b = 10D &= (10)\left(\frac{20\text{ in}}{12\,\frac{\text{in}}{\text{ft}}}\right)\\ &= 16.7\text{ ft}\end{aligned}$$

For $\phi = 35^\circ$, $N_q = 125$.

For a depth of 16 ft into the dense sand,

$$\begin{aligned}q' &= (10\text{ ft})\left(110\,\frac{\text{lbf}}{\text{ft}^3}\right)\\ &\quad + (24\text{ ft})\left(95\,\frac{\text{lbf}}{\text{ft}^3}\right) + (16\text{ ft})\left(125\,\frac{\text{lbf}}{\text{ft}^3}\right)\\ &= 5380\text{ lbf/ft}^2\\ Q_p &= A_p q' N_q\\ &= \left[\left(\frac{20}{12}\right)^2\text{ ft}^2\right]\left(\frac{\pi}{4}\right)\left(5380\,\frac{\text{lbf}}{\text{ft}^2}\right)\left(125\,\frac{\text{lbf}}{\text{ft}^3}\right)\\ &= 1{,}467{,}000\text{ lbf}\\ Q_{\text{allowable}} &= \frac{Q_p}{\text{FS}}\end{aligned}$$

A factor of safety of about 3 should be used.

$$\begin{aligned}Q_{\text{allowable}} &= \frac{1{,}467{,}000\text{ lbf}}{3}\\ &= 489{,}000\text{ lbf}\end{aligned}$$

> $Q_{\text{allowable}} >$ load applied, so the pile end capacity is sufficient.

(Reference: Das, Braja M. *Principles of Foundation Engineering*, 2nd ed., Boston: PWS-Kent Publishing Company, Figs. 8.12 and 8.13, 1990.)

(b) If the clay was overconsolidated, settlement would be negligible and no downdrag would occur. Additional pile capacity would be derived from friction between the pile and the clay. For highly overconsolidated clays, it would be conservative to ignore increased capacity because the clay may crack, which would limit the strength increase. Potential for cracking may be reduced by augering a hole for the pile before driving.

2. (a) Footing A applies 45 kips of load over 36 ft² of footing area. The stress on the bottom of the footing, q, may be computed as

$$q = \left(\frac{45 \text{ kips}}{36 \text{ ft}^2}\right)\left(1000 \, \frac{\text{lbf}}{\text{kip}}\right) = 1250 \text{ lbf/ft}^2$$

Compute effective overburden pressures, $\overline{\sigma}_{v_o}$, and the change in stress, Δp, at the top of the sand layers and the center of the clay. z is the depth from the bottom of the footing. Use stress contour charts or other suitable means to determine the ratio $\Delta p/q$. (The sand settlement is probably negligible, but check it anyway.)

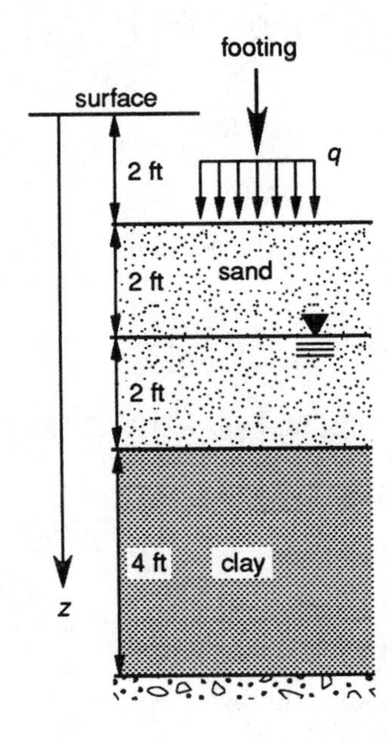

$$\overline{\sigma}_{v_o} = \gamma h$$

$$\overline{\sigma}_{v_o} \text{ at 2 ft} = \left(120 \, \frac{\text{lbf}}{\text{ft}^3}\right)(2 \text{ ft}) = 240 \text{ lbf/ft}^2$$

$$\overline{\sigma}_{v_o} \text{ at 4 ft} = 240 \, \frac{\text{lbf}}{\text{ft}^2} + \left(120 \, \frac{\text{lbf}}{\text{ft}^3}\right)(2 \text{ ft}) = 480 \text{ lbf/ft}^2$$

$$\overline{\sigma}_{v_o} \text{ at 8 ft} = 480 \, \frac{\text{lbf}}{\text{ft}^2} + (2 \text{ ft})\left(130 \, \frac{\text{lbf}}{\text{ft}^3} - 62.4 \, \frac{\text{lbf}}{\text{ft}^3}\right) + (2 \text{ ft})\left(113 \, \frac{\text{lbf}}{\text{ft}^3} - 62.4 \, \frac{\text{lbf}}{\text{ft}^3}\right) = 716.4 \text{ lbf/ft}^2$$

$z+2$ depth (ft)	$\overline{\sigma}_{v_o}$ (lbf/ft²)	$\frac{z}{B}$	$\frac{\Delta p}{q}$	Δp (lbf/ft²)
2	240	0	1	1250
4	480	0.33	0.9	1125
8	716.4	1.0	0.33	412

Use the previous figure to compute the change in stress below the footing at depths of 3 ft, 5 ft, and 8 ft from the bottom of the footing.

$$\text{at 1 ft}, \ \frac{z}{B} = \frac{1 \text{ ft}}{6 \text{ ft}} = 0.17$$

$$\text{at 3 ft}, \ \frac{z}{B} = \frac{3 \text{ ft}}{6 \text{ ft}} = 0.5$$

$$\text{at 5 ft}, \ \frac{z}{B} = \frac{5 \text{ ft}}{6 \text{ ft}} = 0.8$$

$$\text{at 8 ft}, \ \frac{z}{B} = \frac{8 \text{ ft}}{6 \text{ ft}} = 1.3$$

Settlement in the sand layers may be computed using the following formula.

$$S_{\text{sand}} = C_1 C_2 (q - \overline{\sigma}_{v_o}) \sum \left(\frac{I_z}{E_s}\right) \Delta z$$

I_z = strain influence factor

C_1 = correction for embedment $= (1 - 0.5)\left(\frac{\overline{\sigma}_{v_o}}{q - \overline{\sigma}_{v_o}}\right)$

C_2 = correction for creep $= 1 + 0.2 \log \text{(time in yr/0.1)}$

E_s = Young's modulus

(Reference: Das, Braja M. *Principles of Foundation Engineering*. 2nd ed., Boston: PWS-Kent Publishing Company, 1990.)

For a square footing,

$$I_z = 0.1 \text{ at } z = 0$$

$$I_z = 0.5 \text{ at } z = 0.5B = (0.5)(6 \text{ ft}) = 3 \text{ ft}$$

At $z = 1$ ft and 3 ft, $z/B = 0.167$ and 0.5, respectively, and $I_z = 0.23$ and 0.5, respectively.

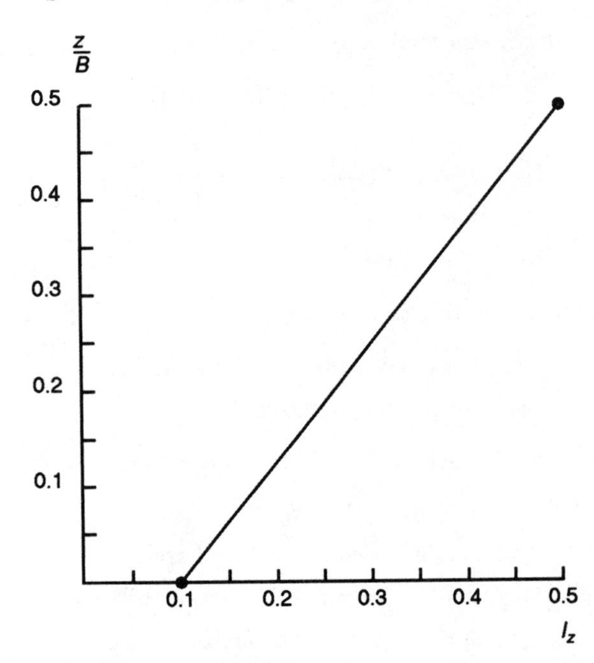

The settlement in the sand 1 yr after the footing is placed is

$$S = \left[1 - (0.5)\left(\frac{240\,\frac{\text{lbf}}{\text{ft}^2}}{1250\,\frac{\text{lbf}}{\text{ft}^2} - 240\,\frac{\text{lbf}}{\text{ft}^2}}\right)\right]$$
$$\times \left[1 + 0.2\log\left(\frac{1\text{ yr}}{0.1\text{ yr}}\right)\right]\left[\left(\frac{0.23}{720{,}000\,\frac{\text{lbf}}{\text{ft}^2}}\right)(24\text{ in})\right.$$
$$\left. + \left(\frac{0.5}{720{,}000\,\frac{\text{lbf}}{\text{ft}^2}}\right)(24\text{ in})\right]\left(1250\,\frac{\text{lbf}}{\text{ft}^2} - 240\,\frac{\text{lbf}}{\text{ft}^2}\right)$$
$$= 0.026\text{ in} \quad \text{[negligible]}$$

Settlement for a normally consolidated clay may be computed as

$$S = \left(\frac{\text{thickness}}{1 + e_o}\right)(C_c)\log\left(\frac{\overline{\sigma}_{v_o} + \Delta p}{\overline{\sigma}_{v_o}}\right)$$
$$= \left(\frac{4\text{ ft}}{1 + 0.80}\right)(0.38)\log\left(\frac{716.4\,\frac{\text{lbf}}{\text{ft}^3} + 412\,\frac{\text{lbf}}{\text{ft}^3}}{716.4\,\frac{\text{lbf}}{\text{ft}^3}}\right)$$
$$= 0.17\text{ ft }(2.0\text{ in})$$

Settlement in the sand layers is negligible, so the total settlement beneath footing A is 2.0 in. Note that this value is excessive for a typical structure.

(b) Assume, based on the previous calculations, that settlement in the sand is negligible for footing B. Footing B should be sized so that 2 in of settlement is induced in the clay layer. In order for this to occur, the change in stress at the center of the clay layer must be the same as for footing A—i.e., 412 lbf/ft^2. Assuming a 2:1 (vertical to horizontal) pressure distribution, the hypothetical loaded area at the top of the clay would be

$$A = \frac{\text{load}}{\text{stress}} = \frac{75{,}000\text{ lbf}}{412\,\frac{\text{lbf}}{\text{ft}^2}}$$
$$= 182\text{ ft}^2$$

A square loaded area of 182 ft^2 would have dimensions of 13.5 ft × 13.5 ft. Assuming the 2:1 stress distribution, project upward to the depth of footing B as shown.

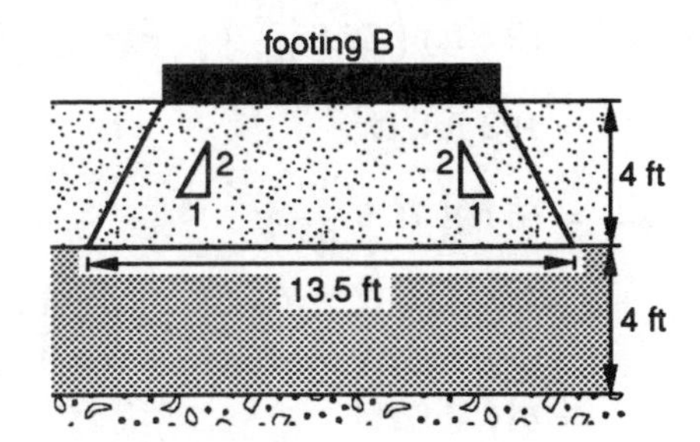

Footing B should have a side dimension of

$$13.5\text{ ft} - (2)(0.5)(4\text{ ft}) = \boxed{9.5\text{ ft}}$$

3. The first step is to determine if the pile will behave as short and rigid or long and flexible. Compute the stiffness factor, R.

$$R = \sqrt[4]{\frac{EI}{kB}}$$
$$k = \frac{k_1}{1.5}$$
k_1 = Terzaghi's subgrade modulus for a one-square-ft plate

B = pile width

I = moment of inertia

$$= \frac{1}{4}\pi r^4 = 0.0491D^4$$

For sand with $\phi = 30°$, density may range from loose to medium. k_1 may be estimated to be from 50–100 tons/ft^3. Solve for R.

$$R = \sqrt[4]{\frac{\left(60{,}000\,\frac{\text{lbf}}{\text{ft}^2}\right)(0.0491)\left[(20)^4\text{ in}^4\right]}{\left(\frac{100{,}000\,\frac{\text{lbf}}{\text{ft}^2}}{1.5\text{ ft}}\right)(20\text{ in})\left(\frac{\text{ft}}{12\text{ in}}\right)}}$$
$$= 8\text{ in}$$

Compute the stiffness factor, T.

$$T = \sqrt[5]{\frac{EI}{n_h}}$$
n_h = coefficient of modulus variation

(Reference: Tomlinson, M.J. *Pile Design and Construction Practice*. London: Cement and Concrete Association, 1981, Table 6.3.)

$$T = \sqrt[5]{\frac{\left(60{,}000\,\frac{\text{lbf}}{\text{ft}^2}\right)(0.0491)\left[(20)^4\text{ in}^4\right]}{\left(40{,}000\,\frac{\text{lbf}}{\text{ft}^3}\right)\left(\frac{\text{ft}}{12\text{ in}}\right)}}$$
$$= 10.7\text{ in}$$

Compare the embedded pile length, L, to R and T.

$$L = (14\text{ ft})\left(\frac{12\text{ in}}{\text{ft}}\right) = 168\text{ in}$$

Since $L \geq 4T$ and $L \geq 3.5R$, the pile is considered to be long and elastic. An elastic solution must be used.

Find the ultimate moment resistance, M_u, for the pile. A 20-in-round (or octagonal) pile will typically have eight #5 or #6 bars. Such piles will have M_u values of 70 kNm and 100 kNm, respectively. (Reference: *Pile Design and Construction Practice*. Figures 6.31 and 7.2f.)

$$k_p = \tan^2\left(45° + \frac{\phi}{2}\right)$$

$$\frac{M_u}{B^4\gamma k_p} = \frac{(70{,}000 \text{ Nm})\left(\frac{1 \text{ lbf}}{4.45 \text{ N}}\right)\left(\frac{3.28 \text{ ft}}{\text{m}}\right)}{\left(\frac{20}{12}\right)^4\left(120\,\frac{\text{lbf}}{\text{ft}^3} - 62.4\,\frac{\text{lbf}}{\text{ft}^3}\right)\tan^2\left(45° + \frac{30°}{2}\right)}$$

$$= 38.7 \quad \text{[say 40]}$$

$$\frac{H_u}{k_p B^3\gamma} = 8 \text{ for } \frac{e}{B} \text{ ratio} = 3.6$$

$$H_u = \text{ultimate lateral load}$$
$$= 8(k_p B^3)\gamma$$
$$= (8)\left[\tan^2\left(45° + \frac{30°}{2}\right)\left(\frac{20 \text{ in}}{12\,\frac{\text{in}}{\text{ft}}}\right)^3 \times \left(120\,\frac{\text{lbf}}{\text{ft}^3} - 62.4\,\frac{\text{lbf}}{\text{ft}^3}\right)\right]$$
$$= 6400 \text{ lbf}$$

The allowable lateral load, $H_{\text{allowable}}$, is

$$H_{\text{allowable}} = \frac{H_u}{\text{FS}} = \frac{6400 \text{ lbf}}{2.5}$$
$$= \boxed{2560 \text{ lbf (minimum)}}$$

If the larger steel is used, then

$$M_u = 100 \text{ kNm}$$
$$\frac{M_u}{B^4\gamma k_p} = 55$$
$$\frac{H_u}{k_p B^3\gamma} = 10$$
$$H_u = 8000 \text{ lbf}$$
$$H_{\text{allowable}} = \frac{8000 \text{ lbf}}{2.5}$$
$$= \boxed{3200 \text{ lbf (maximum)}}$$

4. (a) The applied footing load, q, is

$$q = \frac{\text{load}}{\text{area}} = \frac{50{,}000 \text{ lbf}}{(6 \text{ ft})(6 \text{ ft})}$$
$$= 1389 \text{ lbf/ft}^2$$

Determine the change in stress, Δp, at the center of each soil layer from stress-contour charts or by other means.

center of layer no.	depth (ft)	$\frac{\text{depth}}{B}$ (for $B = 6$ ft)	$\frac{\Delta p}{q}$ $(L/B = 1)$	Δp lbf/ft^2
1	2.5	0.4	0.8	1111
2	7.5	1.3	0.2	278
3	12.5	2.1	0.09	125
4	17.5	2.9	0.04	56

Compute the specific weight of the clay layers. Assume a specific gravity of 2.70, so $\gamma_{\text{solids}} = (2.7)(62.4 \text{ lbf/ft}^3) = 168.5 \text{ lbf/ft}^3$.

$$\gamma_d = \frac{\gamma_{\text{solids}}}{1+e}$$

Clay 1:

$$\gamma_d = \frac{168.5\,\frac{\text{lbf}}{\text{ft}^3}}{1+1.2} = 76.6 \text{ lbf/ft}^3$$

Clay 2:

$$\gamma_d = \frac{168.5\,\frac{\text{lbf}}{\text{ft}^3}}{1+1.0} = 84.3 \text{ lbf/ft}^3$$

Clay 2 is below the water table, so compute the water content, w.

$$w = \frac{Se}{\text{SG}}$$

The degree of saturation, S, is approximately 1 below the water table.

$$w = \frac{(1)(1.0)}{2.70} = 0.37 \quad (37\%)$$

The saturated density of clay 2 is

$$\gamma_{\text{sat}} = \gamma_d(1+w)$$
$$= \left(84.3\,\frac{\text{lbf}}{\text{ft}^3}\right)(1+0.37) = 115.5 \text{ lbf/ft}^3$$

The change in stress at a depth of $2B$ below a footing is usually negligible. Only 4–9% of the applied footing load reaches the sand layers. Settlement in the sand layers is negligible.

The effective overburden pressure, $\overline{\sigma}_{v_o}$, at the center of the clay layers is

$$\overline{\sigma}_{v_o,\text{layer 1}} = (2.5 \text{ ft})\left(76.6\,\frac{\text{lbf}}{\text{ft}^3}\right)$$
$$= 192 \text{ lbf/ft}^2$$

$$\overline{\sigma}_{v_o,\text{layer 2}} = (5 \text{ ft})\left(76.6\,\frac{\text{lbf}}{\text{ft}^3}\right) + (2.5 \text{ ft})\left(115.5\,\frac{\text{lbf}}{\text{ft}^3} - 62.4\,\frac{\text{lbf}}{\text{ft}^3}\right)$$
$$= 516 \text{ lbf/ft}^2$$

Assume the clay is normally consolidated and compute the settlement, S.

$$S = \left(\frac{\text{layer height}}{1 + e_o}\right)\left(C_c \log \frac{\overline{\sigma}_{v_o} + \Delta p}{\overline{\sigma}_{v_o}}\right)$$

$$S_{\text{layer 1}} = \left(\frac{5 \text{ ft}}{1 + 1.2}\right)\left[0.5 \log\left(\frac{192 \frac{\text{lbf}}{\text{ft}^2} + 1111 \frac{\text{lbf}}{\text{ft}^2}}{192 \frac{\text{lbf}}{\text{ft}^2}}\right)\right]$$

$$= 0.95 \text{ ft}$$

$$S_{\text{layer 2}} = \left(\frac{5 \text{ ft}}{1 + 1.0}\right)\left[0.4 \log\left(\frac{516 \frac{\text{lbf}}{\text{ft}^2} + 278 \frac{\text{lbf}}{\text{ft}^2}}{516 \frac{\text{lbf}}{\text{ft}^2}}\right)\right]$$

$$= 0.19 \text{ ft}$$

$$\text{Total settlement} = 0.95 \text{ ft} + 0.19 \text{ ft}$$

$$= \boxed{1.14 \text{ ft } (13.7 \text{ in})}$$

(b) This settlement is not tolerable. Generally, a settlement of 1 in or less is acceptable. For a concrete frame, angular distortion, δ/ℓ of 0.003 is considered acceptable. This results in a settlement of approximately 3/4 in in a column span, ℓ, of 20 ft. The computed settlement would be excessive.

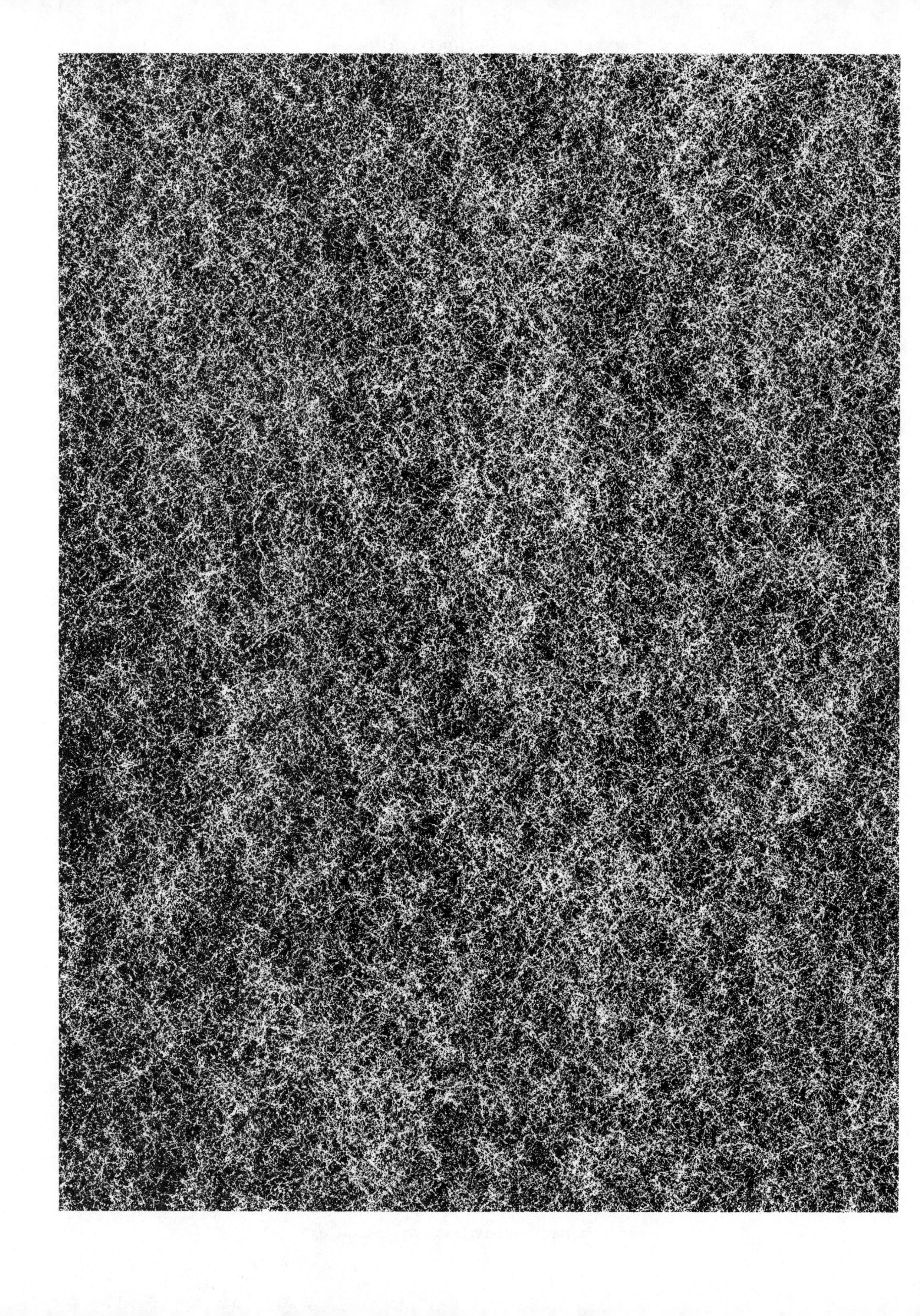

CONCRETE DESIGN

1. (a) Ingredients added to concrete immediately before or during mixing (other than portland cement, aggregates, and water) are known as *admixtures*.

(b) Six types of admixtures are:

1) *Accelerators* (ASTM C494, Type C): Accelerate setting and enhance early strength. Example: calcium chloride (ASTM D98).

2) *Air entraining* (ASTM C260): Improve durability and workability. Example: salts of wood resins (vinsol resins).

3) *Retarders* (ASTM C494, Type B): Retard the setting time to avoid difficulties with placing and finishing (typically used in hot weather). Example: lignins.

4) *Superplasticizers* (ASTM C1017, Type 1): Make high-slump concrete (flowing concrete) from concrete with normal to low water-cement ratios, allow for easy placing, and reduce and sometimes eliminate the need for vibration. Example: lignosulfonates.

5) *Water reducers* (ASTM C494, Type A): Reduce water requirement to produce concrete of a certain slump. Example: lignosulfonates.

6) *Pozzolans*: Improve the properties of concrete by changing the properties of the various types of cement; substituted for certain amounts of cement; reduce temperature rise, alkali-aggregate expansion, and harmful effects of tricalcium aluminate. Examples: fly ash, blast furnace slag, ground pumice.

2. (a) Calculate the volume for one sack of cement.

$$V = \frac{\left(94\ \frac{\text{lbf}}{\text{sack}}\right)(\text{proportion})}{\left(62.4\ \frac{\text{lbf}}{\text{ft}^3}\right)(\text{SG})}$$

SG = SSD specific gravity

Concrete:

$$V = \frac{\left(94\ \frac{\text{lbf}}{\text{sack}}\right)(1)}{\left(62.4\ \frac{\text{lbf}}{\text{ft}^3}\right)(3.15)} = 0.48\ \text{ft}^3/\text{sack}$$

Fine aggregate:

$$V = \frac{\left(94\ \frac{\text{lbf}}{\text{sack}}\right)(1.6)}{\left(62.4\ \frac{\text{lbf}}{\text{ft}^3}\right)(2.62)} = 0.92\ \text{ft}^3/\text{sack}$$

Coarse aggregate:

$$V = \frac{\left(94\ \frac{\text{lbf}}{\text{sack}}\right)(2.6)}{\left(62.4\ \frac{\text{lbf}}{\text{ft}^3}\right)(2.65)} = 1.48\ \text{ft}^3/\text{sack}$$

Water:

$$V = \frac{5.8\ \text{gal}}{7.48\ \frac{\text{gal}}{\text{ft}^3}} = 0.78\ \text{ft}^3/\text{sack}$$

The total volume is

$$0.48\ \text{ft}^3 + 0.92\ \text{ft}^3 + 1.48\ \text{ft}^3 + 0.78\ \text{ft}^3 = 3.66\ \text{ft}^3$$

$$(3.66\ \text{ft}^3)\left(\frac{1}{27}\ \frac{\text{yd}^3}{\text{ft}^3}\right) = 0.136\ \text{yd}^3/\text{sack}$$

For 1 yd^3,

$$\text{cement} = \left(\frac{1}{0.136\ \frac{\text{yd}^3}{\text{sack}}}\right)\left(94\ \frac{\text{lbf}}{\text{sack}}\right) = \boxed{691\ \text{lbf}}$$

$$\text{fine aggregate} = \left(\frac{1\ \text{yd}^3}{0.136\ \frac{\text{yd}^3}{\text{sack}}}\right)(1.6)\left(94\ \frac{\text{lbf}}{\text{sack}}\right) = \boxed{1106\ \text{lbf}}$$

$$\text{coarse aggregate} = \left(\frac{1\ \text{yd}^3}{0.136\ \frac{\text{yd}^3}{\text{sack}}}\right)(2.6)\left(94\ \frac{\text{lbf}}{\text{sack}}\right) = \boxed{1797\ \text{lbf}}$$

$$\text{water} = \left(\frac{1\ \text{yd}^3}{0.136\ \frac{\text{yd}^3}{\text{sack}}}\right)\left(0.78\ \frac{\text{ft}^3}{\text{sack}}\right)\left(62.4\ \frac{\text{lbf}}{\text{ft}^3}\right) = \boxed{358\ \text{lbf}}$$

(b) The ordered material weights are based on SSD density. The water absorbed (−) by sand is

$$(0.016)(1106\ \text{lbf}) = 17.7\ \text{lbf}$$

The excess (+) water in coarse aggregate is

$$(0.032)(1797\ \text{lbf}) = 57.5\ \text{lbf}$$

The adjustment to required water is

$$-57.5\text{ lbf} + 17.7\text{ lbf} = -39.8\text{ lbf (say 40 lbf)}$$

The weight of the water needed is

$$358\text{ lbf} - 40\text{ lbf} = \boxed{318\text{ lbf}}$$

(c) Three techniques that can be used to obtain proper curing are:

1) ponding, which uses earth or sand dikes around the perimeter to retain a pond of water. It requires considerable labor and is generally restricted to small jobs.
2) covering with plastic sheets, which provides an effective moisture barrier and is easily applied.
3) spraying or fogging, which is effective and can be carried out using ordinary lawn sprinklers. This method requires an ample water supply and may be costly.

(d) One way to install traverse joints is to saw a continuous straight slot in the top of the slab. Another method is to place strips of wood, plastic, or metal in the fresh concrete.

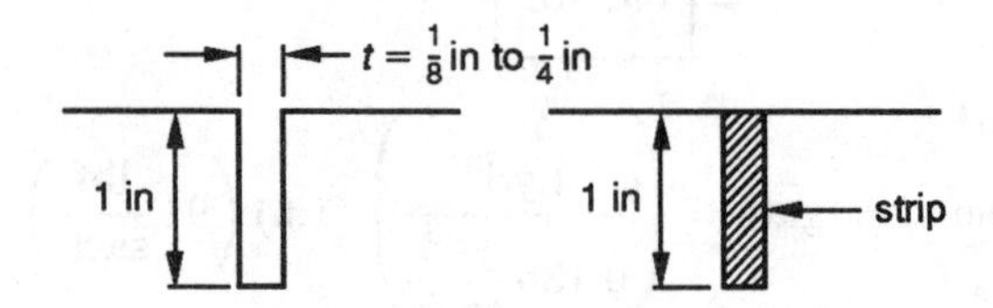

3. (a) Calculate the volumes based on SSD density.

Cement:

$$\text{weight} = (34\text{ sacks})\left(94\,\frac{\text{lbf}}{\text{sack}}\right) = 3196\text{ lbf}$$

$$V = \frac{3196\text{ lbf}}{(3.15)\left(62.4\,\frac{\text{lbf}}{\text{ft}^3}\right)} = 16.3\text{ ft}^3$$

Fine aggregate:

$$V = \frac{6500\text{ lbf}}{(2.67)\left(62.4\,\frac{\text{lbf}}{\text{ft}^3}\right)} = 39.0\text{ ft}^3$$

Water deficit (−) of fine aggregate:

$$(0.02)(6500\text{ lbf}) = 130.0\text{ lbf}$$

Coarse aggregate:

$$V = \frac{11{,}500\text{ lbf}}{(2.64)\left(62.4\,\frac{\text{lbf}}{\text{ft}^3}\right)} = 69.8\text{ ft}^3$$

The excess (+) water in coarse aggregate is

$$(0.015)(11{,}500\text{ lbf}) = 172.5\text{ lbf}$$

Water:

$$V = \frac{142\text{ gal}}{7.48\,\frac{\text{gal}}{\text{ft}^3}} = 19.0\text{ ft}^3$$

The correction from aggregate deviations from SSD is

$$\frac{-130\text{ lbf} + 172.5\text{ lbf}}{62.4\,\frac{\text{lbf}}{\text{ft}^3}} = 0.7\text{ ft}^3$$

The total volume is

$$V = 19.0\text{ ft}^3 + 0.7\text{ ft}^3 = 19.7\text{ ft}^3$$

$$\begin{aligned}\text{yield} &= \frac{16.3\text{ ft}^3 + 39.0\text{ ft}^3 + 69.8\text{ ft}^3 + 19.7\text{ ft}^3}{1 - 0.04}\\ &= 150.8\text{ ft}^3\\ &= (150.8\text{ ft}^3)\left(\frac{1}{27\,\frac{\text{ft}^3}{\text{yd}^3}}\right)\\ &= \boxed{5.6\text{ yd}^3}\end{aligned}$$

$$\begin{aligned}\text{(b) water-cement ratio} &= (19.7\text{ ft}^3)\left(7.48\,\frac{\text{gal}}{\text{ft}^3}\right)\\ &\quad\times\left(\frac{1}{34\text{ sack}}\right)\\ &= \boxed{4.3\text{ gal/sack}}\end{aligned}$$

$$\text{(c) cement factor} = \frac{34\text{ sacks}}{5.6\text{ yd}^3} = \boxed{6.1\text{ sacks/yd}^3}$$

4. (a) Determine the design load, P_u. From ACI 318-89, Eq. 9-1,

$$\begin{aligned}P_u &= 1.4P_{\text{dead}} + 1.7P_{\text{live}}\\ &= (1.4)(280\text{ kips}) + (1.7)(190\text{ kips})\\ &= 715\text{ kips}\end{aligned}$$

Determine the column capacity, $\phi P_{n,\max}$. From ACI 318-89, Eq. 10-2,

$$\phi P_{n,\max} = 0.80\phi\,[0.85f_c'(A_g - A_{\text{st}}) + f_y A_{\text{st}}]$$

$$A_g = (16\text{ in})(16\text{ in}) = 256\text{ in}^2$$

For a tied column, $\phi = 0.7$. Substituting numerical values, the capacity is

$$\phi P_{n,\max} = [(0.80)(0.7)]\left[(0.85)\left(4\ \frac{\text{kips}}{\text{in}^2}\right)(256\ \text{in}^2 - A_{st}) + \left(60\ \frac{\text{kips}}{\text{in}^2}\right)A_{st}\right]$$
$$= 487.4\ \text{kips} + \left(31.7\ \frac{\text{kips}}{\text{in}^2}\right)A_{st}$$

The design criterion is

$$\phi P_{n,\max} \geq P_u$$
$$715\ \text{kips} = 487.4\ \text{kips} + \left(31.7\ \frac{\text{kips}}{\text{in}^2}\right)A_{st}$$
$$A_{st} = 7.18\ \text{in}^2$$

Check the limits.

$$\rho_{\min} < \rho < \rho_{\max}$$
$$\rho = \frac{A_{st}}{A_g}$$

From ACI 318-89, Sec. 10.9.1,

$$\rho_{\min} = 0.01$$
$$\rho_{\max} = 0.08$$
$$\rho = \frac{7.18\ \text{in}^2}{256\ \text{in}^2} = 0.028 \quad \text{[within limits—ok]}$$

Select the longitudinal reinforcement. Various combinations are possible—for example, using #9 bars,

$$\text{eight \#9 bars: } A_{s,\text{provided}} = 8\ \text{in}^2$$

Select the ties. The tie spacing is governed by ACI 318-89, Sec. 7.10.5.

$$16d_b = (16)(1.125\ \text{in}) = 18\ \text{in}$$
$$48d_t = (48)(0.375\ \text{in}) = 18\ \text{in} \quad \text{[for \#3 tie]}$$
$$\text{least column dimension} = 16\ \text{in} \quad \text{[controls]}$$

In accordance with ACI 318-89, Sec. 7.10.5.3, if clear distance dimension A in the figure is larger than 6 in, supplementary ties would be needed. Calculate A.

(b)

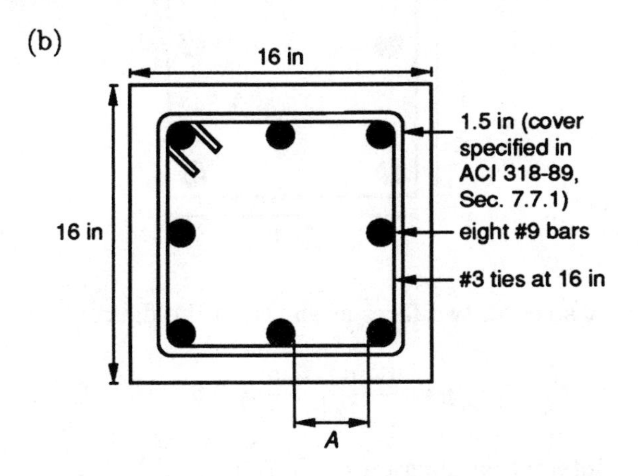

$$A = \frac{16\ \text{in} - (2)(1.5\ \text{in} + 0.375\ \text{in}) - (3)(1.125\ \text{in})}{2}$$
$$= 4.44\ \text{in} < 6\ \text{in}$$

Supplementary ties are not needed.

5. (a) Since information about the framing in the direction perpendicular to the figure is not given, assume (conservatively) that the beam of the frame shown carries the full load. Determine the design load, w_u.

$$w_{\text{dead}} = 110\ \text{lbf/ft}^2$$
$$w_{\text{live}} = 45\ \text{lbf/ft}^2$$

From ACI 318-89, Eq. 9-1, the factored load is

$$w_u = (20\ \text{ft})(1.4w_{\text{dead}} + 1.7w_{\text{live}})$$
$$= (20\ \text{ft})\left[(1.4)\left(110\ \frac{\text{lbf}}{\text{ft}^2}\right) + (1.7)\left(45\ \frac{\text{lbf}}{\text{ft}^2}\right)\right]$$
$$= 4610\ \text{lbf/ft}\ (4.61\ \text{kips/ft})$$

Determine the forces on the column using moment distribution.

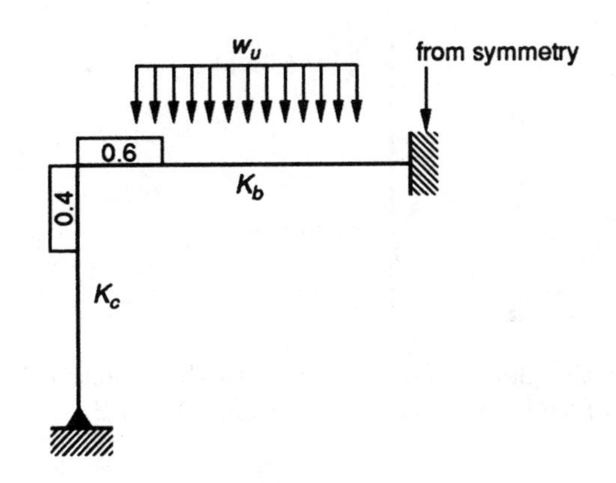

$$K_b = \frac{3EI}{32} = 0.09375EI$$
$$K_c = \left(\frac{EI}{12}\right)(0.75) = 0.0625EI \quad \text{[far end pinned]}$$

$$\text{column distribution factor} = \frac{K_c}{K_c + K_b}$$
$$= 0.4 \quad \text{[0.6 for girder]}$$
$$\text{fixed-end moment} = \frac{1}{12}w_u l^2$$
$$= \left(\frac{1}{12}\right)\left(4.61\ \frac{\text{kips}}{\text{ft}}\right)(32\ \text{ft})^2$$
$$= 393.4\ \text{kips-ft}$$

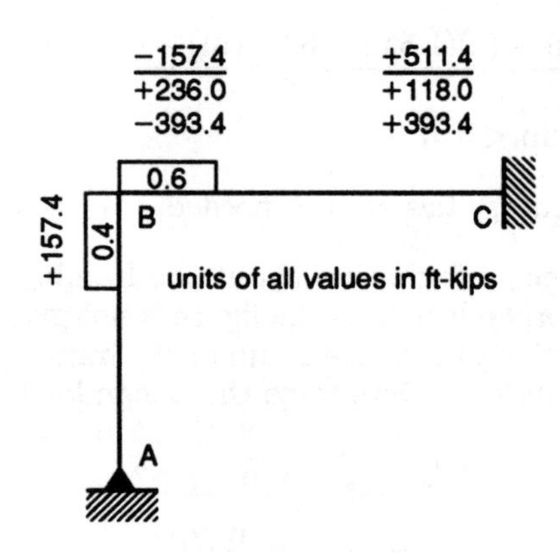

$$R = \frac{\left(4.61\ \frac{\text{kips}}{\text{ft}}\right)(32\ \text{ft})}{2} - \frac{(511.4\ \text{kip-ft} - 157.4\ \text{kip-ft})}{32}$$
$$= 62.70\ \text{kips}$$

The forces on the column are then

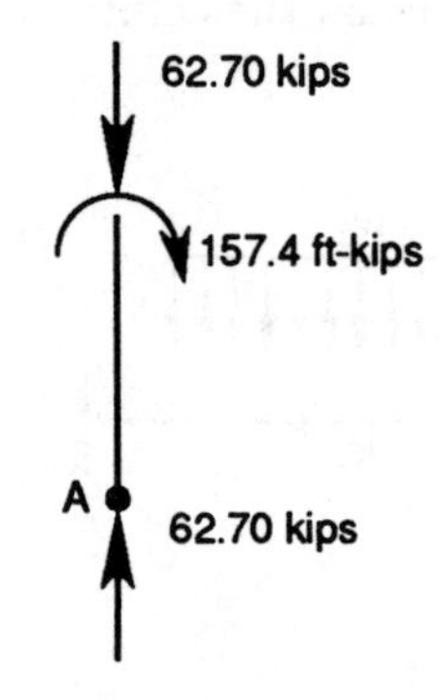

Check the slenderness. Determine the effective length in accordance with ACI 318-89, Sec. 10.11.2.

$$\Psi = \frac{\sum \frac{EI_c}{l_c}}{\sum \frac{EI_b}{l_b}}$$

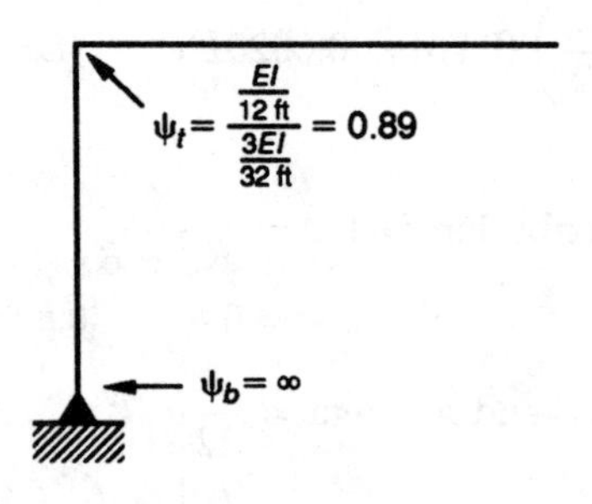

From ACI 318-89, Fig. 10.11.2,

$$k = 0.86$$

In accordance with ACI 318-89, Sec. 10.11.3, the radius of gyration, r, can be taken as

$$r = (0.3)(12\ \text{in}) = 3.6\ \text{in}$$

The column slenderness is defined as

$$\frac{kl_u}{r}$$

l_u is the clear height of the column. Recognizing that the beam depth is 24 in,

$$l_u = 144\ \text{in} - \frac{24\ \text{in}}{2} = 132\ \text{in}$$

The slenderness is

$$\frac{kl_u}{r} = \frac{(0.86)(132\ \text{in})}{3.6\ \text{in}} = 31.5\ \text{in}$$

The limit below which slenderness can be neglected is given in ACI 318-89, Sec. 10.11.4.1.

$$\left(\frac{kl_u}{r}\right)_{\text{limit}} = 34 - (12)\left(\frac{M_{1b}}{M_{2b}}\right)$$
$$= 34\ (\text{because } M_{1b} = 0)$$

Since the actual slenderness (31.5 in) is smaller than the limit, slenderness can be neglected.

Column cross section:

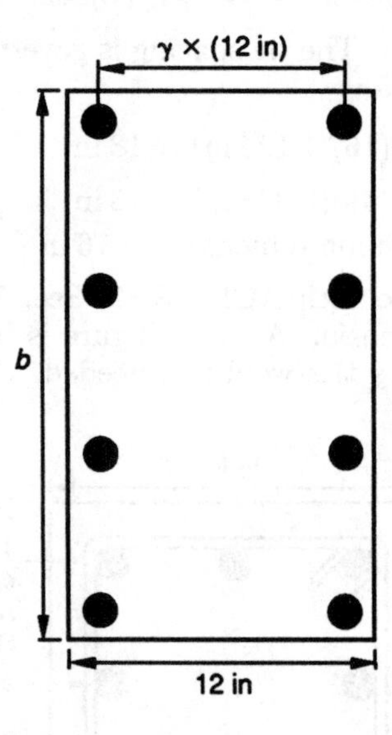

Select steel on two faces as shown in the figure.

$$y \approx \frac{12\ \text{in} - 5\ \text{in}}{12\ \text{in}} = 0.6$$

Use interaction diagram E4-60.75 (ACI SP–17).

Try $b = 24$ in.

$$\frac{\phi P_u}{A_g} = \frac{62.70 \text{ kips}}{(12 \text{ in})(24 \text{ in})} = 0.2177 \text{ kips/in}^2$$

$$\frac{\phi M_u}{A_g h} = \frac{(157.4 \text{ ft-kips})\left(12 \frac{\text{in}}{\text{ft}}\right)}{(12 \text{ in})(24 \text{ in})(12 \text{ in})} = 0.547 \text{ kips/in}^2$$

From the diagram,

$$\rho_{\text{required}} \approx 0.025 \qquad \text{[reasonable]}$$

$$\begin{aligned} A_{st} &= \rho A_g \\ &= (0.025)(12 \text{ in})(24 \text{ in}) \\ &= 7.2 \text{ in}^2 \end{aligned}$$

$$b = \boxed{24 \text{ in}}$$

(b) Select longitudinal steel and ties. One possible arrangement is shown.

Tie spacing:

$$16d_b = (16)(1.125 \text{ in}) = 18 \text{ in}$$

$$48d_t = (48)(0.375 \text{ in}) = 18 \text{ in}$$

$$\text{least column dimension} = 12 \text{ in} \qquad \text{[controls]}$$

Check the assumed y of 0.6.

$$\begin{aligned} y &= \frac{12 \text{ in} - (2)(1.5 \text{ in} + 0.375 \text{ in} + 0.563 \text{ in})}{12 \text{ in}} \\ &= 0.59 \qquad \text{[ok]} \end{aligned}$$

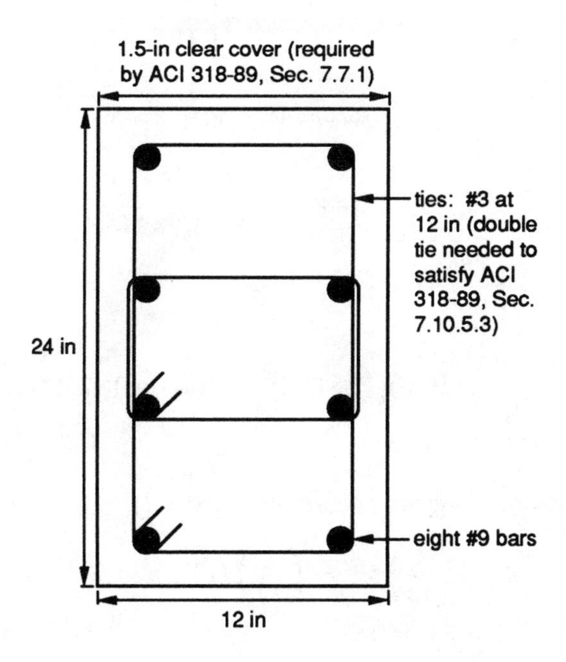

6. Select the beam depth. To control deflection, select a depth that satisfies the requirement of ACI 318-89, Table 9.5.

$$\begin{aligned} h &\geq \left(\frac{l}{16}\right)\left(0.4 + \frac{f_y}{100}\right) \\ &\geq \frac{(22 \text{ ft})\left(12 \frac{\text{in}}{\text{ft}}\right)}{16}\left(0.4 + \frac{36 \frac{\text{kips}}{\text{in}^2}}{100}\right) \\ &\geq 12.54 \text{ in} \end{aligned}$$

The nominal moment of capacity for a singly-reinforced section is given by

$$M_n = (f'_c b d^2 w)(1 - 0.59w)$$

From the minimum and maximum limits of steel allowed in ACI 318-89,

$$\frac{200}{f'_c} \leq w \leq (0.64\beta_1)\left(\frac{87}{87 + f_y}\right)$$

β_1 is as defined in ACI 318-89, Sec. 10.2.7.3.

$$\beta_1 = 0.85 \text{ for } f'_c \leq 4000 \frac{\text{lbf}}{\text{in}^2}$$

Substituting numerical values,

$$\frac{200}{f'_c} = \frac{200}{3000 \frac{\text{lbf}}{\text{in}^2}} = 0.67$$

$$\begin{aligned} (0.64\beta_1)\left(\frac{87}{87 + f_y}\right) &= (0.64)(0.85)\left(\frac{87}{87 + 36 \frac{\text{kips}}{\text{in}^2}}\right) \\ &= 0.38 \end{aligned}$$

$$0.067 \leq w \leq 0.38$$

Large w values will result in small depths and heavy reinforcement, while the opposite occurs when w is small. Use the common intermediate value of $w = 0.18$.

$$\begin{aligned} M_n &= \left(3 \frac{\text{kips}}{\text{in}^2}\right)(16 \text{ in})d^2(0.18)[1 - (0.59)(0.18)] \\ &= 7.72d^2 \end{aligned}$$

Obtain a first estimate of the applied moment by neglecting self-weight.

$$\begin{aligned} w_u &= 1.4w_{\text{dead}} + 1.74w_{\text{live}} \\ &= (1.4)\left(1.3 \frac{\text{kips}}{\text{ft}}\right) + (1.7)\left(3.3 \frac{\text{kips}}{\text{ft}}\right) \\ &= 7.43 \text{ kips/ft} \end{aligned}$$

$$\begin{aligned} M_u &= \frac{1}{8} w_u l^2 \\ &= \left(\frac{1}{8}\right)\left(7.43 \frac{\text{kips}}{\text{ft}}\right)(22 \text{ ft})^2 \\ &= 449.52 \text{ ft-kips} \end{aligned}$$

The design requirement is

$$\phi M_n \geq M_u$$

$$(0.9)(7.72d^2) = (449.52 \text{ ft-kips})\left(12\,\frac{\text{in}}{\text{ft}}\right)$$

$$d = 27.86 \text{ in}$$

On the basis of the previous calculations, select a depth, d, of 30 in.

Calculate the actual value of M_u (including self-weight).

$$\text{self-weight} = \left(0.15\,\frac{\text{kips}}{\text{ft}^3}\right)(16 \text{ in})(30 \text{ in})\left(\frac{1}{144\,\frac{\text{in}^2}{\text{ft}^2}}\right)$$

$$= 0.50 \text{ kips/ft}$$

$$w_u = (1.4)\left(1.3\,\frac{\text{kips}}{\text{ft}} + 0.5\,\frac{\text{kips}}{\text{ft}}\right) + (1.7)\left(3.3\,\frac{\text{kips}}{\text{ft}}\right)$$

$$= 8.13 \text{ kips/ft}$$

$$M_u = \left(\frac{1}{8}\right)\left(8.13\,\frac{\text{kips}}{\text{ft}}\right)(22 \text{ ft})^2$$

$$= 491.9 \text{ ft-kips}$$

Using the actual moment and a depth, d, of 27.5 in, solve for w.

$$M_u = \phi M_n$$

$$(491.9 \text{ ft-kips})\left(12\,\frac{\text{in}}{\text{ft}}\right) = (0.9)\left(3\,\frac{\text{kips}}{\text{in}^2}\right)(16 \text{ in}) \times (27.5 \text{ in})^2 w(1 - 0.59w)$$

$$w(1 - 0.59w) = 0.181$$

$$w = 0.206$$

From this value of w, the area of steel is

$$A_s = w\left(\frac{f'_c}{f_y}\right)bd$$

$$= (0.206)\left(\frac{3\,\frac{\text{kips}}{\text{in}^2}}{36\,\frac{\text{kips}}{\text{in}^2}}\right)(16 \text{ in})(27.5 \text{ in})$$

$$= 7.55 \text{ in}^2$$

Various arrangements and bar selections are possible. Using #11 bars,

$$\text{no. of bars} = \frac{A_s}{A_{\text{one bar}}} = \frac{7.55 \text{ in}^2}{1.56 \text{ in}^2}$$

$$= 4.84 \qquad \text{[Use five \#11 bars]}$$

In order to fit these bars in the 16-in width available and satisfy the spacing requirements, it is necessary to use bundles as shown in the figure.

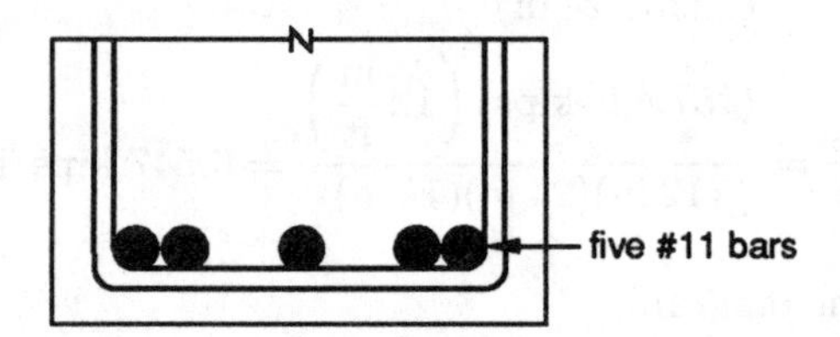

$$\text{beam size} = \boxed{16 \text{ in} \times 30 \text{ in}}$$

7. Determine the cracked moment of inertia, I_{cr}. From ACI 318-89, Sec. 8.5.1,

$$E_c = 57{,}000\sqrt{f'_c} = 57{,}000\sqrt{3000}$$

$$= 3{,}122{,}019\,\frac{\text{lbf}}{\text{in}^2}\ (3122 \text{ kips/in}^2)$$

$$n = \frac{E_s}{E_c} = \frac{29{,}000\,\frac{\text{kips}}{\text{in}^2}}{3122\,\frac{\text{kips}}{\text{in}^2}}$$

$$= 9.3$$

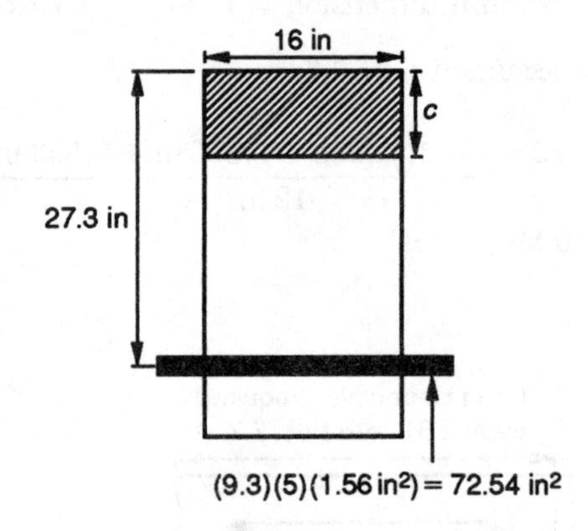

$$\frac{16c^2}{2} = (72.54)(27.3 - c)$$

$$c^2 + 9.1c - 247.5 = 0$$

$$c = 11.83 \text{ in}$$

$$I_{cr} = \left(\frac{1}{3}\right)(16 \text{ in})(11.83 \text{ in})^3 + (72.54 \text{ in}^2)(15.47 \text{ in})^2$$

$$= 26{,}190 \text{ in}^4$$

Calculate the gross moment of inertia, I_g.

$$I_g = \left(\frac{1}{12}\right)bh^3 = \left(\frac{1}{12}\right)(16 \text{ in})(30 \text{ in})^3$$

$$= 36{,}000 \text{ in}^4$$

Determine the cracking moment, M_{cr}. From ACI 318-89, Eq. 9-9,

$$f_r = 7.5\sqrt{f'_c} = (7.5)\sqrt{3000} = 410.8 \text{ lbf/in}^2$$

The cracking moment is given by ACI 318-89, Eq. 9-8.

$$M_{cr} = \frac{f_r I_g}{y_t} = \frac{\left(410.8 \frac{\text{lbf}}{\text{in}^2}\right)(36{,}000 \text{ in}^4)}{(30 \text{ in} - 11.83 \text{ in})\left(1000 \frac{\text{lbf}}{\text{kip}}\right)} = 813.9 \text{ in-kips (67.8 ft-kips)}$$

Obtain the maximum service load moment, M_a.

$$\text{service load} = 3.3 \frac{\text{kips}}{\text{ft}} + 1.3 \frac{\text{kips}}{\text{ft}} + 0.5 \frac{\text{kips}}{\text{ft}} = 5.1 \text{ kips/ft}$$

$$M_a = \frac{1}{8} w_{\text{service}} l^2 = \left(\frac{1}{8}\right)\left(5.1 \frac{\text{kips}}{\text{ft}}\right)(22 \text{ ft})^2 = 308.6 \text{ ft-kips}$$

Calculate the effective moment of inertia, I_e, using ACI 318-89, Eq. 9-7.

$$I_e = \left(\frac{M_{cr}}{M_a}\right)^3 I_g + \left[1 - \left(\frac{M_{cr}}{M_a}\right)^3\right] I_{cr}$$

$$= \left(\frac{67.8 \text{ ft-kips}}{308.6 \text{ ft-kips}}\right)^3 (36{,}000 \text{ in}^4) + \left[1 - \left(\frac{67.8 \text{ ft-kips}}{308.6 \text{ ft-kips}}\right)^3\right](26{,}190 \text{ in}^4) = 26{,}294 \text{ in}^4$$

Calculate the immediate deflections.

$$\Delta_{\text{immediate}} = \frac{5 w_{\text{service}} l^4}{384 E I_e} = \frac{(5)\left(5.1 \frac{\text{kips}}{\text{ft}}\right)\left(\frac{1}{12 \frac{\text{in}}{\text{ft}}}\right)(264 \text{ in})^4}{(384)\left(3122 \frac{\text{kips}}{\text{in}^2}\right)(26{,}294 \text{ in}^4)} = 0.33 \text{ in}$$

Find the long-term additonal deflection using ACI 318-89, Eq. 9-10.

$$\Delta(\text{creep} + \text{shrinkage}) = \lambda \Delta_{\text{immediate}}$$

$$\lambda = \frac{\xi}{1 + 50\rho'} = \frac{2}{1 + (50)(0)} = 2$$

$$\Delta(\text{creep} + \text{shrinkage}) = (2)(0.33 \text{ in}) = 0.66 \text{ in}$$

$$\text{total deflection} = 0.33 \text{ in} + 0.66 \text{ in} = \boxed{0.99 \text{ in} \quad [\text{say 1 in}]}$$

8. (a)

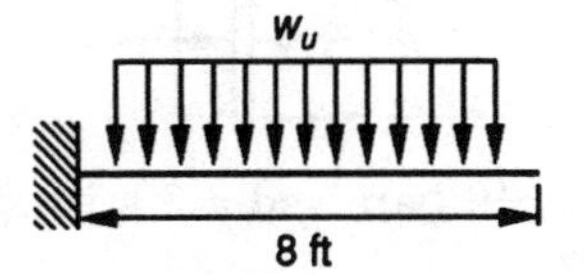

Select the preliminary beam depth based on deflection. From ACI 318-89, Table 9.5,

$$h_{\min} = \frac{l}{10} = \frac{(8)(12 \text{ in})}{10} = 9.6 \text{ in}$$

Estimate the beam weight as that of a 10-in × 10-in beam.

$$w_d = (1.4)\left(0.1 \frac{\text{kips}}{\text{ft}}\right) + \frac{(1.4)(100 \text{ in}^2)\left(0.15 \frac{\text{kips}}{\text{ft}^3}\right)}{\left(144 \frac{\text{in}^2}{\text{ft}^2}\right)} = 0.14 \text{ kips/ft} + 0.15 \text{ kips/ft}$$

$$w_l = (1.7)\left(0.3 \frac{\text{kips}}{\text{ft}}\right) = 0.51 \text{ kips/ft}$$

$$w_u = 0.14 \frac{\text{kips}}{\text{ft}} + 0.15 \frac{\text{kips}}{\text{ft}} + 0.51 \frac{\text{kips}}{\text{ft}} = 0.80 \text{ kips/ft}$$

$$M_u = \frac{w_u l^2}{2} = \left(\frac{1}{2}\right)\left(0.8 \frac{\text{kips}}{\text{ft}}\right)(8 \text{ ft})^2 = 25.6 \text{ ft-kips (307.2 in-kips)}$$

Calculate the required bd^2.

$$M_u = \phi\left[f'_c b d^2 w (1 - 0.59w)\right]$$

Use $w = 0.18$ to define the section size.

$$\rho = w\left(\frac{f'_c}{f_y}\right) = (0.18)\left(\frac{4 \frac{\text{kips}}{\text{in}^2}}{60 \frac{\text{kips}}{\text{in}^2}}\right) = 0.012$$

$$bd^2 = \frac{M_u}{\phi f'_c w (1 - 0.59w)} = \frac{307.2 \text{ in-kips}}{(0.9)\left(4 \frac{\text{kips}}{\text{in}^2}\right)(0.18)[1 - (0.59)(0.18)]} = 530.4 \text{ in}^3$$

$$d = \sqrt{\frac{bd^2}{b}} = \sqrt{\frac{527.6 \text{ in}^3}{10 \text{ in}}} = 7.28 \text{ in}$$

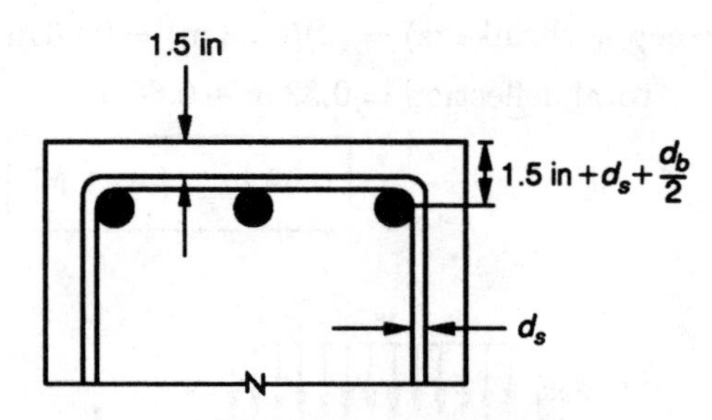

(b) Assuming #6 bars and #3 stirrups,

$$h \geq 7.28\text{ in} + 1.5\text{ in} + \tfrac{3}{8}\text{ in} + \tfrac{3}{8}\text{ in} = 9.53\text{ in}$$

Use $h = 10$ in.

Refine the value of w by using the exact selected dimensions.

$$\begin{aligned}
d &= 10\text{ in} - 1.5\text{ in} - \tfrac{3}{8}\text{ in} - \tfrac{3}{8}\text{ in} = 7.75\text{ in} \\
bd^2 &= (10\text{ in})(7.75\text{ in})^2 \\
&= 600.6\text{ in}^3 \\
(1 - 0.59w)w &= \frac{M_u}{\phi f'_c bd^2} \\
&= \frac{307.2\text{ in-kips}}{(0.9)\left(4\,\frac{\text{kips}}{\text{in}^2}\right)(600.6\text{ in}^3)} \\
&= 0.142 \\
w &= 0.155 \\
\rho &= w\left(\frac{f'_c}{f_y}\right) \\
&= (0.155)\left(\frac{4\,\frac{\text{kips}}{\text{in}^2}}{60\,\frac{\text{kips}}{\text{in}^2}}\right) \\
&= 0.010 \\
A_s &= \rho bd = (0.010\text{ in})(10\text{ in})(7.75\text{ in}) \\
&= 0.78\text{ in}^2
\end{aligned}$$

Two #6 bars provide 0.88 in^2.

Use two #6 bars.

Shear:

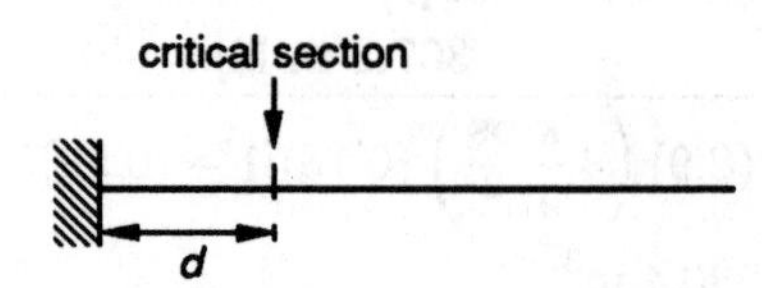

$$V_u = \left(0.8\,\frac{\text{kips}}{\text{ft}}\right)\left(8\text{ ft} - \frac{7.75\text{ in}}{12\,\frac{\text{in}}{\text{ft}}}\right) = 5.88\text{ kips}$$

From ACI 318-89, Eq. 11-3,

$$\begin{aligned}
V_c &= 2\sqrt{f'_c}bwd = 2\sqrt{4000}(10)(7.75) \\
&= 9803\text{ lbf } (9.80\text{ kips}) \\
\phi V_c &= (0.85)(9.80\text{ kips}) \\
&= 8.33\text{ kips}
\end{aligned}$$

Since $\phi V_c > V_u$ and the beam depth is 10 in, no shear reinforcement is needed (ACI 318-89, Sec. 11.5.5.1).

(c)

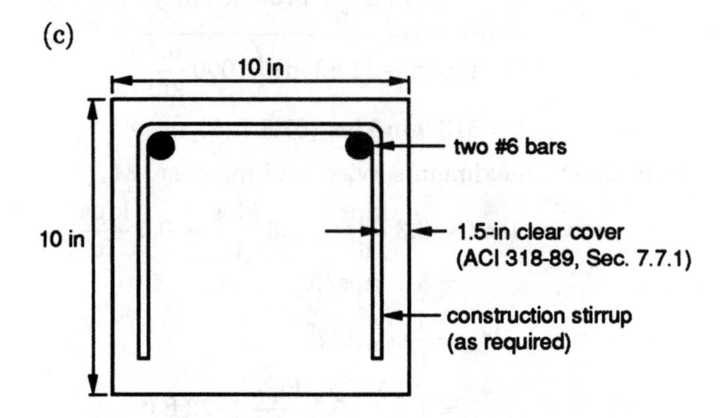

9. (a) Size the footing to attain a uniform pressure distribution under service loads. Locate the resultant of applied service loads.

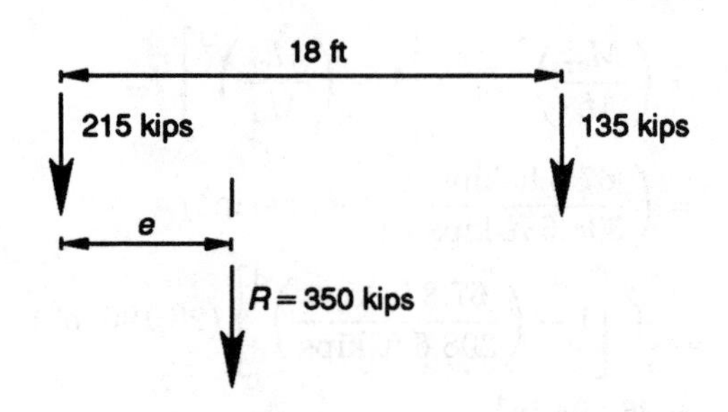

$$\begin{aligned}
(350\text{ kips})e &= (135\text{ kips})(18\text{ ft}) \\
e &= 6.94\text{ ft} \\
\text{footing length} &= [(18\text{ ft} - e) + 2\text{ ft}](2) \\
&= [(18\text{ ft} - 6.94\text{ ft}) + 2\text{ ft}](2) \\
&= 26.1\text{ ft}
\end{aligned}$$

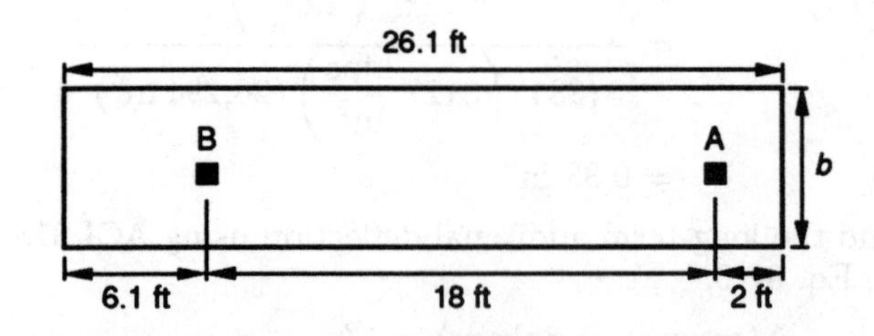

The footing thickness is likely to be governed by punching shear around the columns. Because this problem does not require checking punching shear, assume the footing thickness is the maximum allowed—i.e., 2 ft.

Determine the dimension, b. Equating the total load to the soil capacity times the bearing area gives

$$350 \text{ kips} + \left(0.15 \frac{\text{kips}}{\text{ft}^3}\right)(26.1 \text{ ft})(2 \text{ ft})b = \left(3.5 \frac{\text{kips}}{\text{ft}^2}\right)(26.1 \text{ ft})b$$

$$b = \boxed{4.2 \text{ ft}}$$

(b) Calculate the shear and moment diagrams. The design of the footing uses factored loads.

$$\begin{aligned} \text{column A} &= (1.4)(80 \text{ kips}) + (1.7)(55 \text{ kips}) \\ &= 205.5 \text{ kips} \\ \text{column B} &= (1.4)(110 \text{ kips}) + (1.7)(105 \text{ kips}) \\ &= 332.5 \text{ kips} \\ \text{total} &= 205.5 \text{ kips} + 332.5 \text{ kips} \\ &= 538 \text{ kips} \end{aligned}$$

(The footing weight does not induce shear or moment, so these are not included.)

The resultant of the factored load is not exactly in the same position as the resultant for the service load, but this shift is typically neglected and the pressure is assumed uniform. For this reason, the reactions on the inverted footing are not exactly equal to the factored column loads.

$$q_u = \frac{538 \text{ kips}}{26.1 \text{ ft}} = 20.61 \text{ kips/ft}$$

(c) Determine the steel for the maximum moment. Using a cover of 3 in as specified in ACI 318-89, Sec. 7.7.1, and assuming a bar diameter of approximately 1.25 in,

$$d = h - 3 \text{ in} - \frac{d_b}{2} = 24 \text{ in} - 3 \text{ in} - \frac{1.25 \text{ in}}{2}$$
$$= 20.4 \text{ in}$$

Solve for w.

$$\begin{aligned} (1 - 0.59w)w &= \frac{M_u}{\phi f_c' b d^2} \\ &= \frac{(632.5 \text{ ft-kips})\left(12 \frac{\text{in}}{\text{ft}}\right)}{(0.9)\left(3 \frac{\text{kips}}{\text{in}^2}\right)(50.28 \text{ in})(20.4 \text{ in})^2} \\ &= 0.134 \\ w &= 0.147 \end{aligned}$$

The steel ratio is

$$\begin{aligned} \rho &= w\left(\frac{f_c'}{f_y}\right) = (0.147)\left(\frac{3 \frac{\text{kips}}{\text{in}^2}}{60 \frac{\text{kips}}{\text{in}^2}}\right) \\ &= 0.0074 \qquad [> \rho_{\min} < \rho_{\max}; \text{ ok}] \\ A_s &= \rho b d \\ &= (0.0074)(50.28 \text{ in})(20.4 \text{ in}) \\ &= 7.6 \text{ in}^2 \end{aligned}$$

Use eight #9 bars. (d is slightly conservative.)

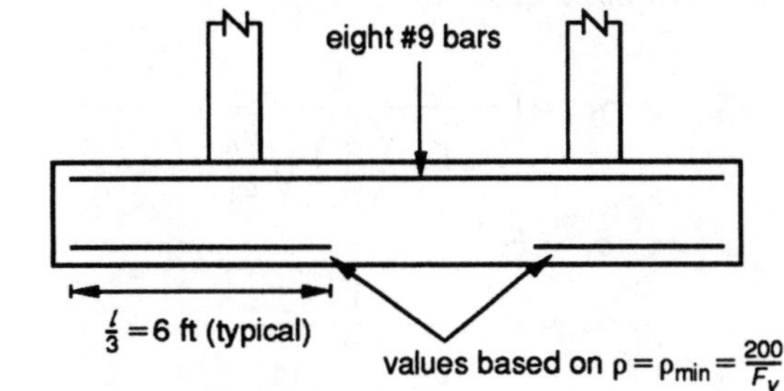

10. (a)

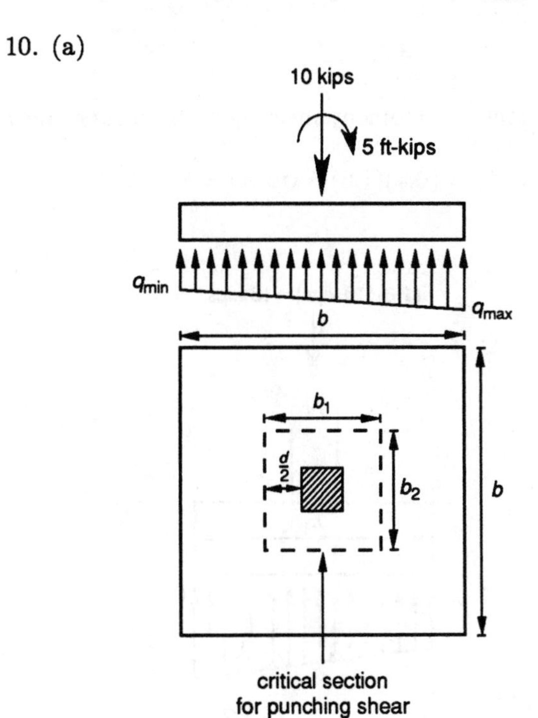

$$I = \frac{1}{12}b^4$$

$$q_{max} = \frac{10}{b^2} + \frac{(5)(0.5b)}{I}$$
$$= \frac{10}{b^2} + \frac{30}{b^3}$$

Solve by trials. To allow for a 1-ft-thick footing, limit q_{max} to 1.5 kips/ft^2 − 0.15 kips/ft^2 = 1.35 kips/in^2.

b (ft)	q_{max} (kips/ft^2)
2	6.25
3	2.22
3.5	1.52
3.6	1.42
3.7	1.32

$$q_{min} = 0.14 \text{ kips/ft}^2$$

$$\boxed{b = 3.7 \text{ ft (square)}}$$

(b) Check the 1-ft depth for punching shear. Find the fraction of the moment transferred by shear, γ_v, from ACI 318-89, Eq. 11-42.

$$\gamma_v = 1 - \frac{1}{1 + \left(\frac{2}{3}\right)\sqrt{\frac{b_1}{b_2}}}$$

$$b_1 = b_2$$

$$\gamma_v = 1 - \frac{1}{1 + \left(\frac{2}{3}\right)(1)}$$
$$= 0.4$$

Find the factored moment affecting the punching shear.

$$\gamma_v M_u = (0.4)(1.6)(5 \text{ kip-ft}) = 3.2 \text{ kip-ft}$$

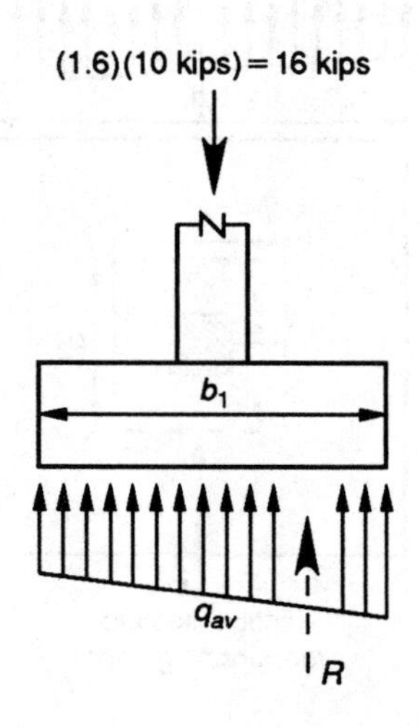

Assuming a single layer of #6 bars and a clear cover of 3 in,

$$d = 12 \text{ in} - 3 \text{ in} - 0.75 \text{ in} = 8.25 \text{ in}$$
$$R = q_{ave} b_1 b_2$$
$$b_1 = b_2 = 10 \text{ in} + d$$
$$= 10 \text{ in} + 8.25 \text{ in}$$
$$= 18.25 \text{ in}$$
$$q_{ave} = \frac{P}{A} = \frac{16 \text{ kips}}{(3.7 \text{ ft})^2}$$
$$= 1.17 \text{ kips/ft}^2$$
$$V_u = 16 \text{ kips} - R$$
$$= 16 \text{ kips} - \left(1.17 \frac{\text{kips}}{\text{ft}^2}\right)\left(\frac{18.25}{12} \text{ ft}\right)^2$$
$$= 13.29 \text{ kips}$$

From ACI 318-89, R11.12.6.2,

$$J_c = \frac{db_1^3}{6} + \frac{d^3 b_1}{6} + \frac{db_2 b_1^2}{2}$$
$$= \frac{(8.25 \text{ in})(18.25 \text{ in})^3}{6} + \frac{(8.25 \text{ in})^3(18.25 \text{ in})}{6} + \frac{(8.25 \text{ in})(18.25 \text{ in})(18.25 \text{ in})^2}{2}$$
$$= 35{,}139 \text{ in}^4$$
$$A_c = 2d(b_1 + b_2) = (2)(8.25 \text{ in})(18.25 \text{ in} + 18.25 \text{ in})$$
$$= 602.25 \text{ in}^4$$

Calculate the maximum punching shear stress using ACI 318-89, R11.12.6.2.

$$v_u = \frac{V_u}{A_c} + \frac{\gamma_v M_u \left(\frac{b_1}{2}\right)}{J_c}$$
$$= \frac{13.29 \text{ kips}}{602.25 \text{ in}^2} + \frac{(3.2 \text{ ft-kips})(9.13 \text{ in})\left(12 \frac{\text{in}}{\text{ft}}\right)}{35{,}139 \text{ in}^4}$$
$$= 0.032 \text{ kips/in}^2 \ (32.0 \text{ lbf/in}^2)$$

Calculate the allowable stress using ACI 318-89, Eq. 11-43.

$$\phi v_n = \frac{\phi V_c}{b_o d}$$

Using ACI 318-89, Eq. 11-36,

$$V_c = \left(2 + \frac{4}{\beta_c}\right)\sqrt{f'_c} b_o d$$

Using ACI 318-89, R11.12.2.1,

$$\beta_c = \frac{\text{long column dimension}}{\text{short column dimension}} = 1$$

From ACI 318-89, Eqs. 11-37 (with $\alpha = 40$) and 11-38, V_c should not exceed the smallest of

$$\left(\frac{40d}{b_o} + 2\right)\sqrt{f'_c}b_o d$$

$$b_o = 2(b_1 + b_2)$$

or

$$4\sqrt{f'_c}b_o d \qquad \text{[controls]}$$

$$\begin{aligned}\phi v_n &= \phi 4\sqrt{f'_c}\\ &= (0.85)(4)\sqrt{3000}\\ &= 186 \text{ lbf/in}^2\end{aligned}$$

Since $\phi v_n > v_u$, the punching shear capacity is adequate.

Check the one-way shear. The capacity is ϕV_c.

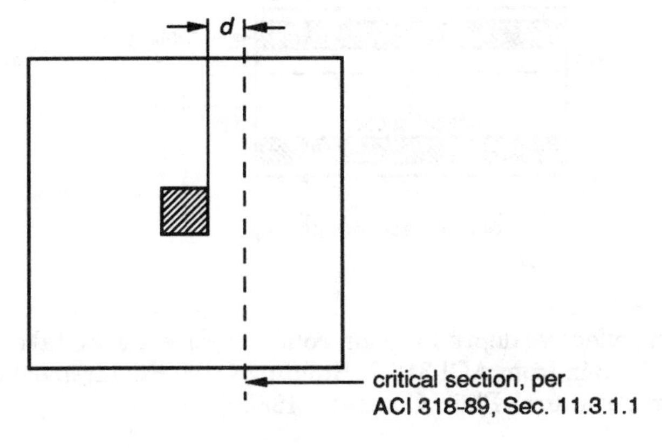

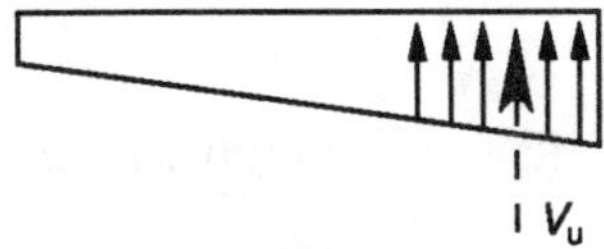

$$\begin{aligned}\phi V_c &= \phi 2\sqrt{f'_c}bd\\ &= (0.85)(2)\sqrt{3000}\,(3.7 \text{ ft})\left(12\,\frac{\text{in}}{\text{ft}}\right)(8.25 \text{ in})\\ &= 34{,}107 \text{ lbf } (34.1 \text{ kips})\end{aligned}$$

ϕV_c is greater than the total factored load, so there is no need to obtain V_u. The one-way shear is adequate.

(c) Calculate the maximum moment.

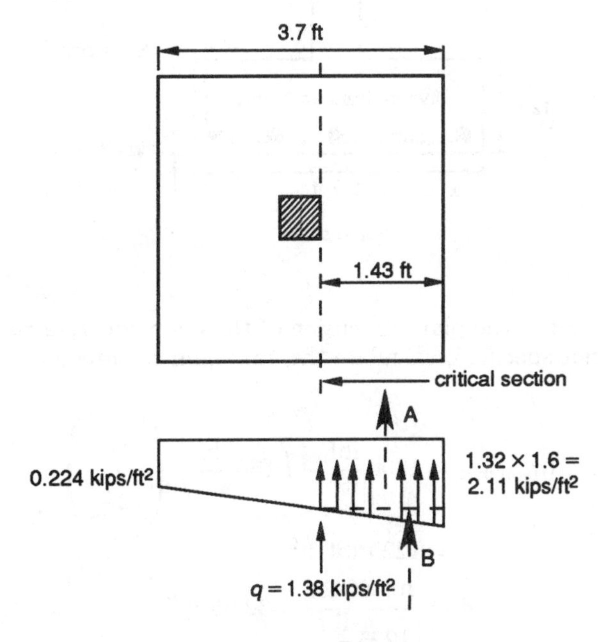

$$\begin{aligned}\text{force } A &= (3.7 \text{ ft})(1.43 \text{ ft})\left(1.38\,\frac{\text{kips}}{\text{ft}^2}\right)\\ &= 7.30 \text{ kips}\end{aligned}$$

$$\begin{aligned}\text{force } B &= \left(2.11\,\frac{\text{kips}}{\text{ft}^2} - 1.38\,\frac{\text{kips}}{\text{ft}^2}\right)\left(\frac{1.43 \text{ ft}}{2}\right)(3.7 \text{ ft})\\ &= 1.93 \text{ kips}\end{aligned}$$

$$\begin{aligned}M_u &= (7.30 \text{ kips})\left(\frac{1.43 \text{ ft}}{2}\right)\\ &\quad + (1.93 \text{ kips})\left(\frac{2}{3}\right)(1.43 \text{ ft})\\ &= 7.06 \text{ ft-kips}\end{aligned}$$

Use the trial approach to compute the required steel area.

$$A_s = \frac{M_u}{(0.9)f_y Jd}$$

Assume $J = 0.95$.

$$\begin{aligned}A_s &= \frac{(7.06 \text{ ft-kips})\left(12\,\frac{\text{in}}{\text{ft}}\right)}{(0.9)\left(60\,\frac{\text{kips}}{\text{in}^2}\right)(0.95)(8.25 \text{ in})}\\ &= 0.19 \text{ in}^2 \qquad \text{[clearly less than minimum]}\end{aligned}$$

Using ACI 318-89, R7.12.2.1,

$$\begin{aligned}(A_s)_{\min} &= \rho_{\min}A\\ &= (0.0018)(44.4 \text{ in})(12 \text{ in})\\ &= 0.96 \text{ in}^2\end{aligned}$$

Use five #4 bars in each direction.

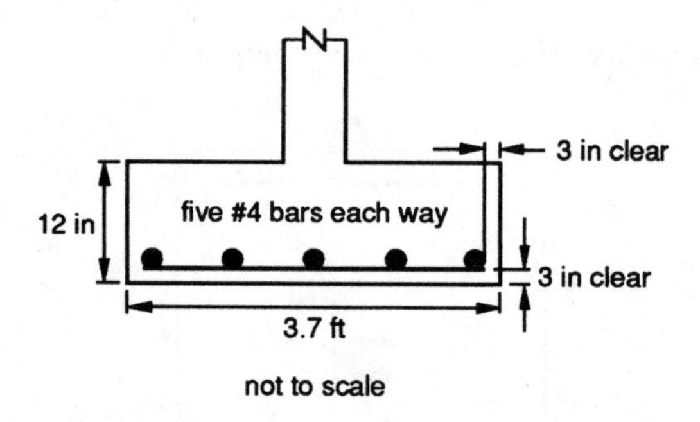

11. Since the plan dimension of the plain concrete slab is not specified, calculate the minimum required.

$$(\sigma_{\text{soil}})_{\text{net}} = 1300\,\frac{\text{lbf}}{\text{ft}^2} - \left(150\,\frac{\text{lbf}}{\text{ft}^3}\right)\left(\frac{6\text{ in}}{12\,\frac{\text{in}}{\text{ft}}}\right)$$

$$= 1225\ \text{lbf/ft}^2$$

$$A = \frac{40{,}000\ \text{lbf}}{1225\,\frac{\text{lbf}}{\text{ft}^2}} = 32.65\ \text{ft}^2$$

Given that the clear distance between the footings is 6 ft and that the length is 10 ft, the minimum area exceeds 60 ft^2.

(a) Assume isolated 2-ft-wide footings and neglect the slab. Assuming all the load is dead, the load factor is 1.4. The shear and moment diagrams for the footings are

$$w_u = \frac{(2)(10\ \text{kips})(1.4)}{10\ \text{ft}}$$

$$= 2.8\ \text{kips/ft}$$

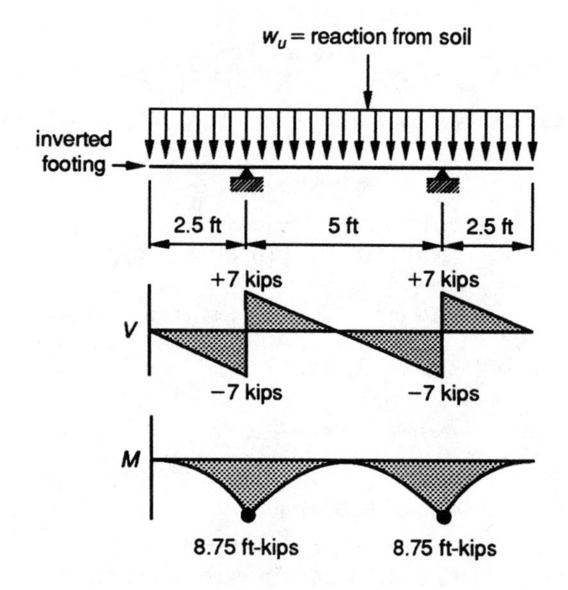

Taking $d \approx (0.8)(20\ \text{in}) = 16$ in,

$$V_c = (2)\left[\frac{\sqrt{3000}(24\ \text{in})d}{1000\,\frac{\text{lbf}}{\text{kip}}}\right]$$

$$= (2)\left[\frac{\sqrt{3000}(24\ \text{in})(16\ \text{in})}{1000\,\frac{\text{lbf}}{\text{kip}}}\right]$$

$$= 42.07\ \text{kips}$$

$$V_u < 7\ \text{kips}$$

$$\phi V_c = (0.85)(42.07\ \text{kips})$$

$$= 35.76\ \text{kips} > V_u$$

The footing size is sufficient for shear. (The moment of 8.75 kips-ft is small and can be easily provided for in the sections of the footing.)

(b) The critical section for shear in the plain slab is at a distance, d, from the face of the footings (see figure).

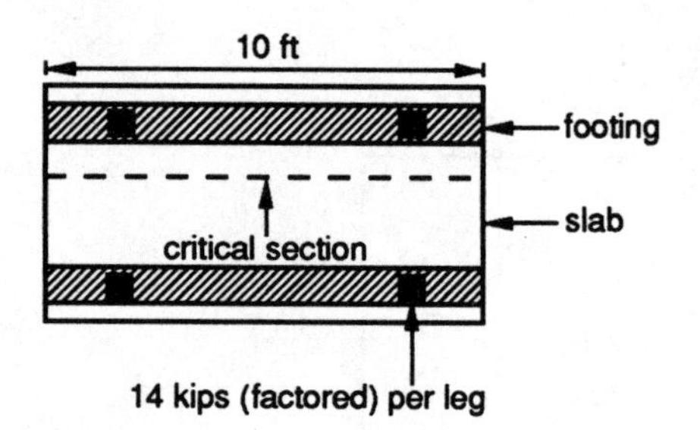

The effective depth for plain concrete slabs can be taken as $h-3$ in from ACI 318.1, Building Code Requirements for Structural Plain Concrete, 1989.

$$d = h - 3\ \text{in} = 3\ \text{in}$$

$$V_c = \frac{2\sqrt{3000}(120\ \text{in})d}{1000\,\frac{\text{lbf}}{\text{kip}}} = \frac{(2)(\sqrt{3000})(120\ \text{in})(3\ \text{in})}{1000\,\frac{\text{lbf}}{\text{kip}}}$$

$$= 39.44\ \text{kips}$$

$$V_u \le 28\ \text{kips}$$

$$\phi V_c = (0.85)(39.44\ \text{kips})$$

$$= 33.52\ \text{kips} > V_u \quad \text{[conservatively estimated]}$$

The plain slab shear strength is sufficient.

12. (a) Calculate the load per foot of ringwall.

$$\text{weight of liquid} = (55{,}000 \text{ gal})\left(0.13368 \frac{\text{ft}^3}{\text{gal}}\right) \times \left(62.4 \frac{\text{lbf}}{\text{ft}^3}\right)$$
$$= 458{,}789 \text{ lbf } (458.8 \text{ kips})$$
$$\text{weight of tank} = 200{,}000 \text{ lbf } (200 \text{ kips})$$
$$\text{total weight} = 458.8 \text{ kips} + 200 \text{ kips} = 658.8 \text{ kips}$$
$$\text{perimeter of centerline of ringwall} = (\pi)(32 \text{ ft}) = 100.53 \text{ ft}$$

The service load per foot of ringwall is

$$q = \frac{658.8 \text{ kips}}{100.53 \text{ ft}} = 6.55 \text{ kips/ft}$$

Neglecting the weight of the ringwall, the required width, b, to keep the soil pressure within allowances is

$$b = \frac{6.55 \frac{\text{kips}}{\text{ft}}}{1.5 \frac{\text{kips}}{\text{ft}^2}} = 4.37 \text{ ft}$$

This dimension is large and suggests the use of a footing under the ringwall. Because of the 18-in-deep frost line, transfer the load at a depth of 24 in. Assuming access to the underside of the tank is unnecessary, the projection of the ringwall above the ground surface is arbitrary. Select 6 in.

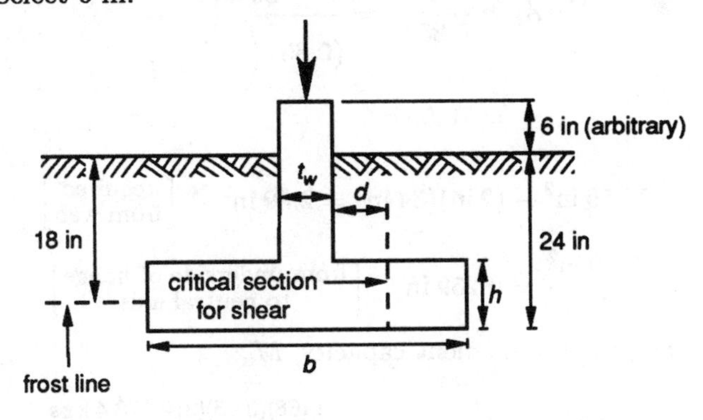

Assuming the full loading is dead load, the factored load per foot is

$$p_u = 1.4q$$

Using ACI 318-89, Sec. 14.5.3.2,

$$t_w \geq 7.5 \text{ in} \quad \text{[say 8 in]}$$

Check the 8-in value for strength.

Estimate h. The shear force at the critical section, V_u, is

$$V_u = P_u \left[\frac{b - (t_w + 2d)}{2b}\right] = P_u\left(\frac{1}{2} - \frac{t_w}{2b} - \frac{d}{b}\right)$$

The shear capacity of a 1 ft-long section is ϕV_c where

$$\phi V_c = (0.85)(2)\sqrt{f'_c}d$$
$$(0.85)(2)\sqrt{3000}(1 \text{ ft})d\left(12 \frac{\text{in}}{\text{ft}}\right) = 1117d \text{ lbf } (1.12d \text{ kips})$$

Taking tentative values of b and t_w as 60 in and 8 in, respectively, and recognizing that $P_u = 9.17$ kips,

$$V_u = 9.17 \text{ kips}\left[\frac{1}{2} - \frac{8 \text{ in}}{(2)(60 \text{ in})} - \frac{d}{60}\right] = 3.97 - 0.153d$$

Equating V_u to ϕV_c,

$$d = 3.12 \text{ in}$$

However, ACI 318-89, Sec. 15.7, presecribes a minimum effective depth of 6 in. With a 3-in clear cover, the total depth is

$$h = d + 3 + d_b \approx d + 4 \text{ in}$$

Try $h = 10$ in. Check the flexure before finalizing h.

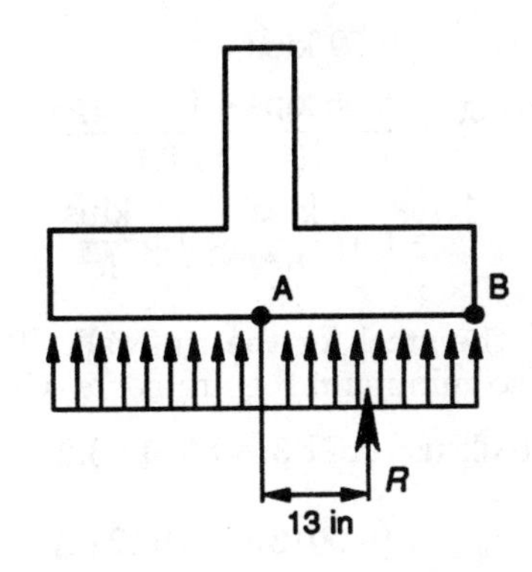

Again taking $b = 60$ in, the load between points A and B in the figure is R.

$$R = \left[\frac{60 \text{ in} - 8 \text{ in}}{(2)(60 \text{ in})}\right](9.17 \text{ kips}) = 3.97 \text{ kips}$$
$$M_u = (3.97 \text{ kips})(13 \text{ in}) = 51.61 \text{ in-kips/ft of ringwall footing}$$

Assume $J = 0.95$.

$$(A_s)_{\text{req}} = \frac{M_u}{\phi f_y J d}$$

$$= \frac{\left(51.61\ \frac{\text{in-kips}}{\text{ft}}\right)}{(0.9)\left(60\ \frac{\text{kips}}{\text{in}^2}\right)(0.95)(6\ \text{in})}$$

$$= 0.17\ \text{in}^2/\text{ft}$$

$$(A_s)_{\min} = (0.0018)(10\ \text{in})(12\ \text{in})$$

$$= 0.216\ \text{in}^2 \quad \text{[controls]}$$

$h = 10$ in can be used.

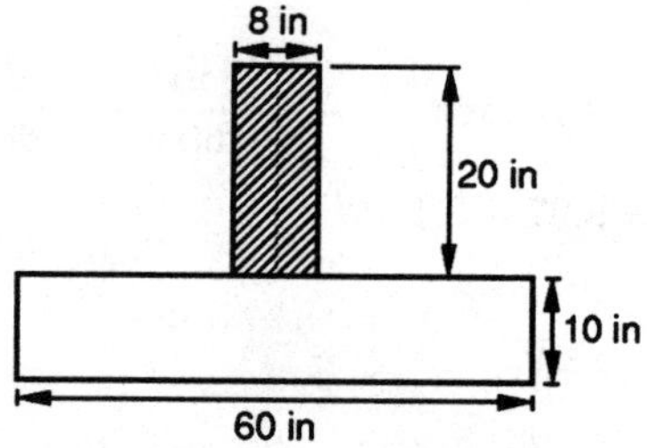

Check the bearing pressure taking into consideration the weight of the wall and footing.

$$\text{added pressure per foot of wall} = \left[\frac{(8\ \text{in})(20\ \text{in}) + (10\ \text{in})(60\ \text{in})}{144\ \frac{\text{in}^2}{\text{ft}^2}}\right] \times \left(0.15\ \frac{\text{kips}}{\text{ft}^2}\right)$$

$$= 0.79\ \text{kips}$$

$$\text{actual bearing pressure} = \frac{6.55\ \text{kips} + 0.79\ \text{kips}}{(5\ \text{ft})(1\ \text{ft})}$$

$$= 1.47\ \frac{\text{kips}}{\text{ft}^2} < 1.5\ \frac{\text{kips}}{\text{ft}^2} \quad \text{[allowable]}$$

(b) Select the steel for the ringwall. The load per foot is small, so minimum requirements will govern.

For vertical steel, use ACI 318-89, 14.3.2.

$$A_{s,v} = (0.0012)(8\ \text{in})(12\ \text{in})$$

$$= 0.12\ \text{in}^2/\text{ft}$$

Use #3 bars at 11 in.

For horizontal steel, use ACI 318-89, 14.3.3.

$$A_{s,h} = (0.002)(8\ \text{in})(20\ \text{in})$$

$$= 0.32\ \text{in}^2/\text{ft} \quad \text{[in shaded area]}$$

Use three #3 or two # 4 bars.

(c) Summary of final design:

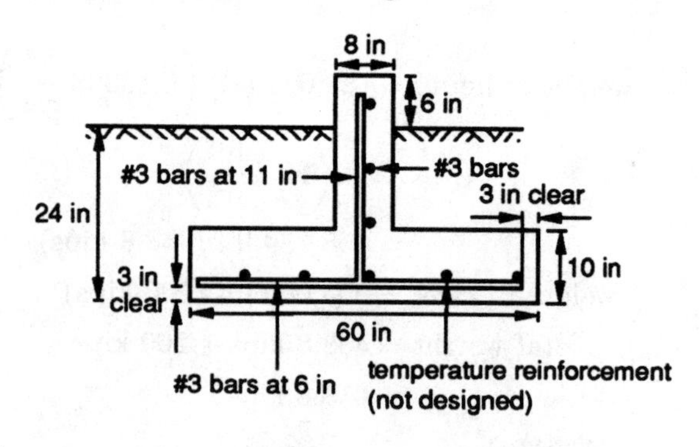

13.

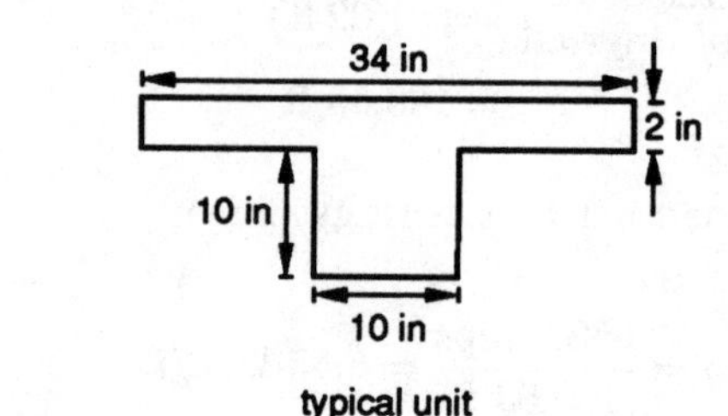

Use ACI 318-89, Sec. 8.10.2 to check the width of the flange.

$$(b_{\text{eff}})_{\max} = (2)(8)(2\ \text{in}) + 10\ \text{in}$$

$$= 42\ \text{in} > 34\ \text{in}$$

The full width is available.

Calculate the area of concrete in compression at ultimate tension.

$$T = f_y A_s = \left(60\ \frac{\text{kips}}{\text{in}^2}\right)(3\ \text{in}^2)$$

$$= 180\ \text{kips}$$

$$A_c = \frac{T}{0.85 f'_c} = \frac{180\ \text{kips}}{(0.85)\left(3\ \frac{\text{kips}}{\text{in}^2}\right)}$$

$$= 70.59\ \text{in}^2$$

$$70.59\ \text{in}^2 - (2\ \text{in})(34\ \text{in}) = 2.59\ \text{in}^2 \quad \left[\begin{array}{c}\text{required}\\ \text{from web}\end{array}\right]$$

$$\frac{2.59\ \text{in}^2}{10\ \text{in}} = 0.259\ \text{in} \quad \left[\begin{array}{c}\text{from underside of flange}\\ \text{to neutral axis}\end{array}\right]$$

Compute the moment capacity, M_u.

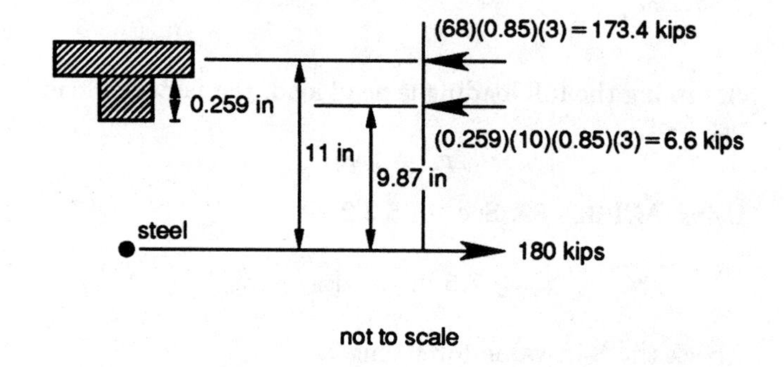

$$M_n = (173.4\ \text{kips})(11\ \text{in}) + (6.6\ \text{kips})(9.87\ \text{in})$$
$$= 1972.5\ \text{in-kips}\ (164.4\ \text{ft-kips})$$
$$\phi M_n = (0.9)(164.4\ \text{ft-kips})$$
$$= \boxed{148.0\ \text{ft-kips}}$$

14. (a) Locate the centroid and calculate the moment of inertia.

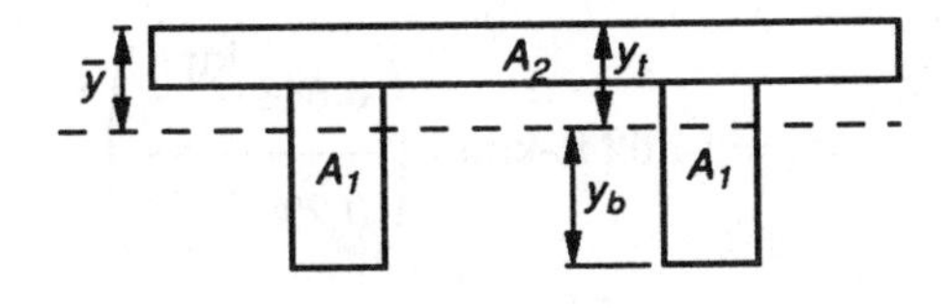

	area (in²)	y (in)	(area)(y) (in³)	(area)$(y-\bar{y})^2$ (in⁴)	I_o (in⁴)
2(A_1)	104	8.5	884	1347.8	1464.7
A_2	96	1.0	96	1460.2	32.0
totals	200		980	2808.0	1496.7

$$\bar{y} = \frac{980\ \text{in}^3}{200\ \text{in}^2} = 4.9\ \text{in}$$
$$I_x = 2808\ \text{in}^4 + 1496.7\ \text{in}^4$$
$$= 4304.7\ \text{in}^4$$
$$y_t = 4.9\ \text{in}$$
$$y_b = 10.1\ \text{in}$$
$$e = 7.1\ \text{in}$$

Check stresses at transfer. Assume tendon stress after transfer, f_{pi}, is the maximum permitted. From ACI 318-89, Sec. 18.5.1,

$$f_{pi} = 0.7f_{pu} = (0.7)\left(250\ \frac{\text{kips}}{\text{in}^2}\right)$$
$$= 175\ \text{kips/in}^2$$
$$\text{area of tendons} = (8)(0.08\ \text{in}^2)$$
$$= 0.64\ \text{in}^2$$

The prestressing force at transfer is

$$P_i = (0.64\ \text{in}^2)\left(175\ \frac{\text{kips}}{\text{in}^2}\right)$$
$$= 112\ \text{kips}$$

The stresses should be checked at 50 diameters of the strand from the ends and at the centerline.

Calculate the bending moments from self-weight at sections 1 and 2.

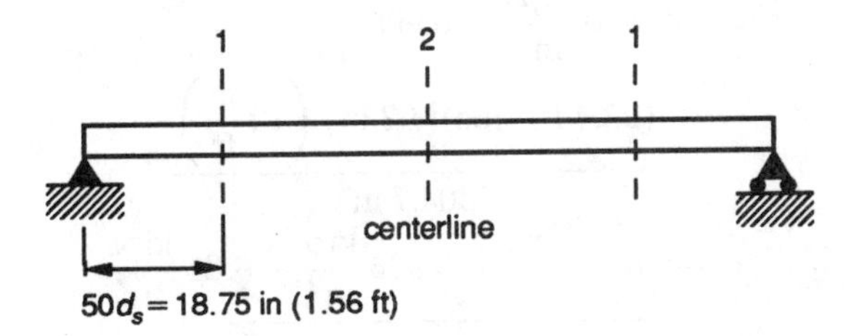

$$w(\text{self weight}) = \frac{(200\ \text{in}^2)\left(0.15\ \frac{\text{kips}}{\text{ft}^3}\right)}{144\ \frac{\text{in}^2}{\text{ft}^2}}$$
$$= 0.208\ \text{kips/ft}$$

At section 1,

$$M_1 = \left(0.208\ \frac{\text{kips}}{\text{ft}}\right)(15\ \text{ft})(1.56\ \text{ft}) - \left(\frac{1}{2}\right)\left(0.208\ \frac{\text{kips}}{\text{ft}}\right)(1.56\ \text{ft})^2$$
$$= 4.62\ \text{ft-kips}$$

At section 2,

$$M_2 = \left(\frac{1}{8}\right)\left(0.208\ \frac{\text{kips}}{\text{ft}}\right)(30\ \text{ft})^2$$
$$= 23.4\ \text{ft-kips}$$

Calculate the stresses at section 1.

$$\sigma_t = -\frac{112\ \text{kips}}{200\ \text{in}^2} + \frac{(112\ \text{kips})(7.1\ \text{in})(4.9\ \text{in})}{4304.7\ \text{in}^4} - \frac{(4.62\ \text{ft-kips})(4.9\ \text{in})\left(12\ \frac{\text{in}}{\text{ft}}\right)}{4304.7\ \text{in}^4}$$
$$= -0.560\ \frac{\text{kips}}{\text{in}^2} + 0.905\ \frac{\text{kips}}{\text{in}^2} - 0.063\ \frac{\text{kip}}{\text{in}^2}$$
$$= \boxed{0.282\ \text{kips/in}^2\ (282\ \text{lbf/in}^2)}$$

$$\sigma_b = -0.560\ \frac{\text{kips}}{\text{in}^2} - \left(0.905\ \frac{\text{kips}}{\text{in}^2}\right)\left(\frac{10.1\ \text{in}}{4.9\ \text{in}}\right) + \left(0.063\ \frac{\text{kip}}{\text{in}^2}\right)\left(\frac{10.1\ \text{in}}{4.9\ \text{in}}\right)$$
$$= -0.560\ \frac{\text{kips}}{\text{in}^2} - 1.865\ \frac{\text{kips}}{\text{in}^2} - 0.130\ \frac{\text{kips}}{\text{in}^2}$$
$$= \boxed{-2.295\ \text{kips/in}^2\ (-2295\ \text{lbf/in}^2)}$$

At section 2,

$$\sigma_t = -0.560\,\frac{\text{kips}}{\text{in}^2} + 0.905\,\frac{\text{kips}}{\text{in}^2} - \frac{(23.4\ \text{ft-kips})(4.7\ \text{in})\left(12\,\frac{\text{in}}{\text{ft}}\right)}{4304.7\ \text{in}^4}$$

$$= -0.560\,\frac{\text{kips}}{\text{in}^2} + 0.905\,\frac{\text{kip}}{\text{in}^2} - 0.31\,\frac{\text{kips}}{\text{in}^2}$$

$$= \boxed{0.035\ \text{kips/in}^2\ (35\ \text{lbf/in}^2)}$$

$$\sigma_b = -0.560\,\frac{\text{kips}}{\text{in}^2} - 1.865\,\frac{\text{kips}}{\text{in}^2} + \left(0.32\,\frac{\text{kips}}{\text{in}^2}\right)\left(\frac{10.1\ \text{in}}{4.9\ \text{in}}\right)$$

$$= -0.560\,\frac{\text{kips}}{\text{in}^2} - 1.865\,\frac{\text{kips}}{\text{in}^2} + 0.660\,\frac{\text{kips}}{\text{in}^2}$$

$$= \boxed{-1.765\ \text{kips/in}^2\ (-1765\ \text{lbf/in}^2)}$$

In summary, at transfer:

$$\text{maximum compression} = \boxed{2295\ \text{lbf/in}^2\quad [\text{not ok}]}$$

According to ACI 318-89, 18.4.1, the allowable compression is

$$0.6f'_{ci} = (0.6)\left(3500\,\frac{\text{lbf}}{\text{in}^2}\right) = 2100\ \text{lbf/in}^2$$

$$\text{maximum tension} = \boxed{282\ \text{psi}\quad [\text{ok}]}$$

According to ACI 318-89, 18.4.1, the allowable tension is

$$6\sqrt{f'_{ci}} = (6)\sqrt{3500\,\frac{\text{lbf}}{\text{in}^2}} = 354\ \text{lbf/in}^2$$

(b) Check stresses at the service load level. Assume total losses = 20% of f_{pi}. (Reference: Zia, et.al. "Estimating Prestress Losses," *Concrete International*, June 1979: 32–38.)

$$f_{pf} = 0.8f_{pi} = (0.8)\left(175\,\frac{\text{kips}}{\text{in}^2}\right) = 140\ \text{kips/in}^2$$

The final tendon force, P_f, is

$$P_f = \left(140\,\frac{\text{kips}}{\text{in}^2}\right)(0.64\ \text{in}^2) = 89.6\ \text{kips}$$

The service load is

$$w_{\text{service}} = \left(\frac{45\,\frac{\text{lbf}}{\text{ft}^2} + 40\,\frac{\text{lbf}}{\text{ft}^2}}{1000\,\frac{\text{lbf}}{\text{kip}}}\right)(4\ \text{ft}) + 0.208\,\frac{\text{kips}}{\text{ft}} = 0.548\ \text{kips/ft}$$

The moments at sections 1 and 2 are

$$M_1 = (4.62\ \text{ft-kips})\left(\frac{0.548\,\frac{\text{kips}}{\text{ft}}}{0.208\,\frac{\text{kips}}{\text{ft}}}\right) = 12.17\ \text{ft-kips}$$

$$M_2 = (23.4\ \text{ft-kips})\left(\frac{0.548\,\frac{\text{kips}}{\text{ft}}}{0.208\,\frac{\text{kips}}{\text{ft}}}\right) = 61.65\ \text{ft-kips}$$

Calculate the stresses at section 1.

$$\sigma_t = -\frac{89.6\ \text{kips}}{200\ \text{in}^2} + \frac{(89.6\ \text{kips})(7.1\ \text{in})(4.9\ \text{in})}{4304.7\ \text{in}^4} - \frac{(12.17\ \text{ft-kips})(12\ \text{in})(4.9\ \text{in})}{4304.7\ \text{in}^4}$$

$$= -0.448\,\frac{\text{kips}}{\text{in}^2} + 0.724\,\frac{\text{kips}}{\text{in}^2} - 0.166\,\frac{\text{kip}}{\text{in}^2}$$

$$= \boxed{0.110\ \text{kips/in}^2\ (110\ \text{lbf/in}^2)}$$

$$\sigma_b = -0.448\,\frac{\text{kips}}{\text{in}^2} - \left(0.724\,\frac{\text{kips}}{\text{in}^2}\right)\left(\frac{10.1\ \text{in}}{4.9\ \text{in}}\right) + \left(0.166\,\frac{\text{kip}}{\text{in}^2}\right)\left(\frac{10.1\ \text{in}}{4.9\ \text{in}}\right)$$

$$= -0.448\,\frac{\text{kips}}{\text{in}^2} - 1.492\,\frac{\text{kips}}{\text{in}^2} + 0.342\,\frac{\text{kips}}{\text{in}^2}$$

$$= \boxed{-1.598\ \text{kips/in}^2\ (-1598\ \text{lbf/in}^2)}$$

At section 2,

$$\sigma_t = -0.448\,\frac{\text{kips}}{\text{in}^2} + 0.724\,\frac{\text{kips}}{\text{in}^2} - \frac{(61.65\ \text{ft-kips})(12\ \text{in})(4.9\ \text{in})}{4304.7\ \text{in}^4}$$

$$= -0.448\,\frac{\text{kips}}{\text{in}^2} + 0.724\,\frac{\text{kip}}{\text{in}^2} - 0.842\,\frac{\text{kips}}{\text{in}^2}$$

$$= \boxed{-0.566\ \text{kips/in}^2\ (-566\ \text{lbf/in}^2)}$$

$$\sigma_b = -0.448\,\frac{\text{kips}}{\text{in}^2} - 1.492\,\frac{\text{kips}}{\text{in}^2} + \left(0.842\,\frac{\text{kips}}{\text{in}^2}\right)\left(\frac{10.1\text{ in}}{4.9\text{ in}}\right)$$

$$= -0.448\,\frac{\text{kips}}{\text{in}^2} - 1.492\,\frac{\text{kips}}{\text{in}^2} + 1.736\,\frac{\text{kips}}{\text{in}^2}$$

$$= \boxed{-0.204\text{ kips/in}^2\ (-204\text{ lbf/in}^2)}$$

In summary, at service:

$$\text{maximum compression} = \boxed{1598\text{ lbf/in}^2\quad[\text{ok}]}$$

According to ACI 318-89, 18.4.2, the allowable compression is

$$0.45f'_c = (0.45)\left(5000\,\frac{\text{lbf}}{\text{in}^2}\right) = 2250\text{ lbf/in}^2$$

$$\text{maximum tension} = \boxed{110\text{ lbf/in}^2\quad[\text{ok}]}$$

According to ACI 318-89, 18.4.2, the allowable tension is

$$6\sqrt{f'_c} = (6)\sqrt{5000\,\frac{\text{lbf}}{\text{in}^2}} = 424\text{ lbf/in}^2$$

(c) Find the section's nominal moment strength. Calculate the stress in the tendon at ultimate, f_{ps}. Using ACI 318-89, Eq. 18.3,

$$f_{ps} = f_{pu}\left[1 - \left(\frac{\gamma_p}{\beta_1}\right)\rho_p\left(\frac{f_{pu}}{f'_c}\right)\right]$$

f_{pu} is the specified tensile strength of the tendons. Using ACI 318-89, Chap. 18 notation,

$$\frac{f_{py}}{f_{pu}} = 0.85$$

$$\gamma_p = 0.4$$

$$f'_c = 5000\text{ lbf/in}^2$$

$$\beta_1 = 0.80$$

$$\rho_p = \frac{0.64\text{ in}^2}{(4\text{ ft})(12\text{ in})\left(12\,\frac{\text{in}}{\text{ft}}\right)} = 0.0011$$

$$f_{ps} = \left(250\,\frac{\text{kips}}{\text{in}^2}\right)\left[1 - \frac{(0.4)(0.0011)\left(250\,\frac{\text{kips}}{\text{in}^2}\right)}{(0.80)\left(5\,\frac{\text{kips}}{\text{in}^2}\right)}\right] = 243.1\text{ kips/in}^2$$

The total tension at ultimate is

$$T = \left(243.1\,\frac{\text{kips}}{\text{in}^2}\right)(0.64\text{ in}^2) = 155.58\text{ kips}$$

The depth of the compression block, a, is

$$a = \frac{T}{0.85f'_c b} = \frac{155.58\text{ kips}}{(0.85)\left(5\,\frac{\text{kips}}{\text{in}^2}\right)(48\text{ in})} = 0.76\text{ in}$$

The flexural capacity is

$$\phi M_n = \frac{(0.9)(155.58\text{ kips})\left(12\text{ in} - \frac{0.76\text{ in}}{2}\right)}{12\,\frac{\text{in}}{\text{ft}}} = \boxed{135.6\text{ ft-kips}}$$

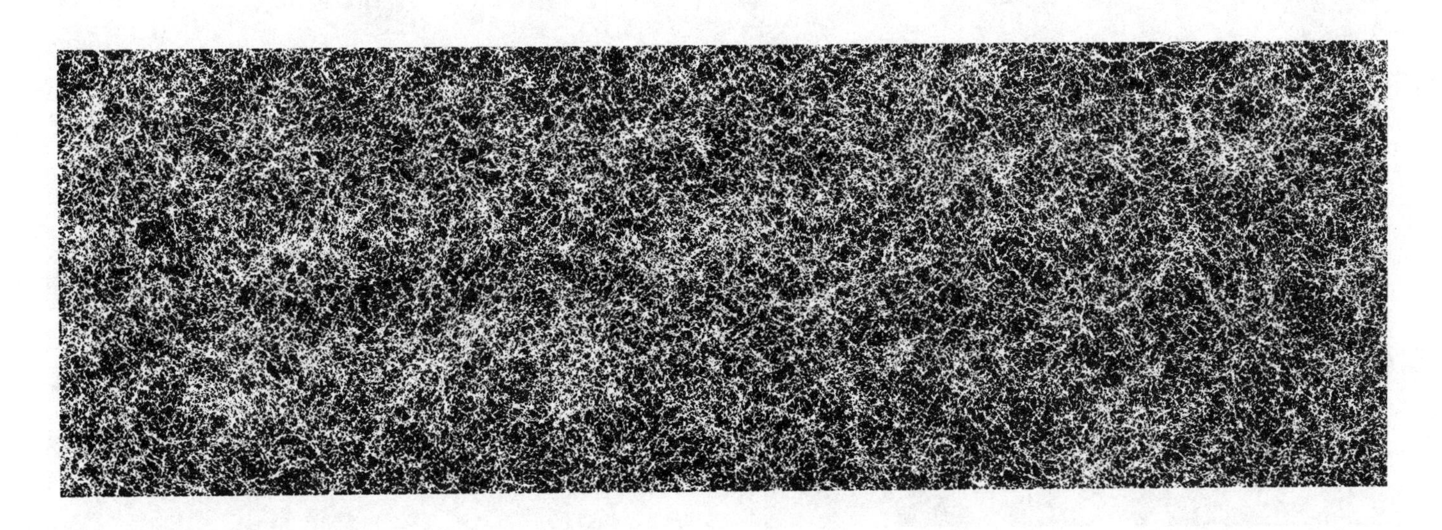

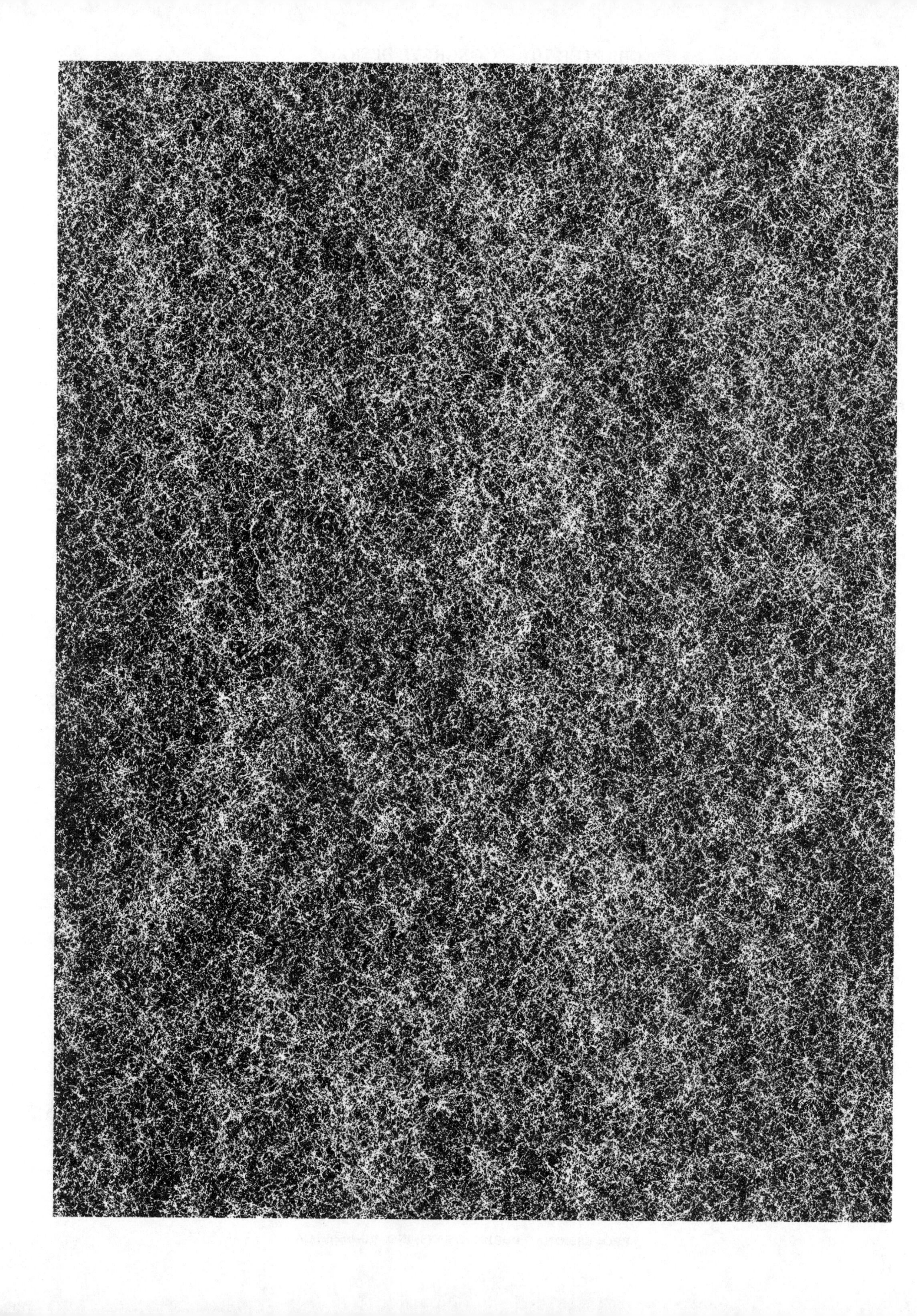

STEEL DESIGN

1. (a) The shear is transferred by the angles connected to the beam web. Use the procedure in the *AISC Manual of Steel Construction*, p. 4-89.

step 1: The horizontal tension and compression forces transmitted by the T sections are

$$T_{\text{all bolts}} = \frac{(140 \text{ ft-kips})\left(12 \frac{\text{in}}{\text{ft}}\right)}{18 \text{ in}} = 93.33 \text{ kips}$$

From AISC, Table I-A (p. 4-3), the capacity of $\frac{7}{8}$-in A325 bolts in tension is

$$B = 26.5 \text{ kips/bolt}$$

The required number of bolts is

$$n \approx \frac{T_{\text{all bolts}}}{B} = \frac{93.33 \text{ kips}}{26.5 \frac{\text{kips}}{\text{bolt}}}$$
$$= 3.52 \quad [\text{try 4 bolts}]$$

The force per bolt is

$$T = T_{\text{one bolt}} = \frac{93.33 \text{ kips}}{4 \text{ bolts}}$$
$$= 23.33 \text{ kips/bolt}$$

(b) *step 2:* Assume $t_w \approx 1.00$ in.

$$b = \frac{5.5 \text{ in} - 1 \text{ in}}{2} = 2.25 \text{ in}$$

The flange width of the W18 × 46 beam is 6.06 in. The load per inch of flange width (i.e., per inch of tee width) is

$$\frac{93.33 \text{ kips}}{6.06 \text{ in}} = 15.4 \text{ kips/in}$$

This exceeds the values in the preliminary selection table (AISC, p. 4-89). Perform a manual analysis.

Use the same length of T section for the top and bottom connectors. Neglect initial tightening and prying action for the initial estimate.

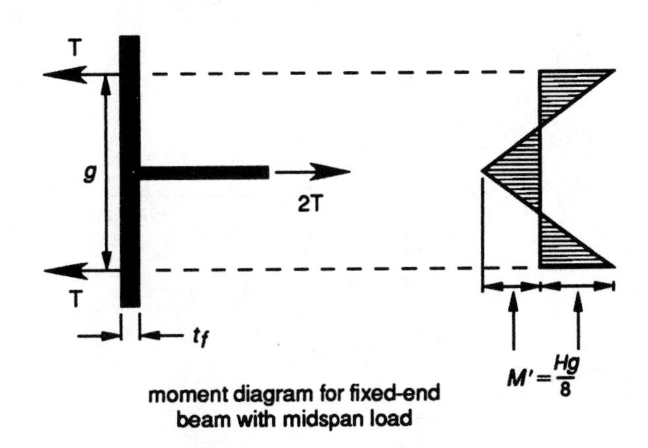

moment diagram for fixed-end beam with midspan load

The moment acting on the T section (considered to be a fixed-end beam) is

$$M' = \frac{Tg}{8} = \frac{(93.33 \text{ kips})(5.5 \text{ in})}{8} = 64.16 \text{ in-kips}$$

From AISC, p. 4-89, the required thickness is

$$t_f = \sqrt{\frac{6M'}{bF_b}} = \sqrt{\frac{(6)(64.16 \text{ in-kips})}{(6 \text{ in})(0.75)\left(36 \frac{\text{kip}}{\text{in}^2}\right)}}$$
$$= 1.54 \text{ in}$$

Try WT7 × 105.5. $[t_f = 1.56 \text{ in}]$

For WT7 × 105.5, $t_w = 0.98$ in and $b_f = 15.8$ in.

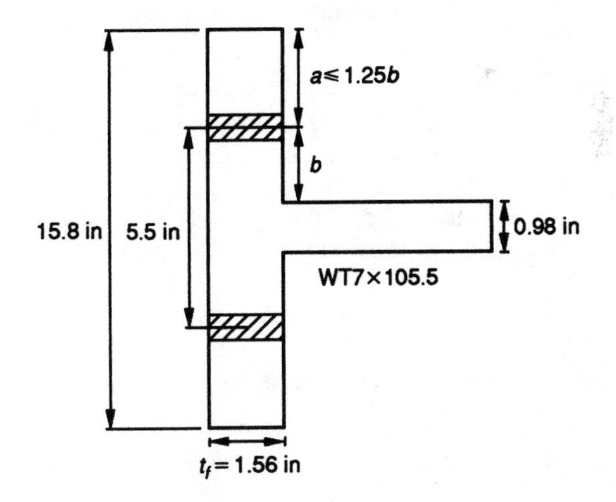

$$p = \frac{6.06 \text{ in}}{2} = 3.03 \text{ in}$$
$$d = \tfrac{7}{8} \text{ in}$$
$$b = \frac{5.5 \text{ in} - 0.98 \text{ in}}{2} = 2.26 \text{ in}$$
$$b' = b - \frac{d}{2} = 2.26 \text{ in} - \frac{\frac{7}{8} \text{ in}}{2}$$
$$= 1.8225 \text{ in}$$

steps 3, 4, and 5: Skipped because Q is to be reduced to an insignificant value.

step 6:

$$t_{f,\text{required}} = \sqrt{\frac{8Tb'}{pF_y}}$$

$$= \sqrt{\frac{(8)(23.33\text{ kips})(1.8225\text{ in})}{(3.03\text{ in})\left(36\,\frac{\text{kips}}{\text{in}^2}\right)}}$$

$$= 1.77\text{ in}$$

Try WT7 × 128.5.

$$t_f = 1.89\text{ in} \quad [> 1.77\text{ in}]$$

$$t_w = 1.175\text{ in}$$

$$b_f = 15.995\text{ in}$$

$$b = \frac{5.5\text{ in} - 1.175\text{ in}}{2} = 2.34\text{ in}$$

$$b' = b - \frac{d}{2} = 2.34\text{ in} - \frac{\frac{7}{8}\text{ in}}{2}$$

$$= 1.9025\text{ in}$$

$$t_{f,\text{required}} = \sqrt{\frac{8Tb'}{pF_y}}$$

$$= \sqrt{\frac{(8)(23.33\text{ kips})(1.9025\text{ in})}{(3.03\text{ in})(36\text{ kips})}}$$

$$\geq 1.80\text{ in} \quad [< 1.89\text{ in, ok}]$$

Use WT7 × 128.5.

(c) From AISC, Table I-A (p. 4-3), the capacity of $\frac{5}{8}$-in A325 bolts in tension is

$$B = 13.5\text{ kips/bolts}$$

The total capacity of four bolts is

$$(4\text{ bolts})\left(13.50\,\frac{\text{kips}}{\text{bolt}}\right) = 54.0\text{ kips}$$

$$d_{\text{min}} = \frac{M}{F}$$

$$= \frac{(140\text{ ft-kips})\left(12\,\frac{\text{in}}{\text{ft}}\right)}{54\text{ kips}}$$

$$= 31.11\text{ in}$$

A W33 beam size is required.

2. (a)

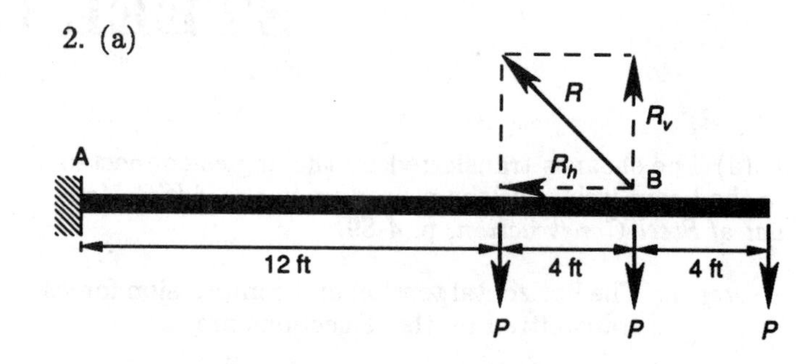

From $\sum M_A = 0$, $R_v = 3P$. From the slope of the tie rod, $R_h = 2R_v = 6P$.

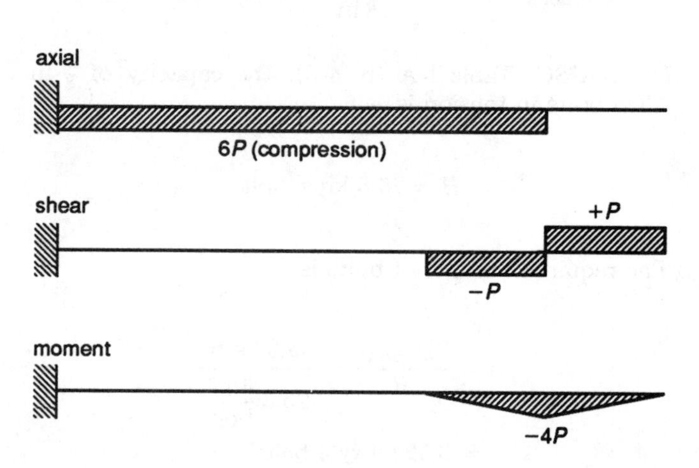

Point B is braced by the combined effect of the lateral wires and the restoring twist provided by the applied loads and the tie rod reaction.

$$P = \frac{3.4\text{ kips}}{3} = 1.13\text{ kips} \quad [\text{neglecting self-weight}]$$

Segment AB will control the selection of the lightest W8 section. Find the self-weight contributions. w is in lbf/ft.

To the moment:

$$\frac{w(4\text{ ft})(2)}{1000\,\frac{\text{lbf}}{\text{kip}}} = 0.008w\text{ ft-kips}$$

To the axial:

$$\left[\frac{w(20\text{ ft})(10\text{ ft})}{\left(1000\,\frac{\text{lbf}}{\text{kip}}\right)(16\text{ ft})}\right](2) = 0.025w\text{ kips}$$

Try W8 × 18.

$$\begin{aligned} A &= 5.26\text{ in}^2 \\ r_T &= 1.39\text{ in} \\ r_y &= 1.23\text{ in} \\ S_x &= 15.2\text{ in}^3 \\ \frac{d}{A_f} &= 4.70\text{ 1/in} \\ r_x &= 3.43\text{ in} \\ w &= 18\text{ lbf/ft} \end{aligned}$$

$$\begin{aligned} M &= (4\text{ ft})\left(\frac{3.4\text{ kips}}{3}\right) + \left(18\,\frac{\text{lbf}}{\text{ft}}\right)\left(0.008\,\frac{\text{kip-ft}^2}{\text{lbf}}\right) \\ &= 4.677\text{ ft-kips}\quad(56.13\text{ in-kips}) \end{aligned}$$

$$\begin{aligned} F &= (6)\left(\frac{3.4\text{ kips}}{3}\right) + (0.025)\left(18\,\frac{\text{lbf}}{\text{ft}}\right) \\ &= 7.25\text{ kips} \end{aligned}$$

$$f_a = \frac{7.25\text{ kips}}{5.26\text{ in}^2} = 1.378\text{ kips/in}^2$$

$$f_b = \frac{56.13\text{ kips}}{15.2\text{ in}^2} = 3.693\text{ kips/in}^2$$

Calculate F_a and F_b. From AISC, Eq. E2-2 or Table C-36,

$$\frac{Kl}{r_y} = \frac{(16\text{ ft})\left(12\,\frac{\text{in}}{\text{ft}}\right)}{1.23\text{ in}} = 156$$

$$\begin{aligned} F_a &= \frac{12\pi^2 E}{(23)\left(\frac{Kl}{r}\right)^2} = \frac{(12)(\pi^2)\left(29{,}000\,\frac{\text{kips}}{\text{in}^2}\right)}{(23)(156)^2} \\ &= 6.14\text{ kips/in}^2 \end{aligned}$$

$$\frac{l}{r_T} = 138.1$$

From AISC, Eq. F1-8,

$$C_b = 1$$

$$\begin{aligned} F_b &= \left[\frac{12{,}000}{l\left(\frac{d}{A_f}\right)}\right]C_b \\ &= \left[\frac{12{,}000}{(16\text{ ft})\left(12\,\frac{\text{in}}{\text{ft}}\right)\left(4.70\,\frac{1}{\text{in}}\right)}\right](1) \\ &= 13.30\text{ kips/in}^2 \end{aligned}$$

From AISC, Sec. H1,

$$F'_{ex} = \frac{12\pi^2 E}{(23)\left(\frac{Kl_b}{r_b}\right)^2}$$

$$\frac{Kl_b}{r_b} = \frac{(16\text{ ft})\left(12\,\frac{\text{in}}{\text{ft}}\right)}{3.43\text{ in}} = 55.98$$

$$F'_{ex} = 47.66\text{ kips/in}^2$$

Take $C_m = 1$ as a conservative estimate. AISC Eq. H1-1 controls.

Stresses should satisfy

$$\frac{f_a}{F_a} + \frac{C_m f_b}{\left(1 - \frac{f_a}{F'_{ex}}\right)F_b} \le 1$$

$$\begin{aligned} &\frac{1.378\,\frac{\text{kips}}{\text{in}^2}}{6.14\,\frac{\text{kips}}{\text{in}^2}} + \frac{3.693\,\frac{\text{kips}}{\text{in}^2}}{\left(1 - \frac{1.378\,\frac{\text{kips}}{\text{in}^2}}{47.66\,\frac{\text{kips}}{\text{in}^2}}\right)\left(13.30\,\frac{\text{kips}}{\text{in}^2}\right)} \\ &\quad = 0.51\quad[\text{ok}] \end{aligned}$$

The next smallest W8 is a W8 × 15, which is likely to be adequate for stress but cannot be used because kl/r_y exceeds 200 (the AISC maximum slenderness ratio permitted in Sec. B7).

Use W8 × 18.

(b) The reaction in each wire is zero. The ratio of the resultant force in the tie rod to the vertical component is

$$\sqrt{(2)^2 + (1)^2} = \sqrt{5}$$

The reaction in the tie rod is

$$\begin{aligned} R &= \sqrt{5}\,(3.4\text{ kips}) + \frac{\left(0.018\,\frac{\text{kips}}{\text{ft}}\right)(20\text{ ft})(10\text{ ft})\sqrt{5}}{6\text{ ft}} \\ &= \boxed{8.94\text{ kips}} \end{aligned}$$

3.

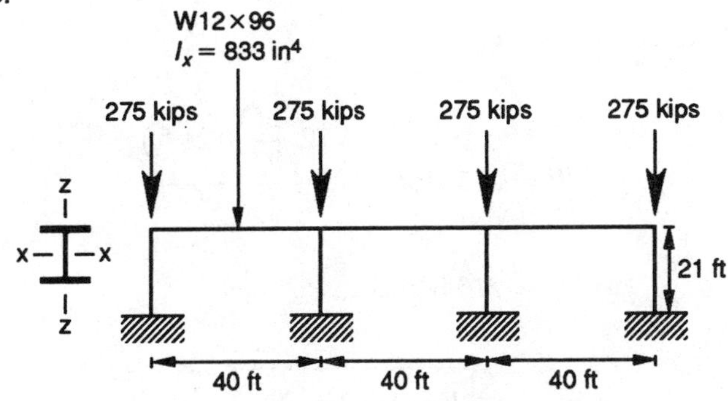

In order to select a section from the AISC tables, it is necessary to know the effective length. Since the effective length is a function of the relative rigidities, assume a value of the effective length factor. Assume $K = 1.5$.

$$Kl = (1.5)(21\ \text{ft}) = 31.5\ \text{ft}$$

From the AISC column tables, select W12 × 76.

$$I_y = 270\ \text{in}^4$$

Calculate the effective length.

$$G_t = \frac{\left(\frac{I}{L}\right)_{\text{column}}}{\left(\frac{I}{L}\right)_{\text{girder}}} = \frac{\frac{270\ \text{in}^4}{21\ \text{ft}}}{\frac{(2)(833\ \text{in}^4)}{40\ \text{ft}}} = 0.309$$

$G_b = 1$ is recommended by AISC for rigid connections to footing. From AISC, Fig. 1 (sidesway uninhibited), p. 3-5 $K \cong 1.2$ and $Kl = 25$ ft.

Try W12 × 72.

$$I_y = 195\ \text{in}^4$$

$$G_t = \frac{\left(\frac{I}{L}\right)_{\text{column}}}{\left(\frac{I}{L}\right)_{\text{girder}}} = \frac{\frac{195\ \text{in}^4}{21\ \text{ft}}}{\frac{(2)(833\ \text{in}^4)}{40\ \text{ft}}} = 0.223$$

$$G_b = 1$$

$$K \cong 1.19$$

$$Kl = 25\ \text{ft} \quad [\text{ok}]$$

Interpolating between 24 ft and 26 ft, the capacity from the AISC table on p. 3-27 is

$$P_{\text{all}} = \frac{288\ \text{kips} + 267\ \text{kips}}{2} = 277.5\ \text{kips}$$

The actual loading is

$$P = 275\ \text{kips} + \frac{\left(72\ \frac{\text{lbf}}{\text{ft}}\right)(21\ \text{ft})}{1000\ \frac{\text{lbf}}{\text{kip}}} = 276.5\ \text{kips} < P_{\text{all}} \quad [\text{ok}]$$

Use W12 × 72.

4. (a) Determine the design forces.

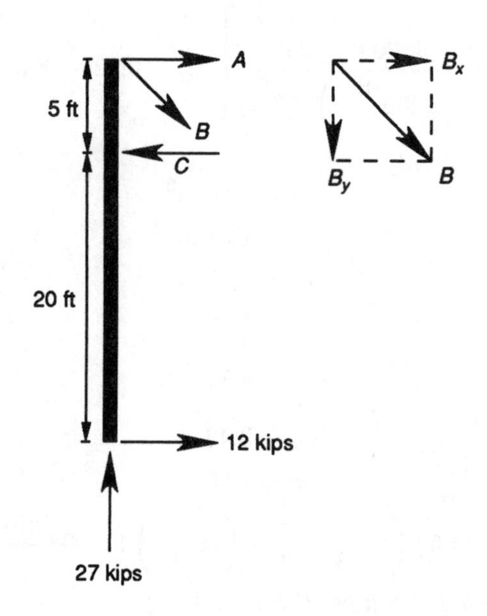

$$\sum F_y = 27 - B_y = 0$$
$$B_y = 27\ \text{kips}$$
$$B = \sqrt{2}(27\ \text{kips}) = 38.18\ \text{kips}$$
$$\sum M_{\text{top}} = -(12\ \text{kips})(25\ \text{ft}) + C(5\ \text{ft}) = 0$$
$$C = 60\ \text{kips}$$
$$\sum F_x = A + B_x - C + 12\ \text{kips} = 0$$
$$A = -12\ \text{kips} + 60\ \text{kips} - 27\ \text{kips} = 21\ \text{kips}$$

The centroids of members A and B intersect at the centerline of the bolted connections to the columns.

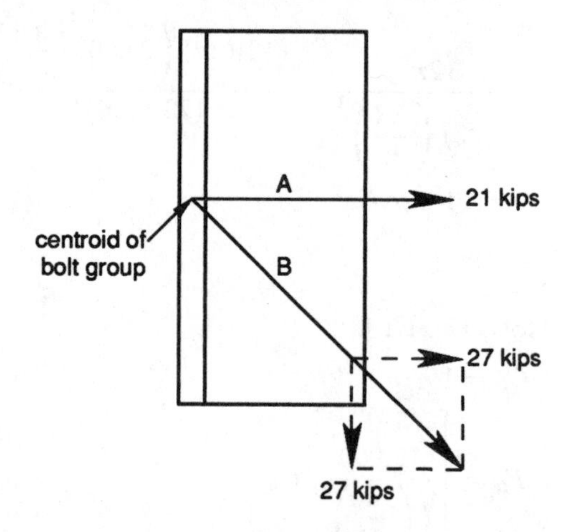

$$\text{tension, } T = 21\ \text{kips} + 27\ \text{kips} = 48\ \text{kips}$$
$$\text{shear, } V = 27\ \text{kips}$$

Find the allowable bolt stresses for A325 (threads not excluded from shear plane), from AISC, Table J3.2.

$$T = 44\ \text{kips/in}^2$$

From AISC, Table J3.2 and Sec. J3.6,

$$V = (17.0)\left(1 - \frac{f_t A_b}{T_b}\right)$$

f_t = average tension stress

A_b = nominal bolt area

T_b = specified pretension load [From AISC, Table J3.7]

Try using $\frac{5}{8}$-in bolts.

$$A_b = 0.306 \text{ in}^2$$

Neglecting the reduction factor for shear, the number of bolts required is determined from

$$(\text{no. of bolts})(0.306 \text{ in}^2)\left(44 \frac{\text{kips}}{\text{in}^2}\right) = 48 \text{ kips}$$

$$\text{no. of bolts} = 3.56$$

Check shear for four bolts.

$$f_v = \frac{27 \text{ kips}}{(4)(0.306 \text{ in}^2)} = 22.06 \text{ kips/in}^2 > 17.0 \text{ kips/in}^2 \quad \text{[not acceptable]}$$

Try six bolts.

$$f_v = \frac{27 \text{ kips}}{(6)(0.306 \text{ in}^2)} = 14.706 \text{ kips/in}^2$$

$$T_b = 19 \text{ kips}$$

$$F_v = \left(17.0 \frac{\text{kips}}{\text{in}^2}\right)\left[1 - \frac{(48 \text{ kips})(0.306 \text{ in}^2)}{(6)(0.306 \text{ in}^2)(19 \text{ kips})}\right] = 9.84 \text{ kips/in}^2 \quad \text{[not acceptable]}$$

Try eight $\frac{5}{8}$-in bolts.

$$f_v = \frac{27 \text{ kips}}{(8)(0.306 \text{ in}^2)} = 11.03 \text{ kips/in}^2$$

$$F_v = \left(17.0 \frac{\text{kips}}{\text{in}^2}\right)\left[1 - \frac{(48 \text{ kips})(0.306 \text{ in}^2)}{(8)(0.306 \text{ in}^2)(19 \text{ kips})}\right] = 11.63 \text{ kips/in}^2 \quad \text{[ok]}$$

Use eight $\frac{5}{8}$-in diameter A325 bolts.

(b)

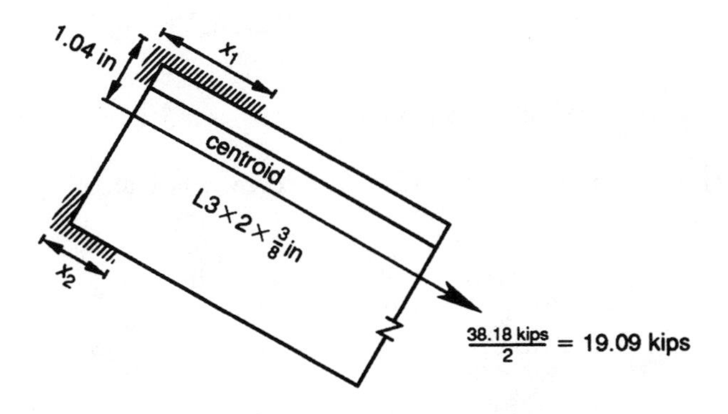

From AISC, Table J2.4,

$$\text{minimum weld size} = \tfrac{3}{16} \text{ in}$$

$$\text{maximum weld size} = \tfrac{3}{8} \text{ in} - \tfrac{1}{16} \text{ in} = \tfrac{5}{16} \text{ in}$$

Select a $\frac{3}{16}$-in fillet.

$$\text{capacity of fillet} = \left(21 \frac{\text{kips}}{\text{in}^2}\right)\left(\frac{3}{16} \text{ in}\right)(0.707) = 2.78 \text{ kips/in}$$

$$\text{length of fillet required} = \frac{19.09 \text{ kips}}{2.78 \frac{\text{kips}}{\text{in}}} = 6.87 \text{ in}$$

Neglecting end returns, the centroid of the weld coincides with the centroid of the member when

$$1.04x_1 = (3 - 1.04)x_2$$

$$x_1 = 1.885x_2$$

$$x_1 + x_2 = 6.87 \text{ in}$$

$$x_2 = 2.38 \text{ in}$$

$$x_1 = 4.49 \text{ in}$$

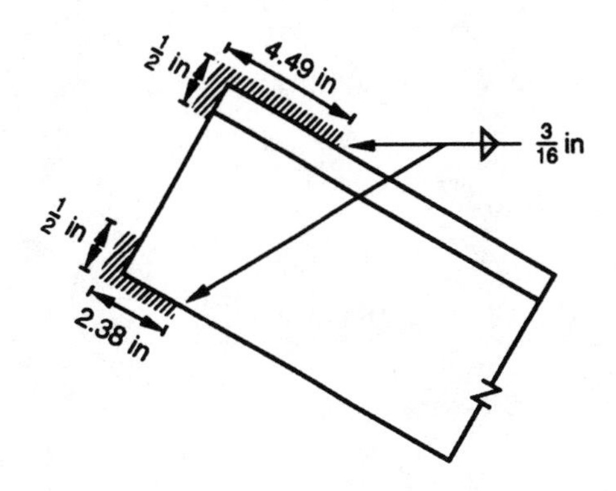

5. (a) For each of the beams, the loading is as a simple beam.

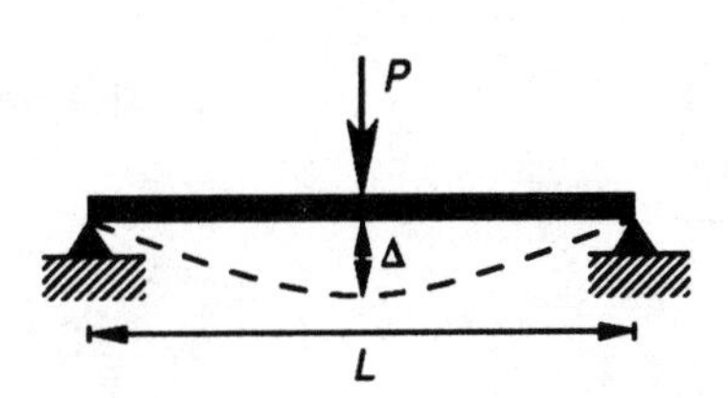

The stiffness is

$$k = \frac{P}{\Delta} = \frac{48EI}{L^3}$$

For the W27 × 146 beam,

$$I = 5630\ \text{in}^4$$
$$L = 216\ \text{in}$$
$$k = \frac{(48E)(5630\ \text{in}^4)}{(216\ \text{in})^3} = 0.0268E$$

For the W24 × 94 beam,

$$I = 2700\ \text{in}^4$$
$$L = 264\ \text{in}$$
$$k = \frac{(48E)(2700\ \text{in}^4)}{(264\ \text{in})^3} = 0.00704E$$

The load on the W27 × 146 beam is

$$F = \left(\frac{0.0268E}{0.0268E + 0.00704E}\right)(180\ \text{kips})$$
$$= \boxed{142.6\ \text{kips}\quad [79\%]}$$

The load on the W24 × 94 beam is

$$F = \left(\frac{0.00704E}{0.0268E + 0.00704E}\right)(180\ \text{kips})$$
$$= \boxed{37.4\ \text{kips}\quad [21\%]}$$

(b) For the W27 × 146 beam,

$$M = \frac{PL}{4} + \frac{wL^2}{8}$$
$$= \frac{(142.6\ \text{kips})(18\ \text{ft})}{4} + \frac{\left(0.146\ \frac{\text{kips}}{\text{ft}}\right)(18\ \text{ft})^2}{8}$$
$$= 647.6\ \text{ft-kips}$$

$$f_b = \frac{M}{S} = \frac{(647.6\ \text{ft-kips})\left(12\ \frac{\text{in}}{\text{ft}}\right)}{411\ \text{in}^3}$$
$$= \boxed{18.91\ \text{kips/in}^2}$$

For the W24 × 94 beam (neglecting self-weight),

$$M = \frac{PL}{4} + \frac{wL^2}{8}$$
$$= \frac{(37.4\ \text{kips})(22\ \text{ft})}{4} + \frac{\left(0.094\ \frac{\text{kips}}{\text{ft}}\right)(22\ \text{ft})^2}{8}$$
$$= 211.4\ \text{ft-kips}$$

$$f_b = \frac{(211.4\ \text{ft-kips})\left(12\ \frac{\text{in}}{\text{ft}}\right)}{222\ \text{in}^3} = \boxed{11.43\ \text{kips/in}^2}$$

$$\text{(c) reaction} = \frac{142.6\ \text{kips}}{2} + \frac{\left(0.146\ \frac{\text{kips}}{\text{ft}}\right)(18\ \text{ft})}{2}$$
$$= 72.61\ \text{kips}$$

For the W27 × 146 beam,

$$b_f = 13.965\ \text{in}$$
$$t_w = 0.605\ \text{in}$$
$$k = 1.6875\ \text{in}$$

Using AISC, Eq. K1-3, the required bearing length is

$$N + 2.5k = \frac{R}{0.66F_y t_w}$$
$$N = \frac{72.61\ \text{kips}}{(0.66)\left(36\ \frac{\text{kips}}{\text{in}^2}\right)(0.605\ \text{in})} - (2.5)(1.6875\ \text{in})$$
$$= 0.83\ \text{in}$$

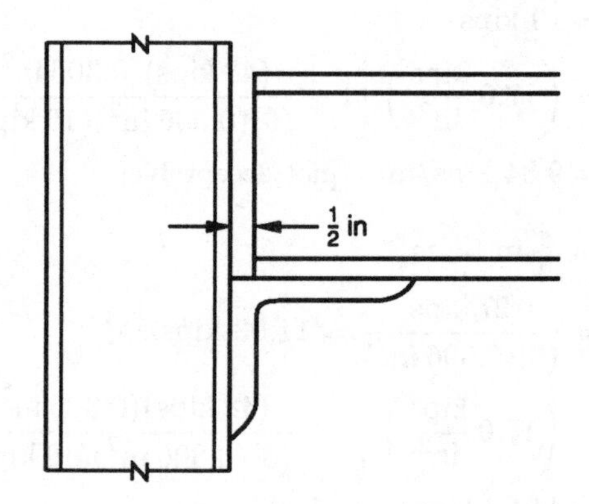

Allow for 0.25-in underrun. The minimum length of the outstanding leg is

$$0.83\ \text{in} + 0.5\ \text{in} = 1.33\ \text{in}$$

The 4-in extended leg is adequate.

Determine thickness based on flexure. $t = 1$ in.

$$e = \frac{N}{2} + 0.5\ \text{in} + 0.25\ \text{in} - t - \frac{3}{8}\ \text{in}$$
$$= \frac{2.76\ \text{in}}{2} + 0.5\ \text{in} + 0.25\ \text{in} - 1\ \text{in} - \frac{3}{8}\ \text{in}$$
$$= 0.755\ \text{in}$$

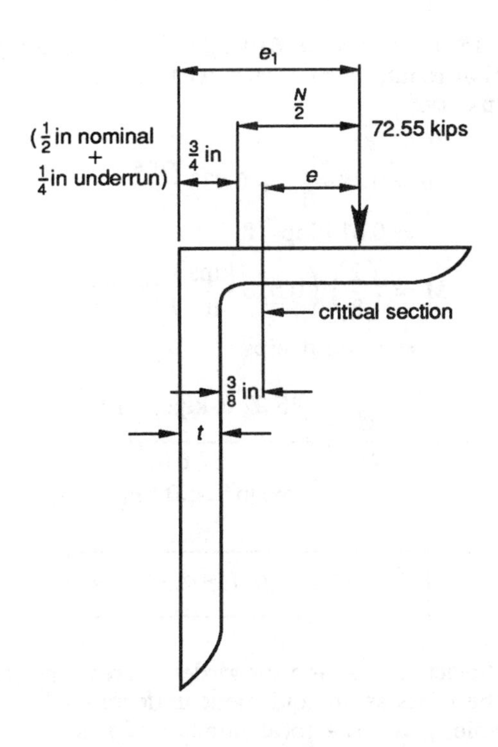

$$M = (72.61 \text{ kips})(0.755 \text{ in}) = 54.82 \text{ in-kips}$$

$$t_{\text{required}} = \sqrt{\frac{6M}{0.75F_yb}}$$

Take $F_y = 36$ kips/in^2 and (based on the flange width of the W27 × 146 beam) $b = 14$ in.

$$t_{\text{required}} = \sqrt{\frac{(6)(54.82 \text{ in-kips})}{(0.75)\left(36 \, \frac{\text{kips}}{\text{in}^2}\right)(14 \text{ in})}} = 0.93 \text{ in}$$

The length of the leg in contact with the column depends on whether the leg is bolted or welded to the column. Select a welded connection to the column.

The maximum weld size is

$$1 \text{ in} - \tfrac{1}{16} \text{ in} = \tfrac{15}{16} \text{ in}$$

From AISC, Table J2.4, the minimum weld size is $\frac{5}{16}$ in.

The eccentricity, e_1, is

$$e_1 = \frac{N}{2} + 0.75 \text{ in} = \frac{2.76 \text{ in}}{2} + 0.75 \text{ in} = 2.13 \text{ in}$$

The moment on one leg of the weld is

$$M = \frac{(2.13 \text{ in})(72.61 \text{ kips})}{2} = 77.3 \text{ in-kips}$$

The shear on one leg of the weld is

$$V = \frac{R}{2} = \frac{72.61 \text{ kips}}{2} = 36.31 \text{ kips}$$

Calculate the weld size. (Neglect returns for simplicity.)

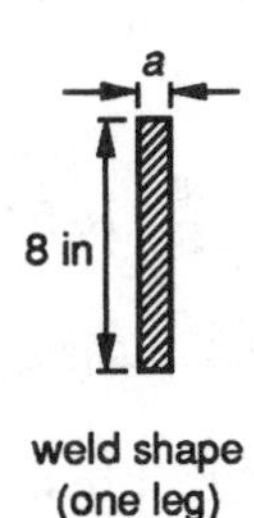

weld shape
(one leg)

The nominal and shear stresses on the weld are

$$S = 10.67a$$
$$A = 8a$$
$$\text{nominal stress} = \frac{77.3 \text{ kips}}{10.67a} = \frac{7.24 \text{ kips}}{a}$$
$$\text{shear stress} = \frac{36.31 \text{ kips}}{8a} = \frac{4.54 \text{ kips}}{a}$$
$$\text{combined stress} = \sqrt{\left(\frac{7.24 \text{ kips}}{a}\right)^2 + \left(\frac{4.54 \text{ kips}}{a}\right)^2} = \frac{8.54 \text{ kips}}{a}$$

For E70 electrodes, the allowable stress is 21.0 kips/in^2.

$$a_{\text{required}} = \frac{8.54 \, \frac{\text{kips}}{\text{in}}}{21 \, \frac{\text{kips}}{\text{in}^2}} = 0.407 \text{ in} \quad \text{[effective]}$$

$$\text{nominal weld size} = \sqrt{2}a = \sqrt{2}(0.407 \text{ in}) = 0.58 \text{ in} \quad \left[\text{say } \tfrac{5}{8} \text{ in}\right]$$

Note: A selection based on AISC, Table XIX (Sec. 4), would lead to a somewhat smaller weld size.

L8 × 4 × 1 is adequate.

The final bracket design is

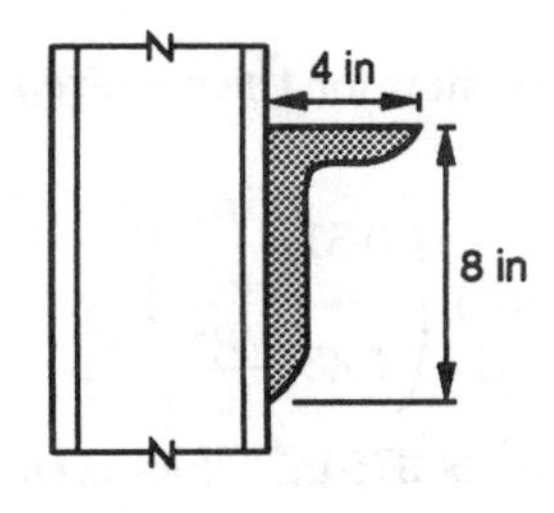

6. (a) Select a W section for the joists.

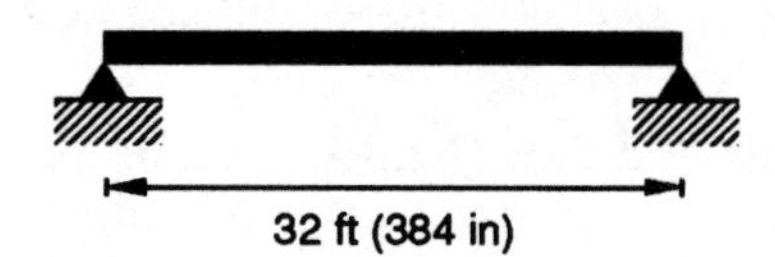

$$w_{\text{dead}} = \left(0.06\,\frac{\text{kips}}{\text{ft}^2}\right)(6\text{ ft}) + \text{self-weight}$$
$$= 0.36\,\frac{\text{kips}}{\text{ft}^2} + \text{self-weight}$$
$$w_{\text{live}} = \left(0.02\,\frac{\text{kips}}{\text{ft}^2}\right)(6\text{ ft})$$
$$= 0.12\text{ kips/ft}$$

For uniform loading,

$$\Delta = \frac{5wl^4}{384EI}$$

The deflection criterion is

$$\frac{\Delta}{l} \leq \frac{1}{360}$$
$$\frac{5wl^3}{384EI} \leq \frac{1}{360}$$

The moment of inertia required to meet the deflection criterion is

$$I \geq \frac{4.69wl^3}{E}$$

Neglecting self-weight,

$$I > \frac{(4.69)\left(0.48\,\frac{\text{kips}}{\text{ft}}\right)(384\text{ in})^3}{\left(29{,}000\,\frac{\text{kips}}{\text{in}^2}\right)\left(12\,\frac{\text{in}}{\text{ft}}\right)}$$
$$\geq 366\text{ in}^4$$

Try W16 × 31. ($I = 375\text{ in}^4$.) Adjust w to account for the self-weight.

$$w = 0.48\,\frac{\text{kips}}{\text{ft}} + 0.031\,\frac{\text{kips}}{\text{ft}} = 0.51\text{ kips/ft}$$

Checking the expression for the required I,

$$I > (366\text{ in}^4)\left(\frac{0.51\,\frac{\text{kips}}{\text{ft}}}{0.48\,\frac{\text{kips}}{\text{ft}}}\right)$$
$$= 389\text{ in}^4 > 375\text{ in}^4 \quad \text{[not acceptable]}$$

Try W18 × 35. ($I = 510\text{ in}^4$; $S = 57.6\text{ in}^3$.) The deflection requirement is met with this section. Check the capacity.

$$w = 0.48\,\frac{\text{kips}}{\text{ft}} + 0.035\,\frac{\text{kips}}{\text{ft}}$$
$$= 0.515\text{ kips/ft}$$
$$M = \left(\frac{1}{8}\right)\left(0.515\,\frac{\text{kips}}{\text{ft}}\right)(32\text{ ft})^2$$
$$= 65.92\text{ ft-kips}$$
$$f_b = \frac{M}{S} = \frac{(65.92\text{ ft-kips})\left(12\,\frac{\text{in}}{\text{ft}}\right)}{57.6\text{ in}^3}$$
$$= 13.73\text{ kips/in}^2 < 0.6F_y \quad \text{[ok]}$$

Use W18 × 35 for the joists.

(b) Select a W shape for girders. Treat the reaction from the joists as an equivalent uniform load (which is justifiable, given the total number of joists). Assume bracing at each joist location.

The deflection criterion is

$$\frac{\Delta}{l} \leq \frac{1}{360}$$

The required moment of inertia is

$$I \geq \frac{4.69wl^3}{E}$$

Calculate w, neglecting self-weight for a first approximation.

$$w = \left(0.06\,\frac{\text{kips}}{\text{ft}^2}\right)(16\text{ ft}) + \left(0.02\,\frac{\text{kips}}{\text{ft}^2}\right)(16\text{ ft})$$
$$= 1.28\text{ kips/ft}$$
$$I = \frac{(4.69)\left(1.28\,\frac{\text{kips}}{\text{ft}}\right)(432\text{ in})^3}{\left(29{,}000\,\frac{\text{kips}}{\text{in}^2}\right)\left(12\,\frac{\text{in}}{\text{ft}}\right)}$$
$$= 1391\text{ in}^4 \quad \text{[required]}$$

Try a W24 × 62 section.

$$I = 1550\text{ in}^4$$
$$S = 131\text{ in}^3$$
$$b_f = 7.04\text{ in}$$
$$\frac{d}{A_f} = 5.72\text{ 1/in}$$

Check the stresses.

$$w = 1.28\,\frac{\text{kips}}{\text{ft}} + 0.062\,\frac{\text{kips}}{\text{ft}}$$
$$= 1.34\ \text{kips/ft}$$
$$M = \frac{1}{8}wl^2 = \left(\frac{1}{8}\right)\left(1.34\,\frac{\text{kips}}{\text{ft}}\right)(36\ \text{ft})^2$$
$$= 217.1\ \text{kip-ft}$$

$$f_b = \frac{M}{S} = \frac{(217.1\ \text{ft-kips})\left(12\,\frac{\text{in}}{\text{ft}}\right)}{131\ \text{in}^3}$$
$$= 19.9\ \text{kips/in}^2$$

Compute the allowable stress.

$$\frac{76b_f}{\sqrt{F_y}} = \frac{(76)(7.04\ \text{in})}{\sqrt{36\,\frac{\text{kips}}{\text{in}^2}}}$$
$$= 89.17\ \text{in}\quad [\text{controls}]$$
$$\frac{20{,}000}{\left(\frac{d}{A_f}\right)F_y} = \frac{20{,}000}{\left(5.72\,\frac{1}{\text{in}}\right)\left(36\,\frac{\text{kips}}{\text{in}^2}\right)}$$
$$= 97.1\ \text{in}$$

The unbraced length is

$$(6\ \text{ft})\left(12\,\frac{\text{in}}{\text{ft}}\right) = 72\ \text{in}$$

Since 72 in < 89.17 in,

$$F_b = 0.66F_y = (0.66)\left(36\,\frac{\text{kips}}{\text{in}^2}\right)$$
$$= 24\ \text{kips/in}^2 > f_b\quad [\text{ok}]$$

Use W24 × 62 for the girders.

(c) The load on the column is

$$P = \frac{\left(1.34\,\frac{\text{kips}}{\text{ft}}\right)(36\ \text{ft})}{2} = 24.12\ \text{kips}$$

Because of the extreme height, the Kl/r limit of 200 is likely to control.

$$\frac{(32\ \text{ft})\left(12\,\frac{\text{in}}{\text{ft}}\right)}{r_y} \geq 200$$
$$r_y > 1.92\ \text{in}$$

From AISC, Table C-36, at $Kl/r = 200$, $F_a = 3.73$ kips/in^2. W14 × 53 has $r_y = 1.92$ and $A = 15.6\ \text{in}^2$, so the allowable load is

$$P_{\text{allowable}} = \left(3.73\,\frac{\text{kips}}{\text{in}^2}\right)(15.6\ \text{in}^2) = 58.19\ \text{kips}$$

This is much larger than the design load.

Use W14 × 53 for the columns.

7. The equations and table references in this solution are from UBC-91.

(a) $V = \boxed{\frac{ZIC}{R_w}W}$ [Eq. 12-1]

(b) $Z = 0.4$ [Table 23-I]
$I = 1$ [Table 23-L]

$$W = 15\ \text{kips} + 30\ \text{kips} + 20\ \text{kips} + 50\ \text{kips}$$
$$= 115\ \text{kips}$$

Since special detailing will not be provided, the frame is an ordinary moment-resisting frame (OMRSF).

$R_w = 6$ [Table 23-O]

$C = \frac{1.25S}{T^{2/3}} \leq 2.75$ [Eq. 12-2]

$S = 1$ [Table 23-J]

$$T = C_t h_n^{3/4}$$
$$C_t = 0.02$$
$$h_n = 14\ \text{ft} + 16\ \text{ft} = 30\ \text{ft}$$
$$T = (0.02)(30\ \text{ft})^{3/4} = 0.256\ \text{sec}$$
$$C = \frac{(1.25)(1)}{(0.256\ \text{sec})^{2/3}} = 3.10\quad [\geq 2.75\ \text{maximum value}]$$
$$C = 2.75$$
$$V = \frac{(0.4)(1)(2.75)(115\ \text{kips})}{6}$$
$$= \boxed{21.08\ \text{kips}}$$

(c) $F_x = (V - F_t)\left(\frac{w_x h_x}{\sum_{i=1}^{n} w_i h_i}\right)$ [Eq. 12-8]

Since $T < 0.7$ sec, $F_t = 0$.

Set up a table to evaluate the F_x.

level	w_i (kips)	h_i (ft)	$w_i h_i$ (ft-kips)	$\frac{w_i h_i}{\sum w_i h_i}$	F_x (kips)
2	45	30	1350	0.58	12.21
1	70	14	980	0.42	8.87
			2330		

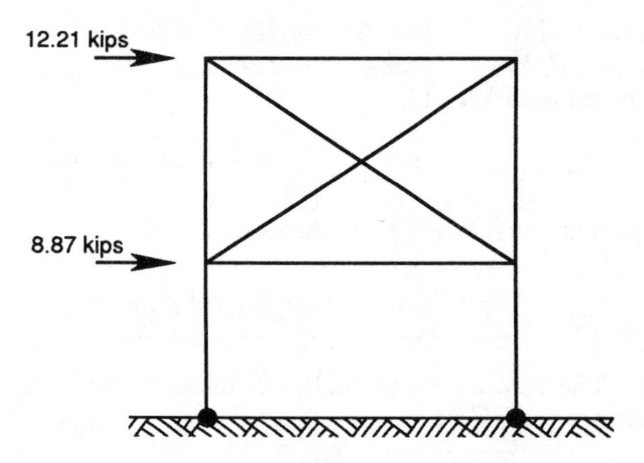

(d) Since the columns of the first level have equal moments of inertia, the shear is equally divided.

$$V_{\text{per column}} = \frac{12.21 \text{ kips} + 8.87 \text{ kips}}{2} = 10.54 \text{ kips}$$

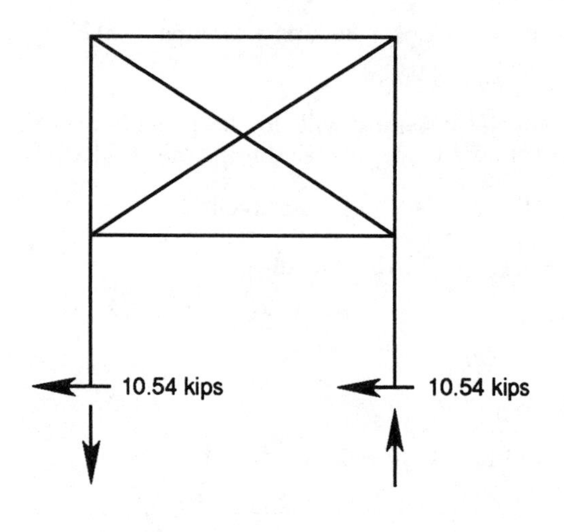

$$V = \boxed{10.54 \text{ kips}}$$

$$M = (10.54 \text{ kips})(14 \text{ ft}) = \boxed{147.6 \text{ ft-kips}}$$

8. Calculate the horizontal reaction, H.

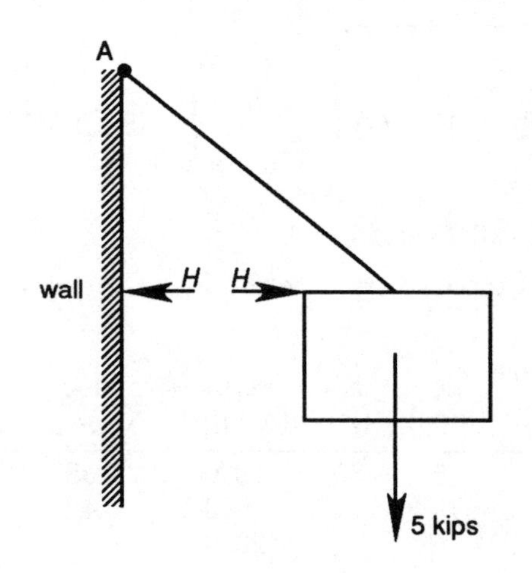

Taking clockwise momennts as positive,

$$\begin{aligned}\sum M_A &= (-H)(6 \text{ ft}) + (5 \text{ kips})(5 \text{ ft}) \\ &= 0 \\ H &= 4.17 \text{ kips}\end{aligned}$$

The loads acting on the column are as illustrated.

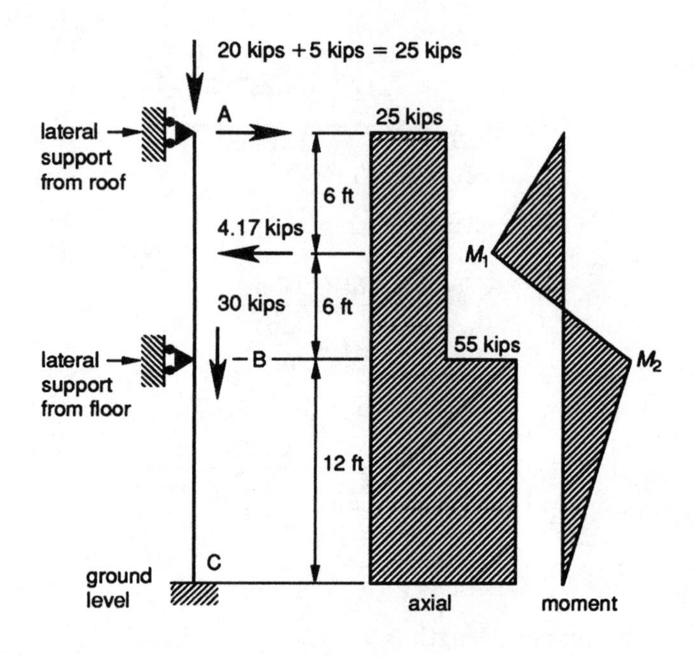

Calculate the moments using moment distribution.

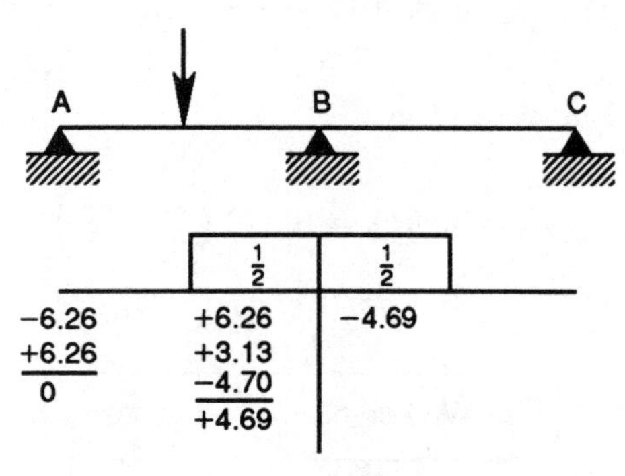

$$\frac{pl}{8} = \frac{(4.17 \text{ kips})(12 \text{ ft})}{8} = 6.26 \text{ ft-kips}$$

$$M_2 = 4.69 \text{ ft-kips}$$

$$\begin{aligned}M_1 &= \frac{(4.17 \text{ kips})(12 \text{ ft})}{4} - \frac{4.69 \text{ ft-kips}}{2} \\ &= 10.17 \text{ ft-kips}\end{aligned}$$

Select a trial column based on the axial force only. From the AISC column tables for $F_y = 36 \text{ kips/in}^2$ and $Kl = 12$ ft (assuming $K = 1$), select W6 × 20, which has an allowable load capacity of 79 kips.

Check the strength about the strong axis including bending.

W6 × 20:

$$S_x = 13.4\ \text{in}^3$$
$$\frac{d}{A_f} = 2.82\ 1/\text{in}$$
$$r_t = 1.64\ \text{in}$$
$$A = 5.87\ \text{in}^2$$
$$r_y = 1.5\ \text{in}$$
$$b_f = 6\ \text{in}$$
$$r_x = 2.66\ \text{in}$$

Segment AB:

$$P = 25\ \text{kips}$$
$$M = 10.17\ \text{ft-kips}$$
$$C_m = 1$$
$$f_a = \frac{25\ \text{kips}}{5.87\ \text{in}^2} = 4.26\ \text{kips/in}^2$$
$$\frac{Kl}{r_y} = \frac{(1)(12\ \text{ft})\left(12\ \frac{\text{in}}{\text{ft}}\right)}{1.5\ \text{in}} = 96$$

From AISC, Table C-36,

$$F_a = 13.48\ \text{kips/in}^2$$
$$F_{bx} = \frac{M}{S} = \frac{(10.17\ \text{kip-ft})\left(12\ \frac{\text{in}}{\text{ft}}\right)}{13.4\ \text{in}^3} = 9.11\ \text{kips/in}^2$$

Calculate the allowable stress in bending.

$$\text{unbraced length} = 12\ \text{ft} \quad (144\ \text{in})$$
$$\frac{l}{r_T} = \frac{144\ \text{in}}{1.64\ \text{in}} = 87.8$$
$$C_b = 1$$

Assume that the sign does not provide support. AISC Eq. F1-6 applies.

$$F_b = \left[\frac{2}{3} - \frac{F_y\left(\frac{l}{r_T}\right)^2}{1530 \times 10^3 C_b}\right] F_y$$
$$= \left[\frac{2}{3} - \frac{\left(36\ \frac{\text{kips}}{\text{in}^2}\right)(87.8)^2}{(1530 \times 10^3)(1)}\right]\left(36\ \frac{\text{kips}}{\text{in}^2}\right)$$
$$= 17.47\ \text{kips/in}^2$$

From AISC, Sec. H1,

$$F'_e = \frac{12\pi^2 E}{(23)\left(\frac{Kl_b}{r_b}\right)^2} = \frac{(12)(\pi^2)\left(29{,}000\ \frac{\text{kips}}{\text{in}^2}\right)}{(23)\left(\frac{144\ \text{in}}{2.66\ \text{in}}\right)^2}$$
$$= 50.96\ \text{kips/in}^2$$

Check the combined stress criteria. From AISC, Eq. H1-1,

$$\frac{f_a}{F_a} + \frac{C_m f_b}{\left(1 - \frac{f_a}{F'_e}\right) F_b} < 1$$

$$\frac{4.26\ \frac{\text{kips}}{\text{in}^2}}{13.48\ \frac{\text{kips}}{\text{in}^2}} + \frac{(1)\left(9.11\ \frac{\text{kips}}{\text{in}^2}\right)}{\left(1 - \frac{4.26\ \frac{\text{kips}}{\text{in}^2}}{50.96\ \frac{\text{kips}}{\text{in}^2}}\right)\left(17.49\ \frac{\text{kips}}{\text{in}^2}\right)}$$
$$= 0.884 \quad \text{[ok]}$$

Segment BC:

$$P = 55\ \text{kips}$$
$$M = 4.69\ \text{ft-kips}$$
$$f_a = \frac{55\ \text{kips}}{5.87} = 9.37\ \text{kips/in}^2$$
$$F_a = 13.48\ \text{kips/in}^2$$
$$f_{bx} = \frac{M}{S} = \frac{(4.69\ \text{ft-kips})\left(12\ \frac{\text{in}}{\text{ft}}\right)}{13.4\ \text{in}^3}$$
$$= 4.2\ \text{kips/in}^2$$
$$C_b = 1.75$$

From AISC, Eq. F1-6,

$$F_b = \left[\frac{2}{3} - \frac{F_y\left(\frac{l}{r_T}\right)^2}{1530 \times 10^3 C_b}\right] F_y$$
$$= \left[\frac{2}{3} - \frac{\left(36\ \frac{\text{kips}}{\text{in}^2}\right)(87.8)^2}{(1530 \times 10^3)(1.75)}\right]\left(36\ \frac{\text{kips}}{\text{in}^2}\right)$$
$$= 20.27\ \text{kips/in}^2$$

$$F'_e = \frac{12\pi^2 E}{(23)\left(\frac{Kl_b}{r_b}\right)^2} = \frac{(12)(\pi^2)\left(29{,}000\ \frac{\text{kips}}{\text{in}^2}\right)}{(23)\left(\frac{144\ \text{in}}{2.66\ \text{in}}\right)^2}$$
$$= 50.96\ \text{kips/in}^2$$

From AISC, Sec H1, since $M_1 = 0$,

$$C_m = 0.6$$

Check the combined stress criteria.

$$\frac{f_a}{F_a} + \frac{C_m f_b}{\left(1 - \frac{f_a}{F'_e}\right) F_b} < 1$$

$$\frac{9.37 \frac{\text{kips}}{\text{in}^2}}{13.48 \frac{\text{kips}}{\text{in}^2}} + \frac{(0.6)\left(4.2 \frac{\text{kips}}{\text{in}^2}\right)}{\left(1 - \frac{9.37 \frac{\text{kips}}{\text{in}^2}}{50.96 \frac{\text{kips}}{\text{in}^2}}\right)\left(20.27 \frac{\text{kips}}{\text{in}^2}\right)} = 0.847 \quad \text{[ok]}$$

Check this value using AISC, Eq. H1-2,

$$\frac{f_a}{0.6F_y} + \frac{f_b}{F_b} < 1$$

$$0.6F_y = (0.6)\left(36 \frac{\text{kips}}{\text{in}^2}\right) = 21.6 \text{ kips/in}^2$$

$$\frac{9.37 \frac{\text{kips}}{\text{in}^2}}{(0.6)\left(3.6 \frac{\text{kips}}{\text{in}^2}\right)} + \frac{4.2 \frac{\text{kips}}{\text{in}^2}}{20.27 \frac{\text{kips}}{\text{in}^2}} = 0.64 \quad \text{[ok]}$$

W6 × 20 is adequate.

9.

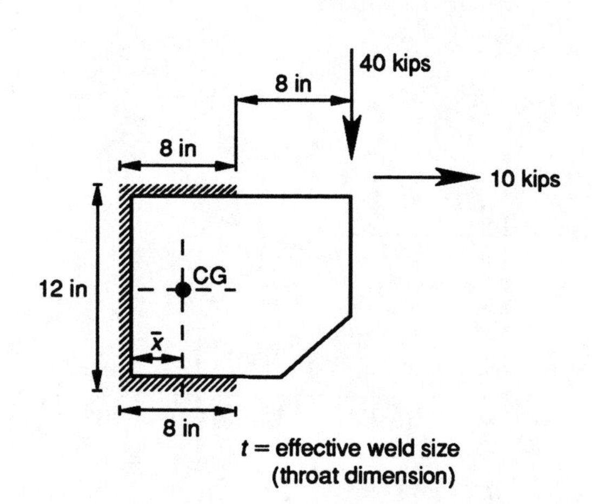

Determine the location of the centroid of the welds.

$$\overline{x} = \frac{(8t)(4)(2)}{[(8)(2) + 12]t} = 2.286 \text{ in}$$

The forces and moment on the weld group are

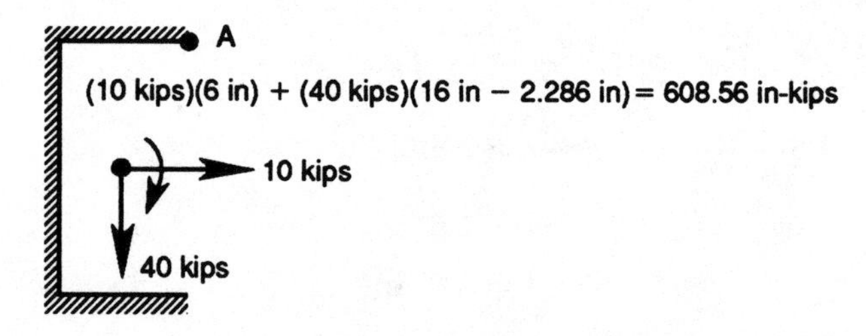

Stresses in the weld are critical at point A.

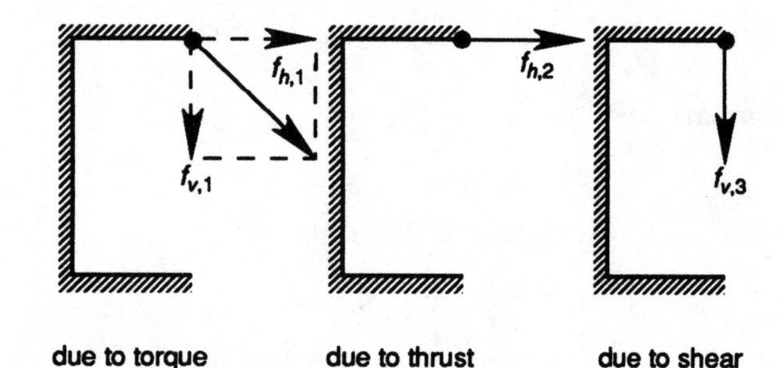

Calculate the polar moment of inertia.

$$I_x = (2)(8 \text{ in})t(6 \text{ in})^2 = 576t \text{ in}^4$$

$$I_y = (12t)(2.286)^2 + \left(\frac{1}{12}t\right)(8)^3(2) + (8t)(2)(4 - 2.286)^2 = 195t \text{ in}^4$$

$$I_p = I_x + I_y = 576t + 195t = 771t \text{ in}^4$$

$$f_{h1} = \frac{Mc}{I} = \frac{\left(608.56 \frac{\text{kips}}{\text{in}^2}\right)(6 \text{ in})}{771t} = \frac{4.736}{t} \frac{\text{kips}}{\text{in}^2}$$

$$f_{h2} = \frac{V}{A} = \frac{10 \text{ kips}}{(8 \text{ in} + 8 \text{ in} + 12 \text{ in})t} = \frac{0.357}{t} \frac{\text{kips}}{\text{in}^2}$$

$$f_{v1} = \frac{Mc}{I} = \left(608.56 \frac{\text{kips}}{\text{in}^2}\right)\left(\frac{8 \text{ in} - 2.286 \text{ in}}{771t}\right) = \frac{4.51}{t} \frac{\text{kips}}{\text{in}^2}$$

$$f_{v3} = \frac{V}{A} = \frac{40 \text{ kips}}{(8 \text{ in} + 8 \text{ in} + 12 \text{ in})t} = \frac{1.429}{t} \frac{\text{kips}}{\text{in}^2}$$

$$F_n = \frac{4.736}{t} + \frac{0.357}{t} = \frac{5.093}{t}$$

$$F_v = \frac{4.51}{t} + \frac{1.429}{t} = \frac{5.939}{t}$$

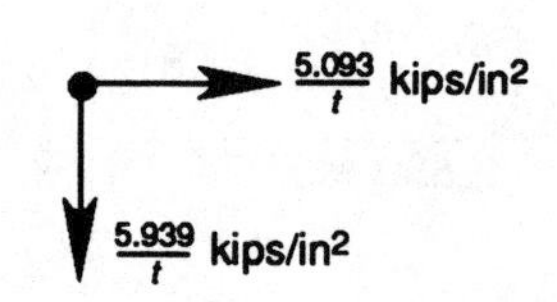

$$\text{resultant} = \sqrt{\left(\frac{5.093}{t}\right)^2 + \left(\frac{5.939}{t}\right)^2}$$
$$= \frac{7.824}{t}\,\frac{\text{kips}}{\text{in}^2}$$

The allowable stress for E70 electrodes is 21 kips/in^2.

$$\frac{7.824}{t}\,\frac{\text{kips}}{\text{in}^2} = 21\ \text{kips/in}^2$$
$$t = 0.373\ \text{in}$$
$$\text{nominal weld size} = \sqrt{2}t = \sqrt{2}\,(0.373\ \text{in})$$
$$= 0.528\ \text{in}$$

Try $\frac{9}{16}$-in fillets.

Check the minimum and maximum limits, using AISC Table J2.4 and Sec. J2.2.b,

$$\text{minimum} = \tfrac{3}{16}\ \text{in}\quad[\text{ok}]$$
$$\text{maximum} = \tfrac{5}{8}\ \text{in} - \tfrac{1}{16}\ \text{in}$$
$$= \tfrac{9}{16}\ \text{in}\quad[\text{ok}]$$

Use a $\frac{9}{16}$-in fillet.

10.

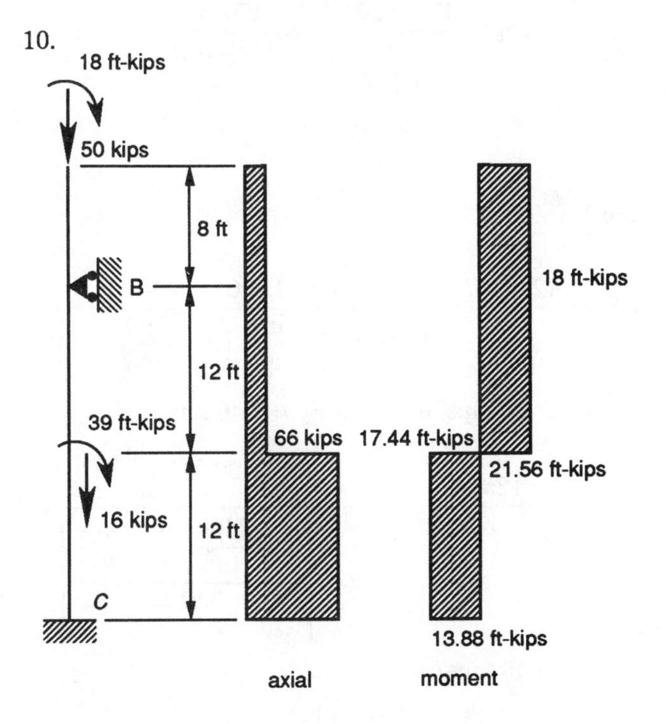

Obtain the bending moments.

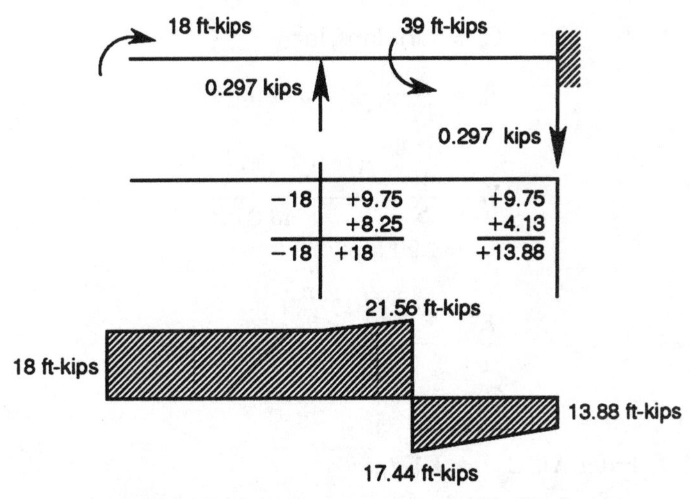

Obtain the effective length for segment AB. Neglecting second-order effects in segment BC.

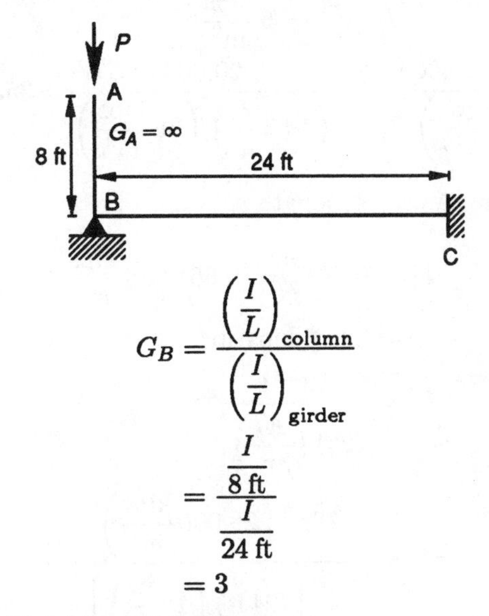

$$G_B = \frac{\left(\frac{I}{L}\right)_{\text{column}}}{\left(\frac{I}{L}\right)_{\text{girder}}} = \frac{\frac{I}{8\ \text{ft}}}{\frac{I}{24\ \text{ft}}} = 3$$

From AISC, Fig. 1 (p. 3-5), $k \approx 3$

$$Kl = (3\ \text{ft})(8\ \text{ft}) = 24\ \text{ft}$$

Try a W8 × 48 beam.

$$A = 14.1\ \text{in}^2$$
$$b_f = 8.11\ \text{in}$$
$$\frac{d}{A_f} = 1.53\ 1/\text{in}$$
$$r_t = 2.23\ \text{in}$$
$$S_x = 43.3\ \text{in}^3$$
$$r_y = 2.08\ \text{in}$$
$$r_x = 3.61\ \text{in}$$
$$\frac{Kl}{r_y} = \frac{(24\ \text{ft})\left(12\,\frac{\text{in}}{\text{ft}}\right)}{2.08\ \text{in}} = 138.46$$

From AISC, Table C-36,

$$F_a = 7.79 \text{ kips/in}^2$$

$$f_a = \frac{50 \text{ kips}}{14.1 \text{ in}^2} = 3.55 \text{ kips/in}^2$$

$$f_b = \frac{M}{S} = \frac{(18 \text{ ft-kips})\left(12 \frac{\text{in}}{\text{ft}}\right)}{43.3 \text{ in}^3} = 4.99 \text{ kips/in}^2$$

$$\frac{l}{r_t} = \frac{(8 \text{ ft})\left(12 \frac{\text{in}}{\text{ft}}\right)}{2.23 \text{ in}} = 43.05 \text{ in}$$

$$C_b = 1$$

From AISC, Eq. F1-2,

$$\frac{76b_f}{\sqrt{F_y}} = \frac{(76)(8.11 \text{ in})}{\sqrt{36 \frac{\text{kips}}{\text{in}^2}}} = 102.73 \text{ in}$$

$$\frac{20{,}000}{\left(\frac{d}{A_f}\right) F_y} = \frac{20{,}000}{\left(1.53 \frac{1}{\text{in}}\right)\left(36 \frac{\text{kips}}{\text{in}^2}\right)} = 363$$

Since the cantilever length is 96 in,

$$F_b = 0.66F_y = (0.66)\left(36 \frac{\text{kips}}{\text{in}^2}\right) = 23.76 \text{ kips/in}^2$$

$$F'_e = \frac{12\pi^2 E}{(23)\left(\frac{Kl}{r}\right)^2} = \frac{12\pi^2\left(29{,}000 \frac{\text{kips}}{\text{in}^2}\right)}{(23)\left[\frac{(24 \text{ ft})\left(12 \frac{\text{in}}{\text{ft}}\right)}{3.61 \text{ in}}\right]^2} = 23.46 \text{ kips/in}^2$$

$$C_m = 0.85$$

Check the combined stress criteria.

$$\frac{f_a}{F_a} + \frac{C_m f_b}{\left(1 - \frac{f_a}{F'_e}\right) F_b} < 1$$

$$\frac{3.55 \frac{\text{kips}}{\text{in}^2}}{7.79 \frac{\text{kips}}{\text{in}^2}} + \frac{(0.85)\left(4.99 \frac{\text{kips}}{\text{in}^2}\right)}{\left(1 - \frac{3.55 \frac{\text{kips}}{\text{in}^2}}{23.76 \frac{\text{kips}}{\text{in}^2}}\right)\left(23.76 \frac{\text{kips}}{\text{in}^2}\right)} = 0.67 \quad \text{[ok]}$$

Segment BC:

$$P = 66 \text{ kips}$$

$$M = 21.56 \text{ ft-kips}$$

$$K = 1$$

$$\frac{Kl}{r_y} = \frac{(1)(24 \text{ ft})\left(12 \frac{\text{in}}{\text{ft}}\right)}{2.08 \text{ in}} = 138.46$$

From AISC, Table C-36,

$$F_a = 7.79 \text{ kips/in}^2$$

$$f_a = \frac{66 \text{ kips}}{14.1 \text{ in}^2} = 4.68 \text{ kips/in}^2$$

$$f_b = \frac{M}{S} = \frac{(21.56 \text{ ft-kips})\left(12 \frac{\text{in}}{\text{ft}}\right)}{43.3 \text{ in}^3} = 5.98 \text{ kips/in}^2$$

$$\frac{l}{r_t} = \frac{(24 \text{ ft})\left(12 \frac{\text{in}}{\text{ft}}\right)}{2.23 \text{ in}} = 129.15 \text{ kips/in}^2$$

$$C_b = 1$$

Use AISC Eq. F1-7.

$$F_b = \frac{170 \times 10^3 C_b}{\left(\frac{l}{r_t}\right)^2} = \frac{170{,}000}{(129.15)^2} = 10.19 \text{ kips/in}^2$$

From Eq. F1-8,

$$F_b = \frac{12{,}000C_b}{\frac{ld}{A_f}} = \frac{12{,}000}{(24 \text{ ft})\left(12 \frac{\text{in}}{\text{ft}}\right)\left(1.53 \frac{1}{\text{in}}\right)} = 27.23 \text{ kips/in}^2 < 0.6F_y = 21.6 \text{ kips/in}^2$$

$$F'_e = \frac{12\pi^2 E}{(23)\left(\frac{Kl}{r}\right)^2} = \frac{12\pi^2\left(29{,}000 \frac{\text{kips}}{\text{in}^2}\right)}{(23)\left[\frac{\left(24 \frac{\text{kips}}{\text{in}^2}\right)\left(12 \frac{\text{in}}{\text{ft}}\right)}{3.61 \text{ in}}\right]^2} = 23.46 \text{ kips/in}^2$$

$$C_m = 1 \quad \text{[conservative value]}$$

Check the combined stress criteria.

$$\frac{f_a}{F_a} + \frac{C_m f_b}{\left(1 - \frac{f_a}{F'_e}\right) F_b} < 1$$

$$\frac{4.68\,\frac{\text{kips}}{\text{in}^2}}{7.79\,\frac{\text{kips}}{\text{in}^2}} + \frac{5.98\,\frac{\text{kips}}{\text{in}^2}}{\left(1 - \frac{4.68\,\frac{\text{kips}}{\text{in}^2}}{23.46\,\frac{\text{kips}}{\text{in}^2}}\right)(21.6\text{ ft-kips})}$$

$$= 0.947 \quad \text{[ok]}$$

Use W8 × 48.

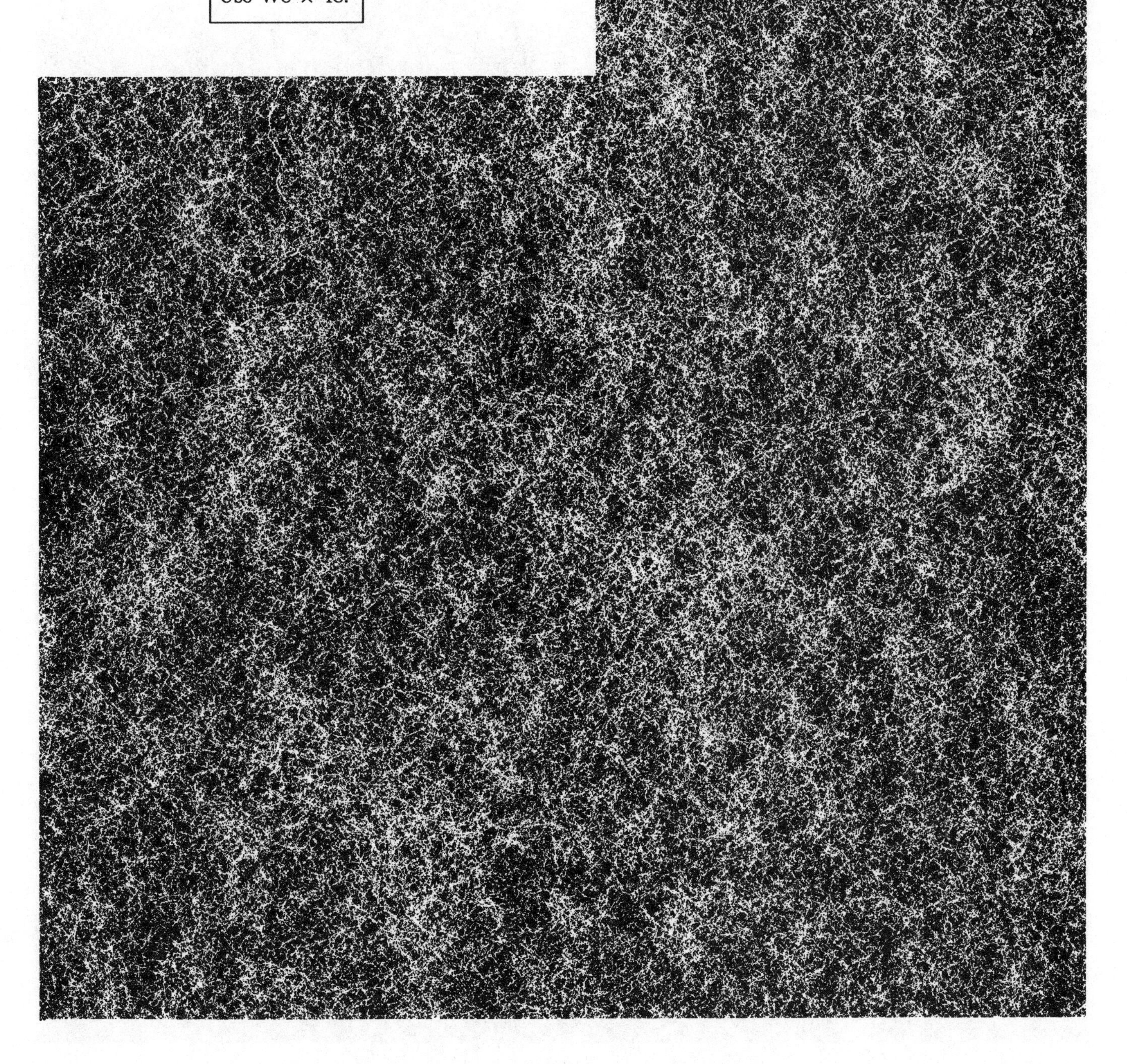

TIMBER DESIGN

1. (a) The uniform load acting on the boards is

$$\text{concrete slab:}\ \left(150\ \frac{\text{lbf}}{\text{ft}^3}\right)\left(\frac{6\ \text{in}}{12\ \frac{\text{in}}{\text{ft}}}\right) = 75\ \text{lbf/ft}^2$$

construction live load: 50 lbf/ft^2

estimated weight of boards: 4 lbf/ft^2

total: 129 lbf/ft^2

The load per foot on a single board, q, is

$$q = \left(\frac{5.5\ \text{in}}{12\ \frac{\text{in}}{\text{ft}}}\right)\left(129\ \frac{\text{lbf}}{\text{ft}^2}\right) = 59.1\ \text{lbf/ft}$$

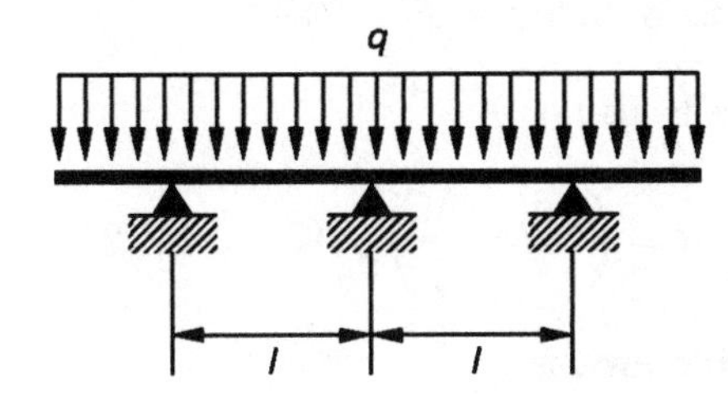

The maximum moment from ACI 318-89, Sec. 8.3.3, is

$$M = \frac{1}{10}ql^2 = \left(\frac{1}{10}\right)\left(59.1\ \frac{\text{lbf}}{\text{ft}}\right)(l^2\ \text{ft}^2)$$
$$= 5.91l^2\ \text{ft-lbf}$$

The section modulus of 1-in × 6-in lumber is

$$S = \frac{1}{6}bh^2 = \left(\frac{1}{6}\right)(5.5\ \text{in})(0.75\ \text{in})^2$$
$$= 0.52\ \text{in}^3$$

The adjustments on F_b from the American Institute of Timber Construction are 0.85 for wet use and 1.25 for load duration. The allowable moment in one board is

$$M_a = F_bS$$
$$= (0.85)(1.25)\left(1200\ \frac{\text{lbf}}{\text{in}^2}\right)(0.52\ \text{in}^3)$$
$$= 663\ \text{in-lbf}\ (55.3\ \text{ft-lbf})$$

Equating the allowable moment to the maximum moment,

$$l = \sqrt{\frac{M_a}{M}} = \sqrt{\frac{55.3\ \text{ft-lbf}}{5.91\ \frac{\text{lbf}}{\text{ft}}}}$$
$$= 3.06\ \text{ft}\quad [\text{for stress}]$$

Check the shear using ACI 318-89, Sec. 8.3.3.

$$V = 1.15\frac{ql}{2}\quad [\text{first interior support}]$$
$$= \left(\frac{1.15}{2}\right)\left(59.1\ \frac{\text{lbf}}{\text{ft}}\right)(3.06\ \text{ft})$$
$$= 104\ \text{lbf}$$

Check the maximum shear stress, f_v.

$$f_v = \frac{3V}{2A}$$
$$= \frac{(3)(104\ \text{lbf})}{(2)(5.5)(0.75\ \text{in}^2)}$$
$$= 37.8\ \text{lbf/in}^2 < F_v\quad [\text{ok}]$$

Check the bearing. The reaction over the first interior joist is the largest.

$$R = 104 + \frac{ql}{2}$$
$$= 104\ \text{lbf} + \left(59.1\ \frac{\text{lbf}}{\text{ft}}\right)\left(\frac{3.06\ \text{ft}}{2}\right)$$
$$= 194.4\ \text{lbf}$$

Assuming the nominal joist width is 2 in, the bearing area is

$$A_b = (5.5\ \text{in})(1.5\ \text{in}) = 8.25\ \text{in}^2$$

$$\text{bearing stress} = \frac{R}{A_b} = \frac{194.4\ \text{lbf}}{8.25\ \text{in}^2}$$
$$= 23.6\ \frac{\text{lbf}}{\text{in}^2} < F_{c\perp}\quad [\text{ok}]$$

Check the deflection.

$$\Delta = \frac{wl^4}{145EI}$$

(Reference: Stalnaker, Judith J. and Ernest C. Harris. *Structural Design in Wood*. New York: Van Nostrand Reinhold Company, 1989.)

Equating the deflection to the allowable value of $l/360$, the limit for l becomes

$$l = \sqrt[3]{\frac{145EI}{360w}}$$
$$= \sqrt[3]{\frac{(145)\left(1\times10^6\ \frac{\text{lbf}}{\text{in}^2}\right)(0.19\ \text{in}^4)\left(12\ \frac{\text{in}}{\text{ft}}\right)}{(360)\left(59.1\ \frac{\text{lbf}}{\text{ft}}\right)}}$$
$$= 24.95\ \text{in}\quad [\text{say 25 in}]$$

The maximum spacing between the joists as limited by the 1-in × 6-in lumber is 25 in.

(b) Calculate the load on the joists. The load from the boards is

$$(1.15)\left(\frac{wl}{2}\right) + \frac{wl}{2} = (2.25)\left(\frac{wl}{2}\right)$$

$$(2.25)\left(129\ \frac{\text{lbf}}{\text{ft}^2}\right)\left(\frac{2.5}{2}\ \text{ft}\right) = 362.8\ \text{lbf/ft}$$

$$\text{assumed self-weight} = \left(64\ \frac{\text{lbf}}{\text{ft}^3}\right)\left[\frac{(1.5\ \text{in})(7.5\ \text{in})}{144\ \frac{\text{in}^2}{\text{ft}^2}}\right]$$
$$= 5.0\ \text{lbf/ft}$$

$$\text{total weight} = 362.8\ \frac{\text{lbf}}{\text{ft}} + 5.0\ \frac{\text{lbf}}{\text{ft}}$$
$$= 367.8\ \text{lbf/ft}$$

Find the maximum moment. (For a conservative estimate, use the total span.)

$$M = \frac{1}{8}wl^2 = \left(\frac{1}{8}\right)\left(367.8\ \frac{\text{lbf}}{\text{ft}^2}\right)(8\ \text{ft})^2$$
$$= 2942.4\ \text{ft-lbf}$$

The required section modulus is

$$S = \frac{M}{F_b} = \frac{(2942.4\ \text{ft-lbf})\left(12\ \frac{\text{in}}{\text{ft}}\right)}{\left(1200\ \frac{\text{lbf}}{\text{in}^2}\right)(1.25)}$$
$$= 23.5\ \text{in}^3$$

For a 2-in × 8-in joist,

$$S_{\text{provided}} = \frac{(1.5\ \text{in})(7.5\ \text{in})^2}{6}$$
$$= 14.06\ \text{in}^3 < \text{required} \quad \text{[unacceptable]}$$

For double joists,

$$S_{\text{provided}} = (2)(14.06\ \text{in}^3)$$
$$= 28.12\ \text{in}^3 > S_{\text{required}} \quad \text{[ok]}$$

Try 2-in × 8-in double joists.

Check the shear.

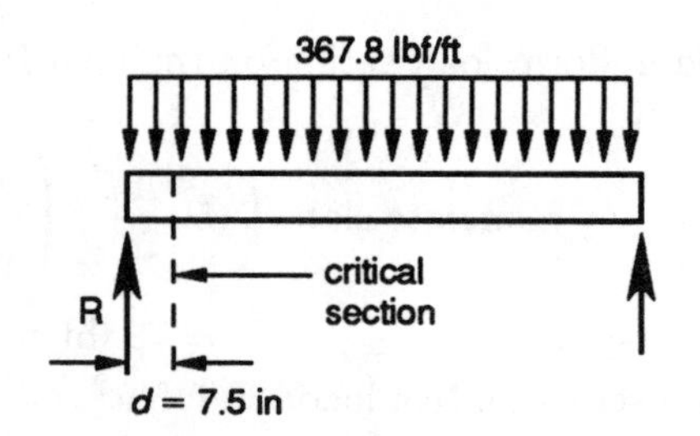

$$V = \left(367.8\ \frac{\text{lbf}}{\text{ft}}\right)\left(4\ \text{ft} - \frac{7.5\ \text{in}}{12\ \frac{\text{in}}{\text{ft}}}\right) = 1241.3\ \text{lbf}$$

$$f_v = \frac{3V}{2A}$$
$$= \frac{(3)(1241.3\ \text{lbf})}{(2)(2)(1.5\ \text{in})(7.5\ \text{in})} \quad \text{[with two joists]}$$
$$= 82.8\ \text{lbf/in}^2 < F_v \quad \text{[ok]}$$

Check the bearing.

$$R = \left(367.8\ \frac{\text{lbf}}{\text{ft}}\right)(4\ \text{ft}) = 1471.2\ \text{lbf}$$

The bearing area is

$$A_b = (2)(1.75\ \text{in})(1.5\ \text{in}) = 5.25\ \text{in}^2$$

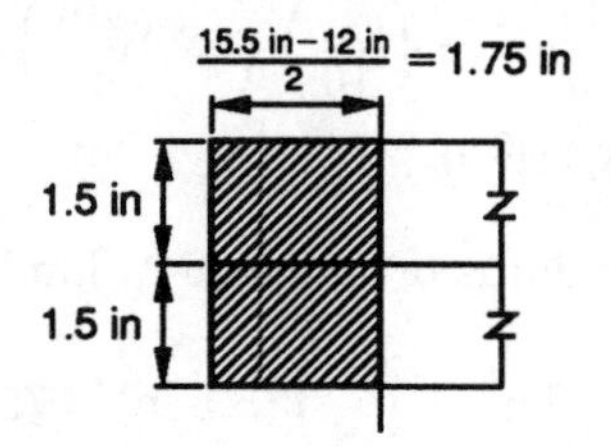

The bearing stress is

$$\frac{1471.2\ \text{lbf}}{5.25\ \text{in}^2} = 280.2\ \frac{\text{lbf}}{\text{in}^2} < F_{c\perp} \quad \text{[ok]}$$

For the 2-ft, 6-in prescribed spacing, use 2-in × 8-in double joists.

(c) (Reference: National Forest Products Association. *National Design Specifications for Wood Construction.* Washington, D.C., 1986.) First, check the limit imposed by the capacity of the post as a column.

$$\text{slenderness} = \frac{l_e}{d} = \frac{(8\ \text{ft})\left(12\ \frac{\text{in}}{\text{ft}}\right)}{3.5\ \text{in}}$$
$$= 27.43\ \text{in}$$

Classify the column by slenderness.

$$K = 0.671\sqrt{\frac{E}{F_c}} = 0.671\sqrt{\frac{1.6\times 10^6\ \frac{\text{lbf}}{\text{in}^2}}{(1.25)\left(1200\ \frac{\text{lbf}}{\text{in}^2}\right)}}$$

$$= 21.91\ \text{lbf/in}^2$$

$$\frac{l_e}{d} > K \quad \text{[column is long]}$$

$$F'_e = \frac{0.3E}{\left(\frac{l_e}{d}\right)^2} = \frac{(0.3)\left(1.6\times 10^6\ \frac{\text{lbf}}{\text{in}^2}\right)}{(27.43\ \text{in})^2}$$

$$= 638\ \text{lbf/in}^2$$

$$P_{\text{allowable}} = AF'_e$$

$$= (3.5\ \text{in})(7.25\ \text{in})\left(638\ \frac{\text{lbf}}{\text{in}^2}\right)$$

$$= 16{,}189\ \text{lbf (16.2 kips)}$$

The maximum post capacity as limited by behavior as a column is 16.2 kips.

Calculate the post capacity as limited by the bearing of the post on the 4-in × 4-in header. Increases in allowable stress for load duration are not permitted for bearing.

$$\text{bearing area} = (4\ \text{in})(8\ \text{in}) = 32\ \text{in}^2$$

$$\text{capacity} = \left(400\ \frac{\text{lbf}}{\text{in}^2}\right)(32\ \text{in}^2)$$

$$= 12{,}800\ \text{lbf (12.8 kips)}$$

Since 12.8 kips is less than the capacity of the post as a column, the bearing on the header controls.

Calculate the spacing, l_p. The reaction on the post from the precast beam is

$$R = \left[\frac{(15\ \text{in})(12\ \text{in})}{144\ \frac{\text{in}^2}{\text{ft}^2}}\right]\left(150\ \frac{\text{lbf}}{\text{ft}^3}\right) l_p$$

$$= 187.5 l_p$$

The total load on one post is

$$\left(129\ \frac{\text{lbf}}{\text{ft}^2}\right)(8\ \text{ft})l_p + \frac{\left(5\ \frac{\text{lbf}}{\text{ft}}\right)(2)(8\ \text{ft})l_p}{2.5\ \text{ft}} + 187.5 l_p$$

$$= 1252 l_p \quad \text{[lbf with } l_p \text{ in ft]}$$

Equating the load to the capacity,

$$12{,}800\ \text{lbf} = 1252 l_p$$

$$l_p = \boxed{10.2\ \text{ft}}$$

(d)

$$R_{\text{joist}} = \left(129\ \frac{\text{lbf}}{\text{ft}^2}\right)(2.5\ \text{ft})(4\ \text{ft}) + \left(10\ \frac{\text{lbf}}{\text{ft}}\right)(4\ \text{ft})$$

$$= 1330\ \text{lbf}$$

$$M = \left(\frac{1}{8}\right)(2R_{\text{joist}})(\text{span}) = \left(\frac{1}{8}\right)(2660\ \text{lbf})(5\ \text{ft})$$

$$= 1662.5\ \text{ft-lbf}$$

The required section modulus is

$$S_{\text{required}} = \frac{M}{F_b}$$

$$= \frac{(1662.5\ \text{ft-lbf})\left(12\ \frac{\text{in}}{\text{ft}}\right)}{(1.25)\left(1200\ \frac{\text{lbf}}{\text{in}^2}\right)}$$

$$= 13.3\ \text{in}^3$$

Assuming the minimum width of 15.5 in,

$$S = 13.3\ \text{in}^3 = \frac{bt^2}{6}$$

$$= \frac{(15.5\ \text{in})t^2}{6}$$

$$t = 2.27\ \text{in}$$

Try a 15.5-in × 3-in soffit beam.

Check the deflection.

$$I = \frac{1}{12}bt^3$$

$$= \left(\frac{1}{12}\right)(15.5\ \text{in})(2.5\ \text{in})^3$$

$$= 20.2\ \text{in}^4$$

$$\Delta = \frac{Pl^3}{192EI}$$

$$= \frac{(2660\ \text{lbf})(60\ \text{in})^3}{(192)\left(1.6\times 10^6\ \frac{\text{lbf}}{\text{in}^2}\right)(20.2\ \text{in}^4)}$$

$$= 0.093\ \text{in}$$

$$\Delta_{\text{allowable}} = \frac{l}{360} = \frac{60\ \text{in}}{360}$$

$$= 0.17\ \text{in}$$

Since $\Delta < \Delta_{\text{allowable}}$, the deflection is ok.

Use a 15.5-in × 3-in soffit beam.

2. (a) Calculate the maximum pressure exerted by the concrete. Use the ACI-recommended equation. (Reference: ACI 347 and Hurd, M.K. *Formwork for Concrete*. American Concrete Institute, 1979.)

For regular (type I) concrete with 4-in slump or less, ordinary workmanship, and internal vibration,

$$R = 2 \text{ ft/hr}$$

$$T = 70°\text{F}$$

$$p = 150 + (9000)\left(\frac{R}{T}\right) \quad [R \leq 7 \text{ ft/hr}]$$

$$= 150 + (9000)\left(\frac{2\,\frac{\text{ft}}{\text{hr}}}{70°\text{F}}\right) = 407.1 \text{ lbf/ft}^2$$

There is no need to account for variations of the pressure with height.

Assume flexure in sheathing controls the stud spacing. Find the load on one board, q_L.

$$q_L = \left(407.1\,\frac{\text{lbf}}{\text{ft}^2}\right)\left(\frac{5.5 \text{ in}}{12\,\frac{\text{in}}{\text{ft}}}\right) = 186.6 \text{ lbf/ft}$$

The maximum moment is

$$M = \left(\frac{1}{10}\right) q_L (\text{stud spacing})^2$$

$$= \left(\frac{1}{10}\right)\left(186.6\,\frac{\text{lbf}}{\text{ft}}\right)(s^2)$$

$$= 18.7 s^2$$

The section modulus is

$$S = \left(\frac{1}{6}\right)(5.5 \text{ in})(0.75 \text{ in})^2$$

$$= 0.52 \text{ in}^3$$

The allowable bending moment is

$$M_a = F_b S = \left(1700\,\frac{\text{lbf}}{\text{in}^2}\right)(0.52 \text{ in}^3)$$

$$= 884 \text{ in-lbf } (73.7 \text{ ft-lbf})$$

Equating to the maximum moment,

$$18.7 s^2 = 73.7 \text{ ft-lbf}$$

$$s = 1.98 \text{ ft} \quad [\text{use 2 ft}]$$

Check the shear.

$$V = \frac{1.15 q s}{2} = \frac{(1.15)\left(186.6\,\frac{\text{lbf}}{\text{ft}}\right)(2 \text{ ft})}{2}$$

$$= 214.6 \text{ lbf}$$

$$f_v = \frac{3V}{2A}$$

$$= \frac{(3)(214.6 \text{ lbf})}{(2)(0.75 \text{ in})(5.5 \text{ in})}$$

$$= 78.0 \text{ lbf/in}^2$$

The allowable shear stress is

$$F_v = (0.85)(1.25)\left(100\,\frac{\text{lbf}}{\text{in}^2}\right)$$

$$= 106.3\,\frac{\text{lbf}}{\text{in}^2} > f_v \quad [\text{ok}]$$

Use studs every 2 ft.

(b) The studs act as continuous beams supported on the wales.

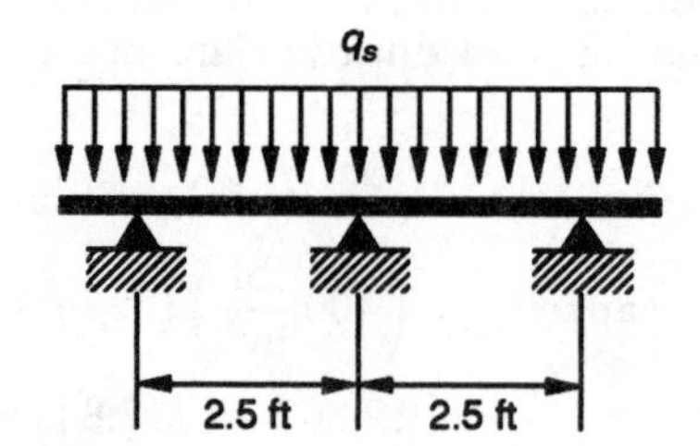

$$q_s = ps$$

$$= \left(407.1\,\frac{\text{lbf}}{\text{ft}^2}\right)(2 \text{ ft}) = 814.2 \text{ lbf/ft}$$

The maximum moment is

$$M \approx \left(\frac{1}{12}\right) q_s (\text{wale spacing})^2$$

$$= \left(\frac{1}{12}\right)\left(814.2\,\frac{\text{lbf}}{\text{ft}}\right)(2.5 \text{ ft})^2$$

$$= 424.1 \text{ lbf-ft}$$

$$S_{\text{required}} = \frac{M}{F_b} = \frac{(424.1 \text{ lbf-ft})\left(12\,\frac{\text{in}}{\text{ft}}\right)}{\left(1700\,\frac{\text{lbf}}{\text{in}^2}\right)(1.25)}$$

$$= 2.40 \text{ in}^3$$

A 2-in × 4-in stud has a section modulus of

$$S_{\text{provided}} = \frac{(1.5 \text{ in})(3.5 \text{ in})^2}{6}$$

$$= 3.06 \text{ in}^3 > S_{\text{required}} \quad [\text{ok}]$$

Check the shear.

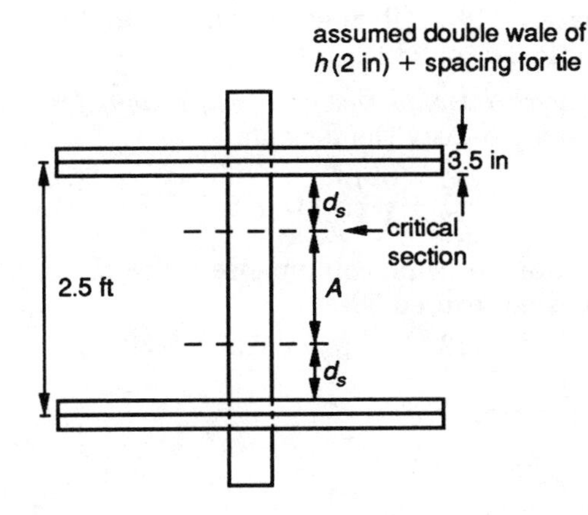

The length of dimension A (the critical section length) in the figure is

$$A = 2.5\text{ ft} - \frac{3.5\text{ in}}{12\,\frac{\text{in}}{\text{ft}}} - d_s$$
$$= 2.5\text{ ft} - \frac{3.5\text{ in}}{12\,\frac{\text{in}}{\text{ft}}} - \frac{3.5\text{ in}}{12\,\frac{\text{in}}{\text{ft}}}$$
$$= 1.917\text{ ft}$$

$$V = \left(814.2\,\frac{\text{lbf}}{\text{ft}}\right)\left(\frac{1.917}{2}\text{ ft}\right) = 780.4\text{ lbf}$$

$$f_v = \frac{3V}{2A} = \frac{(3)(780.4\text{ lbf})}{(2)(1.5\text{ in})(3.5\text{ in})}$$
$$= 222.9\text{ lbf/in}^2$$

Increase F_v by $(1.25)(1.5) = 1.875$. (Reference: *Framework for Concrete.*)

$$F_v = \left(100\,\frac{\text{lbf}}{\text{in}^2}\right)(1.875)$$
$$= 187.5\,\frac{\text{kips}}{\text{in}^2} < f_v \quad \text{[unacceptable]}$$

The 2-in × 4-in studs are overstressed in shear. Try 2-in × 6-in studs.

$$f_v = \frac{(3)\,(780.3\text{ lbf})}{(2)(1.5\text{ in})(5.5\text{ in})}$$
$$= 141.87\text{ kips/in}^2 < F_v \quad \text{[ok]}$$

Use 2-in × 6-in studs.

(c) Assume that the allowable tensile stress in the $\frac{1}{4}$-in-diameter ties is $0.6F_y$.

$$\text{tie capacity} = 0.6F_yA_{\text{tie}}$$
$$= (0.6)\left(36{,}000\,\frac{\text{lbf}}{\text{in}^2}\right)\pi(0.25\text{ in})^2$$
$$= 4241.1\text{ lbf}$$
$$\text{load on a tie} = p(\text{stud spacing})(\text{wale spacing})$$
$$= \left(407.1\,\frac{\text{lbf}}{\text{ft}^2}\right)(2\text{ ft})(2.5\text{ ft})$$
$$= 2035.5\text{ lbf} < 4241.1\text{ lbf}$$

One tie between the studs is sufficient.

(d) Wales are likely to be controlled by shear.

$$V = \frac{\text{force in tie}}{2} = \frac{2035.5\text{ lbf}}{2} = 1017.75\text{ lbf}$$

Assuming a 2-in × 4-in double wale,

$$V_{\text{wale}} = \frac{V}{2} = \frac{1018\text{ lbf}}{2}$$
$$= 508.9\text{ lbf}$$
$$F_v = \frac{(3)(508.9\text{ lbf})}{(2)(1.5\text{ in})(3.5\text{ in})}$$
$$= 145.4\,\frac{\text{lbf}}{\text{in}^2} < 187.5\,\frac{\text{lbf}}{\text{in}^2} \quad \text{[allowable]}$$

Use double 2-in × 4-in wales.

3. (a) A $22F - V3$ beam is visually graded, while a $22F - E3$ beam is machine graded. (The allowable stresses differ for some conditions.)

(b) $$C_F = \left(\frac{12}{d}\right)^{1/9} = \left[\frac{12}{(2\text{ ft})\left(12\,\frac{\text{in}}{\text{ft}}\right)}\right]^{1/9}$$
$$= 0.93 \quad \text{[basic]}$$

The adjustment for the span-to-depth ratio is

$$\frac{l}{d} = \frac{25\text{ ft}}{2\text{ ft}} = 12.5\text{ ft}$$

$$\text{adjustment from AITC table} = 1.03$$

$$C_F = (1.03)(0.93) = \boxed{0.958}$$

(Reference: *Timber Construction Manual*, American Institute of Timber Construction.)

(c)
$$C_c = 1 - (2000)\left(\frac{t}{R_i}\right)^2$$
$$t = 1.5\text{ in}$$
$$R_i = 100\text{ ft}$$
$$C_c = 1 - (2000)\left[\frac{1.5\text{ in}}{(100\text{ ft})\left(12\,\frac{\text{in}}{\text{ft}}\right)}\right]^2$$
$$= \boxed{0.997}$$

(d)
$$\text{self-weight} \approx \left(35\,\frac{\text{lbf}}{\text{ft}^3}\right)\left[\frac{(6.75\text{ in})(24\text{ in})}{144\,\frac{\text{in}^2}{\text{ft}^2}}\right]$$
$$= 39.38\text{ lbf/ft}$$
$$\text{total weight} = 1000\,\frac{\text{lbf}}{\text{ft}} + 39.38\,\frac{\text{lbf}}{\text{ft}}$$
$$= 1039.38\text{ lbf/ft} \quad [\text{use } 1040\text{ lbf/ft}]$$
$$f = \frac{3V}{2A}$$
$$= \frac{(3)\left(1040\,\frac{\text{lbf}}{\text{ft}}\right)(25\text{ ft})}{(2)(2)(6.75\text{ in})(24\text{ in})}$$
$$= \boxed{120\text{ lbf/in}^2}$$

(e) One support is a roller so there is no arch action.

$$M = \frac{1}{8}wl^2 = \left(\frac{1}{8}\right)\left(1040\,\frac{\text{lbf}}{\text{ft}}\right)(25\text{ ft})^2$$
$$= 81{,}250\text{ ft-lbf}$$
$$S = \left(\frac{1}{6}\right)(6.75\text{ in})(24\text{ in})^2$$
$$= 648\text{ in}^3$$
$$\sigma = \frac{M}{S} = \frac{(81{,}250\text{ ft-lbf})\left(12\,\frac{\text{in}}{\text{ft}}\right)}{648\text{ in}^3}$$
$$= \boxed{1504.6\text{ lbf/in}^2}$$

(f)
$$I = \frac{bh^3}{12} = \frac{(6.75\text{ in})(24\text{ in})^3}{12}$$
$$= 7776\text{ in}^4$$
$$\Delta = \frac{5wl^4}{384EI}$$
$$= \frac{(5)\left(1040\,\frac{\text{lbf}}{\text{ft}}\right)(25\text{ ft})^4\left(12\,\frac{\text{in}}{\text{ft}}\right)^4}{(384)\left(1.6\times 10^6\,\frac{\text{lbf}}{\text{in}^2}\right)(7776\text{ in}^4)\left(12\,\frac{\text{in}}{\text{ft}}\right)}$$
$$= \boxed{0.735\text{ in}}$$

(g) The moisture content depends on ambient conditions. For a relative humidity of 60% and a mean temperature of 70°F, the moisture content would be approximately 11%. (Reference: *Structural Design in Wood*. Table 2-4, p. 23.)

4. (Reference: *National Design Specifications for Wood Construction*.) Satisfy the formula

$$\frac{f_c}{F_c'} + \left(\frac{f_b}{F_b'}\right)C_M \le 1$$

Select a tentative beam-column size. The slenderness ratio should not exceed 50.

$$l_e = (2.1)(25\text{ ft}) = 52.5\text{ ft }(630\text{ in})$$
$$\frac{630\text{ in}}{d} \le 50$$
$$d \ge 12.6\text{ in}$$

Try a 14-in × 14-in column.

$$\frac{l_e}{d} = \frac{630\text{ in}}{13.5\text{ in}} = 46.67$$

Calculate F_c'. Include an adjustment factor of 1.33 for wind loading.

$$K = 0.671\sqrt{\frac{E}{F_c}}$$
$$= (0.671)\sqrt{\frac{1.8\times 10^6\,\frac{\text{lbf}}{\text{in}^2}}{(1.33)\left(1200\,\frac{\text{lbf}}{\text{in}^2}\right)}}$$
$$= 22.53\text{ lbf/in}^2$$
$$\frac{l_e}{d} > K$$
$$F_c' = \frac{0.3E}{\left(\frac{l_e}{d}\right)^2} = \frac{(0.3)\left(1.8\times 10^6\,\frac{\text{lbf}}{\text{in}^2}\right)}{(46.67)^2}$$
$$= 247.9\text{ lbf/in}^2$$

Calculate f_c.

$$f_c = \frac{P}{A} = \frac{20{,}000\text{ lbf}}{(13.5\text{ in})^2}$$
$$= 109.7\text{ lbf/in}^2$$

Calculate f_b.

$$M = \frac{wl^2}{2} = \frac{\left(80\,\frac{\text{lbf}}{\text{ft}}\right)(25\text{ ft})^2}{2}$$
$$= 25{,}000\text{ ft-lbf}$$
$$S = \frac{1}{6}bt^2 = \left(\frac{1}{6}\right)(13.5\text{ in})(13.5\text{ in})^2$$
$$= 410.1\text{ in}^3$$
$$f_b = \frac{M}{S} = \frac{(25{,}000\text{ ft-lbf})\left(12\,\frac{\text{in}}{\text{ft}}\right)}{410.1\text{ in}^3}$$
$$= 731.5\text{ lbf/in}^2$$

Calculate F_b'.

$$l_e = 1.84 l_v = (1.84)(25\text{ ft}) = 46\text{ ft }(552\text{ in})$$

$$C_s = \sqrt{\frac{l_e d}{b^2}} = \sqrt{\frac{(552\text{ in})(13.5\text{ in})}{(13.5\text{ in})^2}} = 6.39$$

Since $C_s < 10$, $F_b' = F_b$. With an adjustment for wind loading and wet use,

$$F_b' = (0.91)(1.33)\left(1600\ \frac{\text{lbf}}{\text{in}^2}\right) = 1936\text{ lbf/in}^2$$

Calculate C_M. Since $l_e/d > K$,

$$J = 1$$

$$C_M = \frac{1}{1 - J\left(\frac{F_c}{F_b'}\right)} = \frac{1}{1 - \frac{109.7\ \frac{\text{lbf}}{\text{in}^2}}{1936\ \frac{\text{lbf}}{\text{in}^2}}} = 1.06$$

Check the combined stress criteria.

$$\frac{f_c}{F_c'} + \left(\frac{f_b}{F_b'}\right) C_M \leq 1$$

$$\frac{109.74\ \frac{\text{lbf}}{\text{in}^2}}{247.9\ \frac{\text{lbf}}{\text{in}^2}} + \frac{\left(731.5\ \frac{\text{lbf}}{\text{in}^2}\right)(1.06)}{1936\ \frac{\text{lbf}}{\text{in}^2}} = 0.84 \quad \text{[ok]}$$

Check the shear.

$$V = \left(80\ \frac{\text{lbf}}{\text{ft}}\right)(25\text{ ft}) = 2000\text{ lbf}$$

$$f_v = \frac{3V}{2A} = \frac{(3)(2000\text{ lbf})}{(2)(13.5\text{ in})^2} = 16.46\text{ lbf/in}^2 < F_v \quad \text{[ok]}$$

Use a 14-in × 14-in beam-column.

TRAFFIC

1. (Reference: Transportation Research Board, *Highway Capacity Manual*, 1985, Fig. 9-14(a).)

step 1: Geometry and timing are given. For delay values, see Step 6.

step 2: On the north-to-south streets there are no left turns, so the two lanes in each direction constitute a single lane group. On the east-to-west street there are no turning conflicts, so it can be treated as a single lane group.

step 3: Find the g/C ratio.

$$C = 60 \text{ sec}$$

The effective green time is

green time + amber time − lost time

Assuming the lost time is 3 sec, the effective green ratios are

$$\left(\frac{g}{C}\right)_{\text{WE}} = \frac{32 \text{ sec} - 3 \text{ sec}}{60 \text{ sec}} = 0.483$$

$$\left(\frac{g}{C}\right)_{\text{NS}} = \frac{26 \text{ sec} - 3 \text{ sec}}{60 \text{ sec}} = 0.383$$

step 4: Find the saturation flow rate. The formula is from *Highway Capacity Manual* Eq. 9-8. The area type should be considered a central business district (CBD) because a daytime population of 120,000 is greater than the daytime population of most central cities (e.g., Dallas CBD employment is approximately 80,000). $f_{\text{LT}} = 1.00$ for the west-to-east street (from *Highway Capacity Manual* Table 9-12, Case 4, since $p_{\text{LT}} = 0.05$). For the north-to-south street, assume $N_m = 20$ parking maneuvers/hr (*Highway Capacity Manual* Table 9-3), yielding $f_p = 0.89$. Assume zero grade.

West-to-east:

$$\begin{aligned} f_w &= 0.97 \quad \text{[Table 9-5]} \\ f_{\text{HV}} &= 0.95 \quad \text{[Table 9-6]} \\ f_g = f_p = f_{bb} &= f_{\text{RT}} \\ &= 1 \quad \text{[Tables 9-7, 9-8, 9-9, and 9-11]} \\ f_a &= 0.90 \quad \text{[Table 9-10]} \\ f_{\text{LT}} &= 0.99 \quad \text{[Table 9-12]} \end{aligned}$$

Note that the definition of P_{LT} in *Highway Capacity Manual* Table 9-12, Case 4, should be "proportion of LTs in lane group" as on p. 9-28, items 1.g and 2.e.

$$\begin{aligned} s &= s_o N f_w f_{\text{HV}} f_g f_p f_{bb} f_a f_{\text{RT}} f_{\text{LT}} \\ &= (1800 \text{ vph})(4)(0.97)(0.95)(1)(1)(1)(0.90)(1)(0.99) \\ &= 5912 \text{ vphg} \end{aligned}$$

North-to-south and south-to-north:

$$\begin{aligned} f_w &= 1.00 \quad \text{[Table 9-5]} \\ f_{\text{HV}} &= 0.965 \quad \text{[Table 9-6]} \\ f_g = f_{bb} = f_{\text{RT}} &= f_{\text{LT}} \\ &= 1 \quad \text{[Tables 9-7, 9-9, 9-11, and 9-12]} \\ f_p &= 0.89 \quad \text{[Table 9-8, with } N_m = 20 \text{ from Table 9-3]} \\ f_a &= 0.90 \quad \text{[Table 9-10]} \\ s &= s_o N f_w f_{\text{HV}} f_g f_p f_{bb} f_a f_{\text{RT}} f_{\text{LT}} \\ &= (1800 \text{ vph})(2)(1)(0.965)(1)(0.89)(1)(0.90)(1)(1) \\ &= 2783 \text{ vphg} \end{aligned}$$

step 5: Find the capacity.

West-to-east:

$$\begin{aligned} c &= s\left(\frac{g}{C}\right) \\ &= (5912 \text{ vphg})(0.483) = 2856 \text{ vehicles/hr} \end{aligned}$$

North-to-south and south-to-north:

$$c = (2783 \text{ vphg})(0.383) = 1066 \text{ vehicles/hr}$$

step 6: Find the v/c ratio and the service volume.

The largest permitted average delay for a given LOS is found in *Highway Capacity Manual* Table 9-1. For example, in Case (a), LOS = C, so $d = 25.0$. The volume/capacity ratio, X, is found by trial and error using *Highway Capacity Manual* Eq. 9-18.

$$d = (0.38C)\left(\frac{\left[1-\left(\frac{g}{C}\right)\right]^2}{1-\left(\frac{g}{C}\right)X}\right) + (173)\left[(X-1)^2 + \sqrt{(X-1)^2 + 16\left(\frac{X}{c}\right)}\right]$$

Once X is found, the maximum service flows are

$$\begin{aligned} v_{15} &= \text{maximum service flow in peak 15-min period} \\ &= cX \\ v_{60} &= \text{maximum service volume in an hour} \\ &= (v_{15})(\text{PHF}) \end{aligned}$$

	$\frac{g}{C}$	c	X	d	v_{15} vehicles/hr	PHF	v_{60} vehicles/hr
(a)	0.483	2856	0.917	25	2619	0.8	2095
(b)	0.483	2856	1.073	60	3064	0.8	2452
(c)	0.383	1066	0.162	15	173	0.7	121
(d)	0.383	1066	1.054	60	1124	0.7	786

2. According to Eq. 9-8 of the *Highway Capacity Manual*, there are nine primary factors that influence capacity. They are

1) number of lanes
2) lane width
3) percentage of heavy vehicles in the traffic stream
4) grade of the approach
5) presence of a parking lane and number of parking maneuvers per hour
6) blockage caused by local buses making stops
7) area type (capacity is less in central business districts than elsewhere)
8) percentage of right-turning vehicles in the traffic stream
9) percentage of left-turning vehicles in the traffic stream

Key secondary factors that interact with items 8 and 9 are

1) whether turns are made from an exclusive or shared lane
2) whether turning movements are protected, permitted (unprotected), or both
3) volume of pedestrians using the conflicting crosswalk for right turns
4) opposing flow rate when the signal phasing includes permitted left turns

3. (a) The practical capacity of loop ramps (270° turns) is 800–1200 vehicles/hr (Reference: AASHTO, *A Policy on Geometric Design of Highways and Streets*, 1990, p. 904), regardless of width (loop ramps are not amenable to double lanes). Assuming a k-factor (ratio of peak hour flow to ADT) of between 0.12 (urban) and 0.15 (rural), peak hour flow for the lower volume left turn is at least 1200 vehicles/hr. A directional three-leg interchange is recommended.

(b) The low flows on all except the freeway mainline dictate a simple diamond with stop- or signal-controlled intersections on the north-to-south road. If the river prevents ramps on the east, a half cloverleaf with diagonal and loop ramps in the two western quadrants is recommended, with stop or signal control at the minor street terminals.

(c) The relatively low flows suggest that no interchange will be needed. The high proportion of turning movements suggests a rotary intersection. If an interchange is desired, a loop-connector type is recommended, with the east-to-west street at a separate grade from a loop that carries all other movements. This is an inexpensive type of interchange that works well with low flows. Because it treats the north-to-south flows as turning flows (they must enter the loop and then leave it), it is most appropriate when the through volume is small compared to the turning flows.

(d) The interchange is between two very high-volume roads, presumably freeways. With its high turning volumes, it requires a fully directional or semidirectional interchange that provides the best level of service—i.e., high speeds for all turning flows. The lowest volume left turn movement (east-to-south) will have a peak hour flow of at least 1000–1200 vehicles/hr, so it could possibly be accommodated by a loop ramp; however, including a single loop ramp in a directional interchange is impractical, and the level of service on the loop would be poor.

4. (a) (Reference: *Highway Capacity Manual*, Tables 3-4, 3-5, and 7-4.) On basic freeway sections and on multilane rural highways, E_T for typical trucks can range from 3 to 10, depending on the length of the grade and the precentage of trucks. For light trucks (100 lbm/hp), E_T is 2.

(b) (Reference: *Highway Capacity Manual*, pp. 3-8, 4-9, 5-4, 7-6, 8-5, 9-3, 10-9, and 11-3.)

- On basic freeway segments and multilane highways, density is the primary variable.
- On freeway weaving sections, speed for weaving and nonweaving vehicles is the primary variable.
- On freeway ramp sections, flow in the merge and/or diverge lanes and total flow are the primary variables.
- On two-lane rural highways, percent time delay is the primary variable.
- At signalized intersections, average stopped delay is the primary variable.
- At unsignalized intersections, reserve capacity is the primary variable.
- On urban/suburban arterials, average travel speed is the primary variable.

(c) (Reference: AASHTO, *Roadside Design Guide*, 1989, Chaps. 5, 6, 9.) There are three sets of warrants for roadside guardrails.

1) Embankment slope and height: Various combinations warrant a roadside barrier, as illustrated (from Fig. 5.1). Modifications for traffic and embankment length may also be included in the warrant.

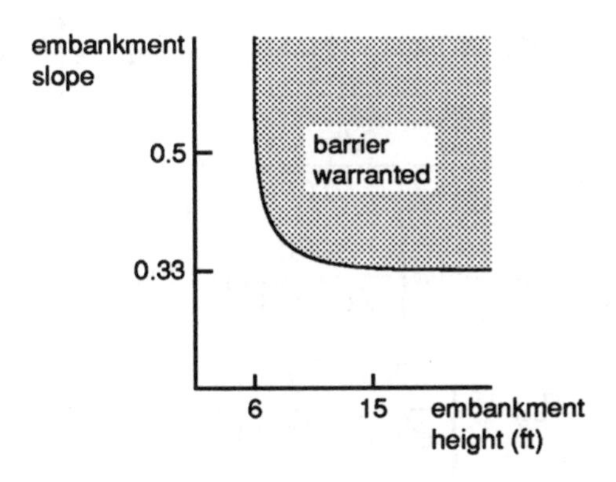

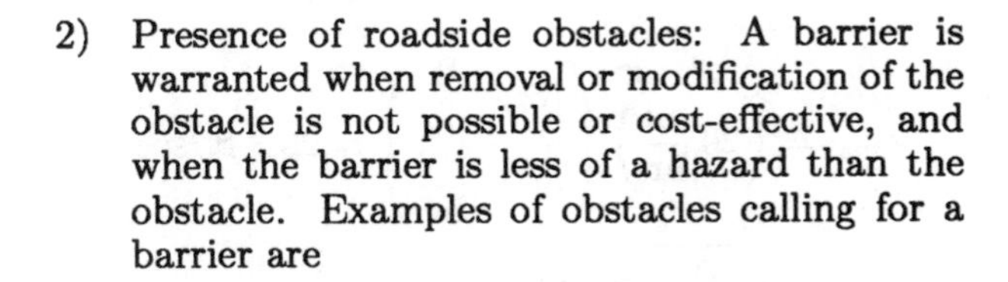

2) Presence of roadside obstacles: A barrier is warranted when removal or modification of the obstacle is not possible or cost-effective, and when the barrier is less of a hazard than the obstacle. Examples of obstacles calling for a barrier are

- wood poles or posts with an area greater than or equal to 50 in^2
- metal shapes with a moment of inertia greater than or equal to 3.0 in^4 for steel and 4.5 in^4 for aluminum
- concrete bases extending 6 in or more above the ground
- sign bridge supports
- breakaway light poles, with breakaway linear impulse greater than or equal to 110 lbf-sec
- trees with diameter greater than or equal to 6 in

3) Protection of bystanders, cyclists, and pedestrians: No simple numerical value is given, but under normal circumstances, a raised curb is a sufficient barrier to protect pedestrians for speeds up to 40 mph. Special circumstances (e.g., accident experience and the presence of schoolchildren) may apply.

4) Median barriers: These may be needed, based on roadside warrants. In addition, there is a median warrant to prevent head-on collisions, based on various combinations of ADT and median width.

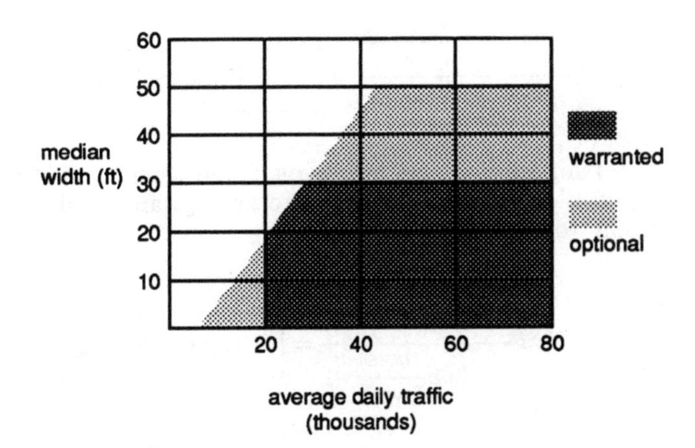

5) Barriers in work areas: No numerical values are given. Work-area barriers are warranted when needed to

- protect traffic from entering unsafe areas
- protect workers
- separate two-way traffic
- protect exposed construction works (e.g., falsework)

(d) The primary variable is the total number of accidents. Other measures used are the accident rate (accidents per 100,000 vehicles) and a weighted sum of personal injury accidents and property damage only (PDO) accidents, in which personal injury accidents count as four or five PDO accidents.

(e) (Reference: Institute of Traffic Engineers, *Transportation and Traffic Engineering Handbook*, 4th ed., 1992, Tables 7-4 and 7-5. Prentice-Hall, Inc., Englewood Cliffs, NJ) For most purposes, a 90° parking stall can be 8.5 ft wide by 17.5 ft long. Recommended width varies from 9 ft for high turnover uses to 8.25 ft for low turnover uses. "Small car only" stalls can, of course, be smaller.

(f) (Reference: *Transportation and Traffic Engineering Handbook*.) A typical parking module is 8.5 ft wide by 61 ft long, including two stalls and an aisle. Adding 10% for losses at corners and such,

$$\text{area per vehicle} = (1.1)(8.5\text{ ft})\left(\frac{61\text{ ft}}{2}\right)$$

$$= \boxed{285\text{ ft}^2}$$

5. (a) There are (24 days)(365 hr) = 8760 hr in a year, so the 30th busiest hour is the $(1 - 30/8760)(100) =$ 99.66 percentile value, which (from the problem statement) is 21,550,000 vehicles/yr. Converting to an hourly volume, the 30th hour volume is

$$\frac{21{,}550{,}000\ \dfrac{\text{vehicles}}{\text{yr}}}{8760\ \dfrac{\text{hr}}{\text{yr}}} = \boxed{2460\text{ vehicles/hr}}$$

(b) $$\frac{28{,}040{,}000\ \frac{\text{vehicles}}{\text{yr}}}{8760\ \frac{\text{hr}}{\text{yr}}} = \boxed{3201\ \text{vehicles/hr}}$$

(c) An approximate average is found using the 50th percentile volume. Converting to average annual daily traffic volume (AADT),

$$\text{AADT} \approx \frac{990{,}000\ \frac{\text{vehicles}}{\text{yr}}}{365\ \frac{\text{days}}{\text{yr}}} = \boxed{2712\ \text{vehicles/day}}$$

(d) Density = volume/speed. Using the volume from (a),

$$\text{density} = \frac{2460\ \frac{\text{vehicles}}{\text{hr}}}{42\ \frac{\text{mi}}{\text{hr}}} = \boxed{58.6\ \text{vehicles/mi}}$$

(e) Assuming a 50–50 directional split, the volume in each direction is

$$\begin{aligned}\text{volume} &= \frac{2460\ \frac{\text{vehicles}}{\text{hr}}}{2} \\ &= 1230\ \text{vehicles/hr}\end{aligned}$$

$$\begin{aligned}\text{headway} &= \frac{1}{\text{volume}} = \frac{1}{\left(1230\ \frac{\text{vehicles}}{\text{hr}}\right)\left(\frac{1}{3600}\ \frac{\text{hr}}{\text{sec}}\right)} \\ &= 2.93\ \text{sec/vehicle} \quad [2.93\ \text{sec}]\end{aligned}$$

(f) Because the road is sometimes congested, its flow reaches capacity. An estimate of the capacity is the maximum hourly flow as determined in (b). At capacity, the average speed is approximately 30 mph.

$$\text{density} \approx \frac{3201\ \frac{\text{vehicles}}{\text{hr}}}{30\ \frac{\text{mi}}{\text{hr}}} = \boxed{107\ \text{vehicles/mi}}$$

(g) Space mean speed differs little from time mean speed, so the given speed can be equated to a space mean speed.

$$\begin{aligned}\text{mean unit travel time} &= \frac{1}{\text{mean speed}} \\ &= \left(\frac{1}{42\ \frac{\text{mi}}{\text{hr}}}\right)\left(60\ \frac{\text{min}}{\text{hr}}\right) \\ &= \boxed{1.43\ \text{min/mi}}\end{aligned}$$

(h) $(1.02)^4 = 1.0824$. The increase will be $\boxed{8.24\%.}$

6. (a) $$F_{\text{gravity}} = \left[\frac{0.05}{\sqrt{1+(0.05)^2}}\right] mg$$

$$F_{\text{friction}} = -\mu\left[\frac{1.0}{\sqrt{1+(0.05)^2}}\right] mg$$

$$F_{\text{net}} = \left[\frac{32.2\ \frac{\text{ft}}{\text{sec}^2}}{\sqrt{1+(0.05)^2}}\right](0.05-\mu)m$$

$$a = \frac{F_{\text{net}}}{m} = \left(32.16\ \frac{\text{ft}}{\text{sec}^2}\right)(0.05-\mu)$$

$$\text{v}_{\text{final}} = \left(20\ \frac{\text{mi}}{\text{hr}}\right)\left(\frac{5280\ \frac{\text{ft}}{\text{sec}}}{3600\ \frac{\text{sec}}{\text{hr}}}\right) = 29.3\ \text{ft/sec}$$

$$\begin{aligned}v_0 &= \sqrt{v_{\text{final}}^2 - 2as} \\ &= \sqrt{\left(29.3\ \frac{\text{ft}}{\text{sec}}\right)^2 - (2)\left[\left(32.16\ \frac{\text{ft}}{\text{sec}^2}\right)(0.05-\mu)\right](180\ \text{ft})} \\ &= \sqrt{280 + 11{,}578\mu}\end{aligned}$$

Since μ must be considered with v_0, solve by trial and error.

assumed μ	resulting v_0	
0.30	61 ft/sec = 42 mph	[inconsistent with μ]
0.32	63 ft/sec = $\boxed{43\ \text{mph}}$	[consistent with μ]

(b) The answer to (a) would not change appreciably because the given coefficients of friction are valid for most wet pavements.

7. (Reference: American Association of State Highway and Transportation Officials, *A Policy on Geometric Design of Highways and Streets*, 1990, pp. 745–760.)

Assumptions:

- stop control exists on the minor street, making this Case III
- design speed is 10 mph greater than posted speed
- design vehicle is passenger car (little truck traffic)
- no median, 12-ft lanes
- AASHTO assumed values are used for variables D, L, and J

Sight distance requirements apply for stopped minor street vehicles crossing the major street, turning left onto the major street, and turning right onto the major street.

(a) Crossing maneuver: On the right, all four lanes of traffic must be cleared, so the distance to clear the intersection is

$$\begin{aligned} S &= D + N(\text{lane width}) + L \\ &= 10\text{ ft} + (4)(12\text{ ft}) + 19\text{ ft} \\ &= 77\text{ ft} \end{aligned}$$

From Fig. IX-33, $t_a = 5.6$ sec.

The sight distance required is

$$\begin{aligned} d &= 1.47V(J + t_a) \\ &= \left(1.47\,\frac{\frac{\text{ft}}{\text{sec}}}{\frac{\text{mi}}{\text{hr}}}\right)\left(60\,\frac{\text{mi}}{\text{hr}}\right)(2.0\text{ sec} + 5.6\text{ sec}) \\ &= 670\text{ ft on the right} \end{aligned}$$

This value could also be read from Fig. IX-39.

Solving for the left (only two lanes need to be cleared),

$$\begin{aligned} S &= D + N(\text{lane width}) + L \\ &= 10\text{ ft} + (2)(12\text{ ft}) + 19\text{ ft} = 53\text{ ft} \end{aligned}$$

From Fig. IX-33,

$$\begin{aligned} t_a &= 4.8\text{ sec} \\ d &= 1.47v(J + t_a) \\ &= \left(1.47\,\frac{\frac{\text{ft}}{\text{sec}}}{\frac{\text{mi}}{\text{hr}}}\right)\left(60\,\frac{\text{mi}}{\text{hr}}\right)(2.0\text{ sec} + 4.8\text{ sec}) \\ &= 600\text{ ft on the left} \end{aligned}$$

(b) Left-turning maneuver: The distance the left-turning car must cover to clear traffic on the left is

$$\begin{aligned} S &= 10\text{ ft} + (2)(12\text{ ft})(1.5) + 19\text{ ft} \\ &= 65\text{ ft} \end{aligned}$$

From Fig. IX-33, $t_a = 5.2$ sec on the left. The required sight distance is

$$\begin{aligned} d &= \left(1.47\,\frac{\frac{\text{ft}}{\text{sec}}}{\frac{\text{mi}}{\text{hr}}}\right)\left(60\,\frac{\text{mi}}{\text{hr}}\right)(2.0\text{ sec} + 5.2\text{ sec}) \\ &= 635\text{ ft on the left} \end{aligned}$$

To safely clear traffic on the right, 85% of the design speed is

$$v_{85\%} = (0.85)\left(60\,\frac{\text{mi}}{\text{hr}}\right) = 51\text{ mi/hr (75 ft/sec)}$$

From Fig. IX-4, the distance covered to reach 51 mph is $P = 800$ ft. From Table IX-7, $t_a = 19.1$ sec.

$$\begin{aligned} Q &= v_{85\%}(t_a + 2\text{ sec}) \\ &= \left(75\,\frac{\text{ft}}{\text{sec}}\right)(19.1\text{ sec} + 2\text{ sec}) = 1580\text{ ft} \\ h &= P - 16 - (\text{width of 2nd lane and median}) \\ &\quad - L - 1.9v_{85\%} \\ &= 800\text{ ft} - 16\text{ ft} - 12\text{ ft} - 19\text{ ft} - (1.9)(75\text{ ft}) \\ &= 610\text{ ft} \\ d &= Q - h = 1580\text{ ft} - 610\text{ ft} \\ &= 970\text{ ft on the right} \end{aligned}$$

Right-turning maneuver: According to AASHTO, the required distance on the left is 1–3 ft less than the distance to the right that is required under (b). Take $d = 970$ ft on the left.

Combining all three cases, the controlling distances, measured from the center of the minor street approach lane, are 970 ft on the left and 970 ft on the right.

The short dimension of the sight triangles for minor street traffic is based on a 10-ft setback from the stopline to the edge of the traveled way, plus 5 ft from the front of the vehicle to the eye of the driver.

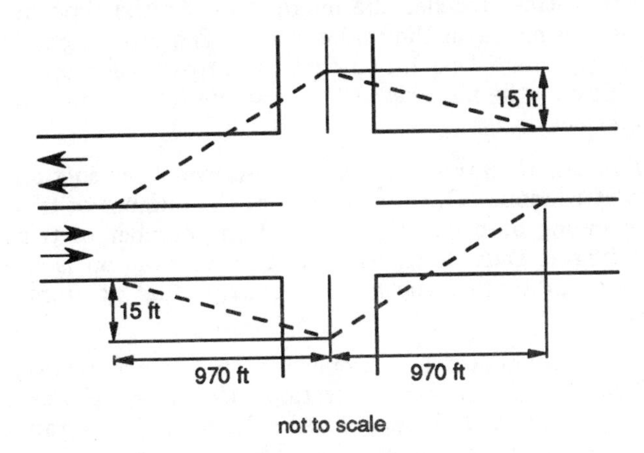

8. (a) Reasonable gradients are 0–15% longitudinal and 1.5% crown traverse.

(b) There is more than one way to lay out two intersecting runways. One way is

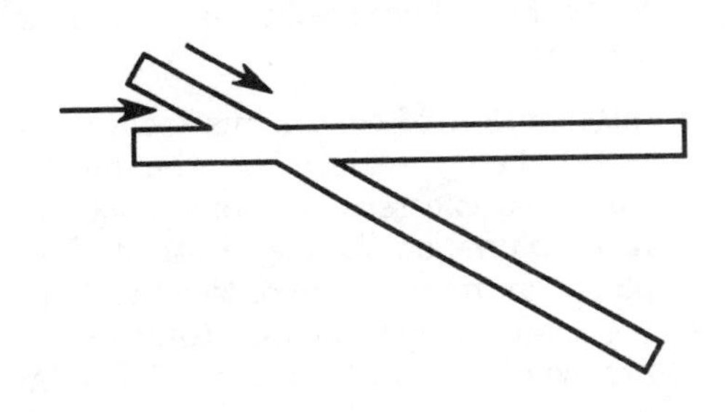

PHOCAP (practical hourly capacity) under VFR (visual flight rules) is between 72 and 98 operations/hr, depending on the mix of aircraft. For IFR (instrument flight rules), PHOCAP is about 60 operations/hr. (Reference: FAA Advisory Circular 150/5060-5, "Airport Capacity and Delay," cited in Haefner, L.E., *Introduction to Transportation Systems*, Holt, Rinehart and Winston, 1986.)

(c) Other factors to consider are the weight of the largest expected aircraft, the strength of the base and subbase, drainage, and frost interaction.

9. (a) (Reference: *Highway Capacity Manual*, pp. 9-7 and 9-8.) A very good progression is arrival type 5. Arrival type = 5 if PVG/PTG > 1.5. PVG is the percentage of all vehicles at an intersection approach that arrive during the green phase. PTG is the green time/cycle length.

(b) (Reference: Institute of Traffic Engineers, *Trip Generation*, 4th ed., 1987, Table 1.) Weekday ADT (trip ends) are 7431. The peak hour volume is 623.

(c) The peak hour factor is

$$4\left(\frac{\text{traffic volume in the peak 15-min period}}{\text{traffic volume in the peak hour}}\right)$$

(d) A *gap* is the time until the next vehicle on the major street reaches the intersection, or the time between vehicles on the major street. The *critical gap* is the minimum length gap that the driver of a stopped vehicle on the minor street finds acceptable to enter the intersection.

Trip distribution is the result of matching productions to attractions. Given estimates of the number of trips beginning at each origin, i, and the number of trips ending at each destination, j, trip distribution is the art of estimating the number of trips that go from i to j.

(e) (Reference: U.S. Department of Transportation, Federal Highway Administration, *Manual on Uniform Traffic Control Devices*.) This book lists many warrants for signalizing an intersection. They include

- Minimum vehicular volume: For example, at an intersection of two two-lane roads, the major street volume (total for both approaches) should be at least 500 vehicles/hr and the minor street volume at least 150 vehicles/hr. These volumes should hold for 8 hours.
- Interruption of continuous traffic: This applies when the delay on the minor street would be excessive without a signal due to high volume on the major street. For example, at an intersection of two two-lane roads, the major street volume (total for both approaches) should be at least 750 vehicles/hr and the minor street volume at least 75 vehicles/hr. These volumes should hold for 8 hours.
- Minimum pedestrian volume: In the simplest case, estimate at least 600 vehicles/hr on the major street and 150 pedestrians/hr at the busiest crosswalk, for 8 hours a day.
- School crossing
- Progressive movement (to keep platoons from dispersing)
- Accident experience: Signalizing may be required if there are at least 5 accidents in 12 months, if less restrictive remedies have failed to reduce accident frequency, and if there are traffic volumes that meet 80% of the first three given warrants.

10. (Reference: Asphalt Institute, *The Asphalt Handbook*, (Manual MS-4), Lexington, KY, 1989.)

(a) The bulk specific gravity is a weighted average of the aggregate and mineral specific gravities.

$$\begin{aligned} G_{sb} &= \frac{\sum x_i G_i}{\sum x_i} \\ &= \frac{(0.52)(2.61) + (0.34)(2.71) + (0.07)(2.70)}{0.52 + 0.34 + 0.07} \\ &= \boxed{2.653} \end{aligned}$$

(b)
$$\begin{aligned} G_{se} &= \frac{1 - x_b}{\frac{1}{G_{mm}} - \frac{x_b}{G_b}} = \frac{1 - 0.07}{\frac{1}{2.455} - \frac{0.07}{1.01}} \\ &= \boxed{2.751} \end{aligned}$$

(c)
$$\begin{aligned} G_{mm} &= \frac{1}{\frac{1 - x_b}{G_{se}} + \frac{x_b}{G_b}} = \frac{1}{\frac{1 - 0.08}{2.751} + \frac{0.08}{1.01}} \\ &= \boxed{2.418} \end{aligned}$$

(d)
$$\begin{aligned} x_{ba} &= G_b\left(\frac{G_{se} - G_{sb}}{G_{sb}G_{se}}\right) = (1.01)\left[\frac{2.751 - 2.653}{(2.751)(2.653)}\right] \\ &= \boxed{0.0136\ (1.36\%)} \end{aligned}$$

(e)
$$\begin{aligned} x_e &= x_b - x_{ba}(1 - x_b) = 0.07 - (0.0136)(1 - 0.07) \\ &= \boxed{0.0574\ (5.74\%)} \end{aligned}$$

(f)
$$\begin{aligned} \text{VMA} &= 1 - \frac{G_{mb}(1 - x_b)}{G_{sb}} = 1 - \frac{(2.360)(1 - 0.07)}{2.653} \\ &= \boxed{0.1727\ (17.27\%)} \end{aligned}$$

(g) $x_a = \dfrac{G_{\text{mm}} - G_{\text{mb}}}{G_{\text{mm}}} = \dfrac{2.455 - 2.360}{2.455}$

$= \boxed{0.0387\ (3.87\%)}$

(h) Bleeding is the migration of free asphalt to the pavement surface, resulting in a loss of skid resistance.

(i) Bleeding typically occurs under loading on hot days in mixtures that have too few air voids.

(j) Air voids are generally 3%–5% of the mixture by volume.

SURVEYING

1. (a)

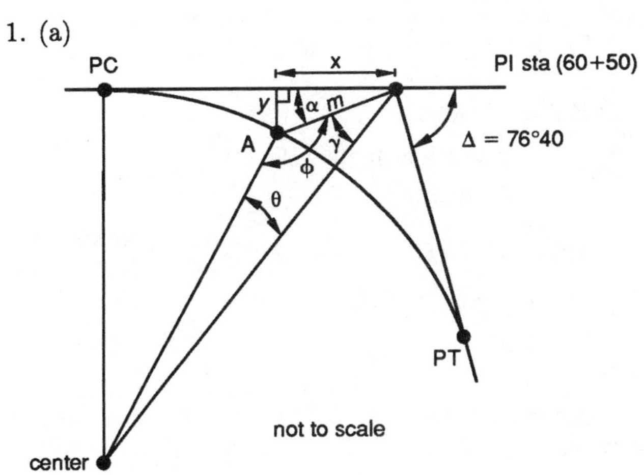

$$\alpha = \arctan \frac{y}{x} = \arctan \frac{72.25 \text{ ft}}{215.21 \text{ ft}}$$
$$= 18.558°$$
$$m = \sqrt{x^2 + y^2}$$
$$= \sqrt{(215.21 \text{ ft})^2 + (72.25 \text{ ft})^2} = 227.01 \text{ ft}$$
$$\Delta = 76°40' = 76.667°$$
$$\gamma = 90° - \frac{\Delta}{2} - \alpha$$
$$= 90° - \frac{76.667°}{2} - 18.558°$$
$$= 33.109°$$
$$\phi = 180° - \arcsin \left(\frac{\sin \gamma}{\cos \frac{\Delta}{2}} \right)$$
$$= 180° - \arcsin \left[\frac{\sin 33.109°}{\cos \left(\frac{76.667°}{2} \right)} \right]$$
$$= 135.864°$$
$$\theta = 180° - \phi - \gamma$$
$$= 180° - 135.864° - 33.109°$$
$$= 11.027°$$

From the law of sines,

$$\frac{\sin \theta}{m} = \frac{\sin \phi}{R \sec \left(\frac{\Delta}{2} \right)} = \frac{\sin \phi \cos \left(\frac{\Delta}{2} \right)}{R}$$
$$R = \frac{(227.01 \text{ ft})(\sin 135.864°) \left(\cos \frac{76.667°}{2} \right)}{\sin 11.027°}$$
$$= \boxed{648.30 \text{ ft}}$$

(b)
$$T = R \tan \left(\frac{\Delta}{2} \right) = (648.30 \text{ ft}) \left(\tan \frac{76.667°}{2} \right)$$
$$= 512.61 \text{ ft}$$
$$\text{PC} = \text{PI} - T = (60+50) - 512.61 \text{ ft}$$
$$= \boxed{55+37.39 \text{ sta}}$$

The length of curve, LC, is

$$LC = R\Delta = (648.30 \text{ ft}) \left[(76.667°) \left(\frac{2\pi}{360°} \right) \right]$$
$$= 867.48 \text{ ft}$$
$$\text{sta PT} = \text{sta PC} + LC = (55+37.39) + 867.48 \text{ ft}$$
$$= \boxed{64+4.87 \text{ sta}}$$

(c) Calculate the minimum radius length. e is the superelevation, and f is the side-friction factor.

$$R_{\text{min}} = \frac{v^2}{15(e+f)}$$

The minimum value of e is 0.04, and for a speed of 40 mph, $f = 0.15$. (Reference: AASHTO, *A Policy on Geometric Design of Highways and Streets*, 1990.)

$$R_{\text{min}} = \frac{(40 \text{ mph})^2}{(15)(0.04 + 0.15)}$$
$$= 561 \text{ ft} < 660 \text{ ft} \quad \text{[ok]}$$

$$\boxed{\text{The radius is acceptable.}}$$

2. (a)
$$\angle \text{BVA} = 180° - \angle \text{VBA} - \angle \text{VAB}$$
$$= 180° - 25° - 21° = 134°$$
$$\Delta = 180° - \angle \text{BVA}$$
$$= 180° - 134° = 46°$$

From the law of sines,

$$\frac{\text{BV}}{\sin 21°} = \frac{\text{BA}}{\sin 134°}$$
$$\text{BV} = (627.13 \text{ ft}) \left(\frac{\sin 21°}{\sin 134°} \right) = 312.43 \text{ ft}$$

The distance between the PC and point B is

$$(120+10) - (115+83) = 427 \text{ ft}$$

$$T = 427 \text{ ft} + \text{BV} = 427 \text{ ft} + 312.43 \text{ ft}$$
$$= 739.43 \text{ ft}$$

So,

$$T = R\tan\left(\frac{\Delta}{2}\right)$$

$$R = \frac{739.43 \text{ ft}}{\tan\left(\frac{46^\circ}{2}\right)} = \boxed{1742.0 \text{ ft}}$$

(b) $D = \frac{5729.6}{R} = \frac{5729.6}{1742 \text{ ft}} = \boxed{3.289^\circ}$

(c) $LC = (100)\left(\frac{\Delta}{D}\right) = (100)\left(\frac{46^\circ}{3.289^\circ}\right)$

$$= \boxed{1398.60 \text{ ft}}$$

(d) $\text{sta PT} = \text{sta PC} + LC$

$$= (115{+}83) + 1398.60 \text{ ft}$$
$$= \boxed{129{+}81.60 \text{ sta}}$$

(e) The deflection angle for the whole curve is

$$\frac{\Delta}{2} = \frac{46^\circ}{2} = 23^\circ$$

The deflection angle per foot of curve is

$$\frac{\left(\frac{\Delta}{2}\right)}{LC} = \frac{\left(\frac{46^\circ}{2}\right)}{1398.60 \text{ ft}} = 0.016445^\circ$$

For station 116+00,

$$\text{deflection angle} = [(116{+}00) - (115{+}83)](0.016445^\circ)$$
$$= \boxed{0.2796^\circ}$$

(f) $\text{deflection angle} = [(117{+}00) - (115{+}83)] \times (0.016445^\circ)$

$$= \boxed{1.9241^\circ}$$

(g)

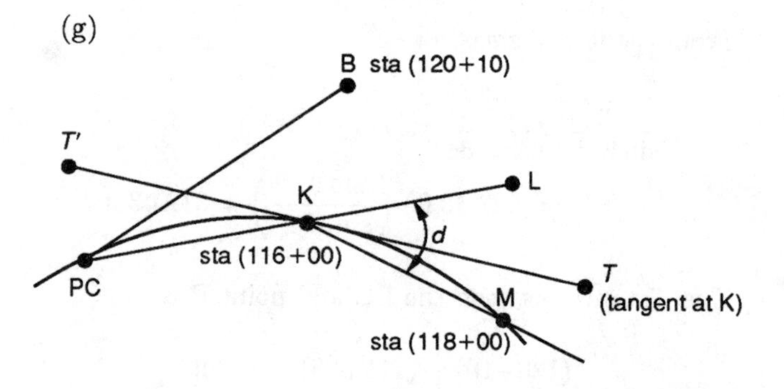

With the instrument at K (sta 116+00), locate direction KL (along line PC-L). The total deflection angle, ∠LKM, will be ∠LKT + ∠TKM. (M is at sta 118+00 and TT' is tangent to the curve at point K.

$$\angle\text{LKT} = \angle\text{T}'\text{KPC}$$
$$= \text{deflection of sta } 116{+}00$$
$$= 0.2796^\circ$$

$$\angle\text{TKM} = [(118{+}00) - (116{+}00)]\left(\begin{array}{c}\text{deflection angle}\\ \text{per ft of curve}\end{array}\right)$$
$$= (200 \text{ ft})(0.016445)$$
$$= 3.289^\circ$$

The deflection angle for M (sta 118+00) as measured from bearing PC-L is

$$\angle\text{LKT} + \angle\text{TKM} = 0.2796^\circ + 3.289^\circ = 3.5686^\circ$$

The deflection angle for M as measured from the tangent to the curve at K is the angle between the tangent and the chord.

$$\angle\text{TKM} = 3.289^\circ \quad \text{[generally not used in practice]}$$

3.

PI sta (37+00)

not to scale

(a) $I = 180^\circ - 50^\circ - 35^\circ = 95^\circ$

$$LC = RI = (800 \text{ ft})(95^\circ)\left(\frac{2\pi}{360^\circ}\right)$$
$$= \boxed{1326.45 \text{ ft}}$$

(b) $T = R\tan\left(\frac{I}{2}\right) = (800 \text{ ft})\left(\tan\frac{95^\circ}{2}\right)$

$$= 873.05 \text{ ft} \quad [8{+}73.05 \text{ sta}]$$
$$\text{PC} = \text{PI} - T = (37{+}00) - (8{+}73.05 \text{ sta})$$
$$= \boxed{28{+}26.95 \text{ sta}}$$

(c) sta PT = sta PC + LC

$$= (28{+}26.95) + (13{+}26.45)$$

$$= \boxed{41{+}53.40 \text{ sta}}$$

(d) The interior angle at PI is

$$\frac{I}{2} = \frac{95^\circ}{2} = \boxed{47.5^\circ}$$

(e) T was calculated in (b).

$$T = \boxed{873.05 \text{ ft}}$$

(f) The long chord distance, C, is

$$C = 2R\sin\left(\frac{I}{2}\right)$$

$$= (2)(800 \text{ ft})\left(\sin\frac{95^\circ}{2}\right)$$

$$= \boxed{1179.64 \text{ ft}}$$

(g) $E = R\tan\left(\frac{I}{2}\right)\tan\left(\frac{I}{4}\right)$

$$= (800 \text{ ft})\left(\tan\frac{95^\circ}{2}\right)\left(\tan\frac{95^\circ}{4}\right)$$

$$= \boxed{384.15 \text{ ft}}$$

(h) arc basis: $D = \frac{5729.6}{R} = \frac{5729.6}{800 \text{ ft}}$

$$= \boxed{7.162^\circ}$$

(i) chord basis: $\sin\frac{D}{2} = \frac{50}{R} = \frac{50}{800 \text{ ft}}$

$$= 0.0625$$

$$D = 2\arcsin(0.0625)$$

$$= \boxed{7.167^\circ}$$

(j) chord length: $2R\sin\frac{D}{2}$

$$= (2)(800 \text{ ft})\left(\sin\frac{7.167^\circ}{2}\right)$$

$$= \boxed{100.00 \text{ ft}}$$

4. (a) Compound curves fit the topography much better than simple curves due to the inequality of their tangent lengths. This flexibility sometimes creates operating disadvantages. Where possible, a compound curve should be replaced by a simple curve.

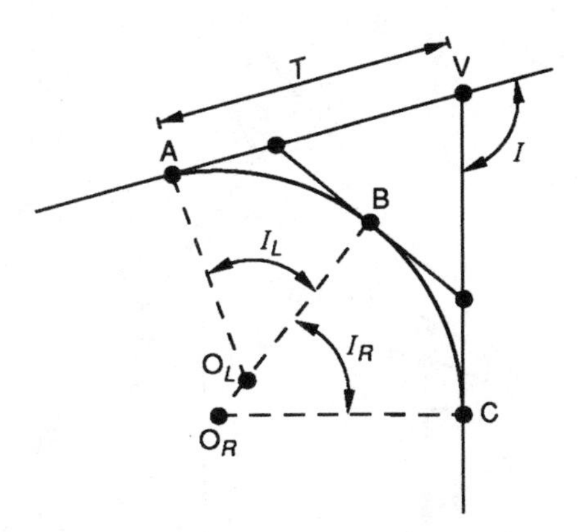

The objective, as shown in the figure, is to fit a simple circular curve between two tangents: AV and CV. These are the relationships governing circular curves.

$$I = I_L + I_R$$

$$T = R\tan\left(\frac{I}{2}\right)$$

$$E = R\left[\sec\left(\frac{I}{2}\right) - 1\right]$$

$$L = 100\left(\frac{I}{D}\right)$$

$$D = \frac{5729.6}{R}$$

$$M = R\left[1 - \cos\left(\frac{I}{2}\right)\right]$$

$$\text{LC} = 2R\sin\left(\frac{I}{2}\right)$$

(b) Using spirals instead of simple circular curves increases operating safety and performance. The most common way is to replace a simple circle with two spirals and one circle. The original circle has radius R and angle I. The replacement circle with equal spirals has two symmetrical spirals with angles I_s and lengths L_s.

$$L_s = 1.6\left(\frac{v}{R}\right)^3$$

$$I_s = \frac{L_s D}{200}$$

$$D = \frac{5729.6}{R}$$

$$L_C = 100\left(\frac{I - 2I_s}{D}\right)$$

L_C is the length of the circular curve with the degree of curve, D.

(c) Elimination of the tangent between curves is a common practice. If topography and clearances allow, one simple curve can be inserted between the two outer tangents. One possible solution is given in the following figure.

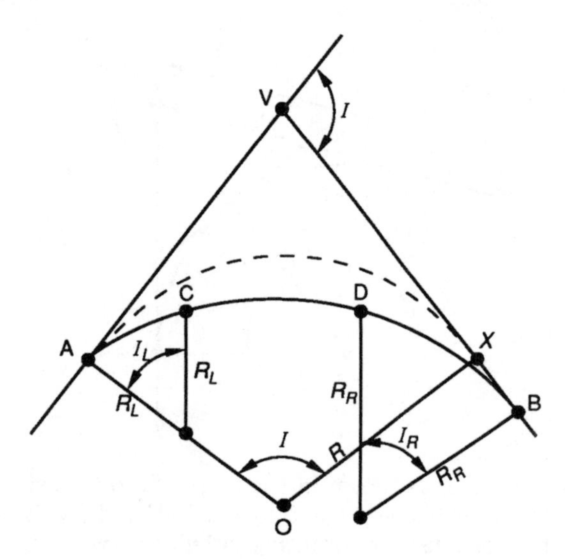

The new curve starts at point A (the starting point of the original shorter-radius curve) and finishes at point X on the existing tangent. The radius of the new single curve, R, and the distance BX (B is the ending point of the original longer-radius curve) is given by the following equation. (Reference: Meyer and Gibson, *Route Surveying and Design*, Harper & Row, 1980.)

$$R = \frac{R_L(\cos I_R - \cos I) + \text{CD} \sin I_R + R_R(1 - \cos I_R)}{1 - \cos I}$$

$$\text{BX} = (R_R - R_L)\sin I_R - (R - R_L)\sin I + \text{CD} \cos I_R$$

The offset, BX, between the original and proposed layouts may be so large that it prohibits the replacement.

5.

BVC elev 562 ft sta (60+00) — −1.5% — PVI sta (64+00) — +2.5% — EVC sta (68+00)

(a) $LC = (68{+}00) - (60{+}00) = 800 \text{ ft}$

$$y = \left(\frac{g_2 - g_1}{2LC}\right)x^2 + g_1 x + y_{\text{BVC}}$$

$$= \left[\frac{2.5 - (-1.5)}{(2)(8 \text{ sta})}\right]x^2 - 1.5x + 562 \text{ ft}$$

$$= 0.25x^2 - 1.5x + 562 \text{ ft}$$

Elevations of the road at 50-ft intervals can be calculated by inserting values of $x = 0.5, 1.0, 1.5, \ldots, 8.0$ stations in the preceding equation.

$$y_{64+50} = (0.25)(4.5 \text{ sta})^2 - (1.5)(4.5 \text{ sta}) + 562 \text{ ft} = \boxed{560.31 \text{ ft}}$$

$$y_{65+00} = (0.25)(5 \text{ sta})^2 - (1.5)(5.0 \text{ sta}) + 562 \text{ ft} = \boxed{560.75 \text{ ft}}$$

$$y_{65+50} = (0.25)(5.5 \text{ sta})^2 - (1.5)(5.5 \text{ sta}) + 562 \text{ ft} = \boxed{561.31 \text{ ft}}$$

$$y_{66+00} = (0.25)(6 \text{ sta})^2 - (1.5)(6 \text{ sta}) + 562 \text{ ft} = \boxed{562.0 \text{ ft}}$$

(b)

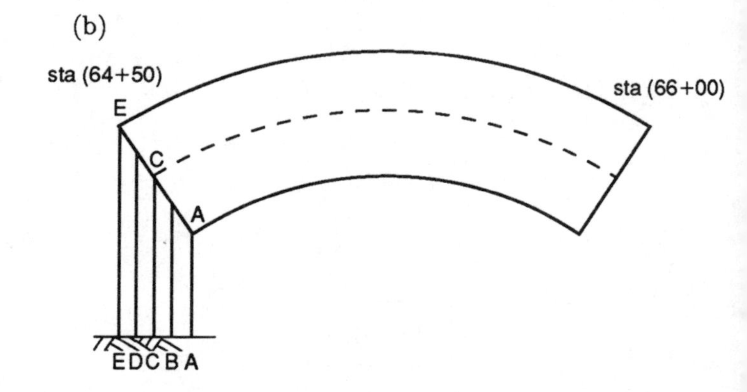

The elevation of the centerline pile (pile C) is found from (a), less 3 ft. There is a 12% superelevation, so for each 10-ft run, the rise is 1.2 ft.

10(.12) = 1.2 ft

For the first bent at 64+50,

pile C top = 560.31 ft − 3.0 ft = 557.31 ft
pile B top = 557.31 ft − 1.2 ft = 556.11 ft
pile A top = 557.31 ft − (2)(1.2 ft) = 554.91 ft
pile D top = 557.31 ft + 1.2 ft = 558.51 ft
pile E top = 557.31 ft + (2)(1.2 ft) = 559.71 ft

bent at station	A (ft)	B (ft)	C (ft)	D (ft)	E (ft)
64+50	554.91	556.11	557.31	558.51	559.71
65+00	555.35	556.55	557.75	558.95	560.15
65+50	555.91	557.11	558.31	559.51	560.71
66+00	556.6	557.8	559.0	560.2	561.4

6. (a) $\text{sta}_{\text{BVC}} = \text{sta}_{\text{V}} - \frac{L}{2}$

$$= (7{+}20) - \frac{1400 \text{ ft}}{2} = 0{+}20 \text{ sta}$$

$$\text{sta}_{\text{EVC}} = (0{+}20) + L = (0{+}20) + 1400 \text{ ft} = 14{+}20 \text{ sta}$$

$$\text{elevation}_{BVC} = \text{elevation}_V + g_1\left(\frac{L}{2}\right)$$
$$= 226.88 \text{ ft} + (0.02)\left(\frac{1400 \text{ ft}}{2}\right)$$
$$= 240.88 \text{ ft}$$
$$y = \left(\frac{g_2 - g_1}{2LC}\right)x^2 + g_1x + \text{elevation}_{BVC}$$
$$= \left[\frac{4-(-2)}{(2)(14 \text{ sta})}\right]x^2 - 2x + 240.88 \text{ ft}$$
$$= 0.21429x^2 - 2x + 240.88 \text{ ft}$$

At $y = 240$ ft, calculate x.

$$0.21429x^2 - 2x + 240.88 = 240$$
$$x_1 = 8.871 \text{ sta } (887.1 \text{ ft})$$
$$x_2 = 0.462 \text{ sta } (462. \text{ ft})$$
$$\text{sta}_1 = (0{+}20) + 46.2 \text{ ft}$$
$$= \boxed{0{+}66.2 \text{ sta}}$$
$$\text{sta}_2 = (0{+}20) + 887.1 \text{ ft}$$
$$= \boxed{9{+}07.1 \text{ sta}}$$

Between stations 0+66.2 and 9+07.1, the curve drops below the flood plain. The stations dropping below this 50-year flood plain are

1+00, 2+00, 3+00, 4+00, 5+00, 6+00, 7+00, and 8+00

(b) $907.1 \text{ ft} - 66.2 \text{ ft} = \boxed{840.9 \text{ ft}}$

(c) Increasing the length of curve will decrease the length of the portion that will be submerged because elevations on the curve will generally increase.

(d) Decreasing the length of curve will increase the length of the portion that will submerge.

(e) When the road is flooded, the embankment and the subgrade may deteriorate, which in turn degrades the pavement, regardless of pavement material. Depending on its importance, the highway may have to be rerouted. In addition, providing an adequate drainage system, paving the shoulders, and using high-quality, plant-mixed hot bituminous pavement material can help.

7. (a) The curve passes through point E at sta 75+20 and elevation 314 ft.

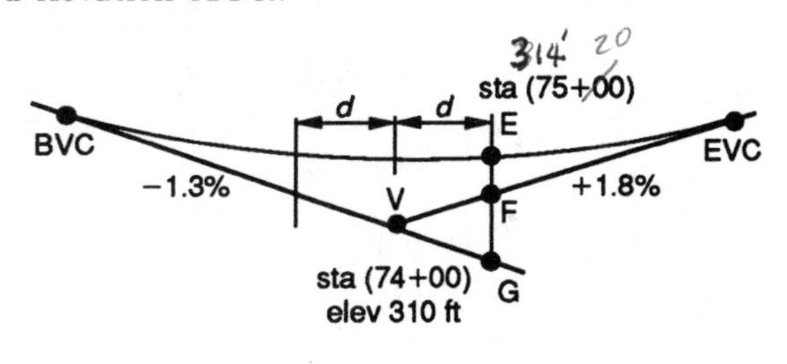

$$\text{EG} = \text{elevation}_E - \text{elevation}_V + |dg_1|$$
$$= 314 \text{ ft} - 310 \text{ ft} + |(7520 \text{ ft} - 7400 \text{ ft})(-0.013)|$$
$$= 5.56 \text{ ft}$$
$$\text{EF} = \text{elevation}_E - \text{elevation}_V - dg_2$$
$$= 314 \text{ ft} - 310 \text{ ft} - [(7520 - 7400)(0.018)]$$
$$= 1.84 \text{ ft}$$

The length of curve is

$$\boxed{LC = 2d\left(\frac{\sqrt{\frac{\text{EG}}{\text{EF}}}+1}{\sqrt{\frac{\text{EG}}{\text{EF}}}-1}\right)}$$
$$= (2)(120 \text{ ft})\left(\frac{\sqrt{\frac{5.56 \text{ ft}}{1.84 \text{ ft}}}+1}{\sqrt{\frac{5.56 \text{ ft}}{1.84 \text{ ft}}}-1}\right)$$
$$= \boxed{890.13 \text{ ft}}$$

(b) $LC = 1300$ ft.

$$\text{sta}_{BVC} = \text{sta}_V + \frac{LC}{2}$$
$$= (74{+}00) - \frac{1300 \text{ ft}}{2}$$
$$= 67{+}50 \text{ sta}$$

$$\text{elevation}_{BVC} = \text{sta}_V + \left(\frac{LC}{2}\right)g_1$$
$$= 310 \text{ ft} + \left(\frac{1300 \text{ ft}}{2}\right)(0.013)$$
$$= 318.45 \text{ ft}$$
$$y = \left(\frac{g_2 - g_1}{2LC}\right)x^2 + g_1x + \text{elevation}_{BVC}$$
$$= \left[\frac{1.8-(-1.3)}{(2)(13 \text{ sta})}\right]x^2 - 1.3x + 318.45 \text{ ft}$$
$$= 0.1192x^2 - 1.3x + 318.45 \text{ ft}$$

At point E,

$$x_E = (75{+}20) - (67{+}50) = 7.70 \text{ sta}$$
$$y_E = (0.1192)(7.7 \text{ sta})^2 - (1.3)(7.7 \text{ sta}) + 318.45 \text{ ft} = 315.51 \text{ ft}$$

The difference in elevation is

$$315.51 \text{ ft} - 314 \text{ ft} = \boxed{1.51 \text{ ft}}$$

(c) The grate inlet should be placed at the low point on the curve.

$$y = 0.1192x^2 - 1.3x + 318.45 \text{ ft}$$

At the low point, $dy/dx = 0$.

$$\frac{dy}{dx} = 0.2384x - 1.3 = 0$$
$$x = 5.453 \text{ sta } (543.3 \text{ ft})$$
$$\text{sta}_{\text{grate inlet}} = \text{sta}_{\text{BVC}} + x = (67{+}50) + (5{+}45.3) = \boxed{72{+}95.3 \text{ sta}}$$

$$\text{elevation}_{\text{grate inlet}} = (0.1192)(5.453 \text{ sta})^2 - (1.3)(5.453 \text{ sta}) + 318.45 \text{ ft} = \boxed{314.91 \text{ ft}}$$

8.(a)

AB: departure $= (-905.21 \text{ ft})(\sin 35.15°) = -521.15 \text{ ft}$
latitude $= (905.21 \text{ ft})(\cos 35.15°) = 740.14 \text{ ft}$

BC: departure $= (1135.76 \text{ ft})(\sin 81.28°) = 1122.63 \text{ ft}$
latitude $= (1135.76 \text{ ft})(\cos 81.28°) = 172.19 \text{ ft}$

CD: departure $= (1207.92 \text{ ft})(\sin 7.19°) = 151.18 \text{ ft}$
latitude $= (-1207.92 \text{ ft})(\cos 7.19°) = -1198.42 \text{ ft}$

DE: departure $= (-800.25 \text{ ft})(\sin 15.25°) = -210.49 \text{ ft}$
latitude $= (-800.25 \text{ ft})(\cos 15.25°) = -772.07 \text{ ft}$

EF: departure $= (-1100.85 \text{ ft})(\sin 48.17°) = -820.27 \text{ ft}$
latitude $= (-1100.85 \text{ ft})(\cos 48.17°) = 734.18 \text{ ft}$

FA: departure $= (429.53 \text{ ft})(\sin 40.73°) = 280.27 \text{ ft}$
latitude $= (429.53 \text{ ft})(\cos 40.73°) = 325.49 \text{ ft}$

$$\sum \text{departure} = -521.15 \text{ ft} + 1122.63 \text{ ft} + 151.18 \text{ ft} - 210.49 \text{ ft} - 820.27 \text{ ft} + 280.27 \text{ ft} = 2.17 \text{ ft}$$

$$\sum \text{latitude} = 740.14 \text{ ft} + 172.19 \text{ ft} - 1198.42 \text{ ft} - 772.07 \text{ ft} + 734.18 \text{ ft} + 325.49 \text{ ft} = 1.51 \text{ ft}$$

The length of the traverse closure is

$$\sqrt{(2.17 \text{ ft})^2 + (1.51 \text{ ft})^2} = \boxed{2.64 \text{ ft}}$$

(b) The total traverse length is

$$905.21 \text{ ft} + 1135.76 \text{ ft} + 1207.92 \text{ ft} + 800.25 \text{ ft} + 1100.85 \text{ ft} + 429.53 \text{ ft} = 5579.52 \text{ ft}$$

Use the compass rule.

$$\frac{\text{leg departure correction}}{\text{closure in departure}} = -\frac{\text{leg length}}{\text{total traverse length}} \quad \text{[Eq. 1]}$$

$$\frac{\text{leg latitude correction}}{\text{closure in latitude}} = -\frac{\text{leg length}}{\text{total traverse length}} \quad \text{[Eq. 2]}$$

Using Eqs. 1 and 2 for each leg,

$$\frac{\text{AB departure correction}}{2.17 \text{ ft}} = -\frac{905.21 \text{ ft}}{5579.52 \text{ ft}}$$

$$\text{AB departure correction} = -\left(\frac{905.21 \text{ ft}}{5579.52 \text{ ft}}\right)(2.17 \text{ ft}) = -0.35 \text{ ft}$$

$$\text{AB latitude correction} = -\left(\frac{905.21 \text{ ft}}{5579.52 \text{ ft}}\right)(1.51 \text{ ft}) = -0.24 \text{ ft}$$

$$\text{BC departure correction} = -\left(\frac{1135.76 \text{ ft}}{5579.52 \text{ ft}}\right)(2.17 \text{ ft}) = -0.44 \text{ ft}$$

$$\text{BC latitude correction} = -\left(\frac{1135.76 \text{ ft}}{5579.52 \text{ ft}}\right)(1.51 \text{ ft}) = -0.31 \text{ ft}$$

$$\text{CD departure correction} = -\left(\frac{1207.92 \text{ ft}}{5579.52 \text{ ft}}\right)(2.17 \text{ ft}) = -0.47 \text{ ft}$$

$$\text{CD latitude correction} = \left(\frac{1207.92 \text{ ft}}{5579.52 \text{ ft}}\right)(1.51 \text{ ft}) = -0.33 \text{ ft}$$

$$\text{DE departure correction} = -\left(\frac{800.25\text{ ft}}{5579.52\text{ ft}}\right)(2.17\text{ ft})$$
$$= -0.31\text{ ft}$$

$$\text{DE latitude correction} = -\left(\frac{800.25\text{ ft}}{5579.52\text{ ft}}\right)(1.51\text{ ft})$$
$$= -0.22\text{ ft}$$

$$\text{EF departure correction} = -\left(\frac{1100.85\text{ ft}}{5579.52\text{ ft}}\right)(2.17\text{ ft})$$
$$= -0.43\text{ ft}$$

$$\text{EF latitude correction} = -\left(\frac{1100.85\text{ ft}}{5579.52\text{ ft}}\right)(1.51\text{ ft})$$
$$= -0.30\text{ ft}$$

$$\text{FA departure correction} = -\left(\frac{429.53\text{ ft}}{5579.52\text{ ft}}\right)(2.17\text{ ft})$$
$$= -0.17\text{ ft}$$

$$\text{FA latitude correction} = -\left(\frac{429.53\text{ ft}}{5579.52\text{ ft}}\right)(1.51\text{ ft})$$
$$= -0.12\text{ ft}$$

Find the adjusted lengths.

AB: departure = −521.15 ft − 0.35 ft = −521.50 ft
latitude = 740.14 ft − 0.24 ft = 739.90 ft

BC: departure = 1122.63 ft − 0.44 ft = 1122.19 ft
latitude = 172.19 ft − 0.31 ft = 171.88 ft

CD: departure = 151.18 ft − 0.47 ft = 150.71 ft
latitude = −1198.42 ft − 0.33 ft = −1198.75 ft

DE: departure = −210.49 ft − 0.31 ft = −210.80 ft
latitude = −772.07 ft − 0.22 ft = −772.29 ft

EF: departure = −820.27 ft − 0.43 ft = −820.70 ft
latitude = 734.18 ft − 0.30 ft = 733.88 ft

FA: departure = 280.27 ft − 0.17 ft = 280.10 ft
latitude = 325.49 ft − 0.12 ft = 325.37 ft

(c) Use the double meridian distance (DMD) method.

leg	latitude (ft)	departure (ft)	DMD (ft)	latitude × DMD (ft²)
AB	739.90	−521.50	−521.50	−385,857.85
BC	171.88	1122.19	−521.50 − 521.50 + 1122.19 = 79.19	13,611.18
CD	−1198.75	150.71	79.19 + 1122.19 + 150.71 = 1352.09	−1,620,817.89
DE	−772.29	−210.80	1352.09 + 150.71 − 210.80 = 1292.00	−997,798.68
EF	733.88	−820.70	1292.00 − 210.80 − 820.70 = 260.50	191,175.74
FA	325.37	280.10	−280.10	− 91,136.14
			total =	−2,890,823.64

$$A = \left(\frac{1}{2}\right)(2{,}890{,}823.64\text{ ft}^2)$$
$$= \boxed{1{,}445{,}411.82\text{ ft}^2}$$

(d) $\text{BC} = \sqrt{(171.88\text{ ft})^2 + (1122.19\text{ ft})^2} = 1135.28\text{ ft}$

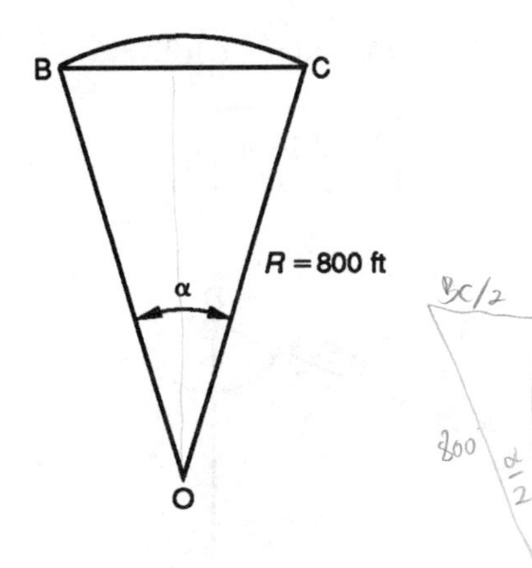

$$\sin\left(\frac{\alpha}{2}\right) = \frac{\frac{\text{BC}}{2}}{R} = \frac{\frac{1135.28\text{ ft}}{2}}{800\text{ ft}}$$
$$= 0.70955$$
$$\alpha = (2)(\arcsin 0.70955) = 90.397°$$

The area of circular region BC is

$$\left(\frac{90.397°}{360°}\right)(\pi R^2) = \left(\frac{90.397°}{360°}\right)\pi(800\text{ ft})^2$$
$$= 504{,}872.1\text{ ft}^2$$

The area of triangle OBC is

$$\frac{(1135.28\text{ ft})(800\text{ ft})\cos\left(\frac{\alpha}{2}\right)}{2} = \frac{(1135.28\text{ ft})(800\text{ ft})\cos\left(\frac{90.397°}{2}\right)}{2}$$
$$= 319{,}991.3\text{ ft}^2$$

The area of the circular segment bounded by chord and arc BC is

$$504{,}870.1 \text{ ft} - 319{,}991.3 \text{ ft} = \boxed{184{,}878.8 \text{ ft}^2}$$

(e) The length of curve BC is

$$BC = R\alpha = (800 \text{ ft})(90.397°)\left(\frac{2\pi}{360°}\right)$$

$$= \boxed{1262.18 \text{ ft}}$$

(a)

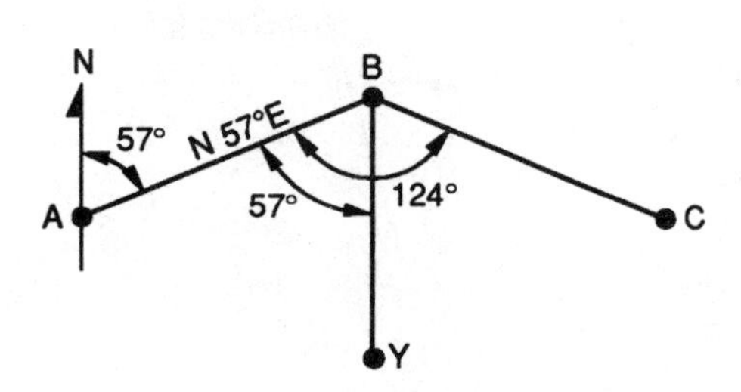

$$\angle ABY = 57°$$

$$\angle YBC = 124° - 57° = 67°$$

$$\text{bearing}_{BC} = \boxed{\text{S}67°\text{E}}$$

(b)

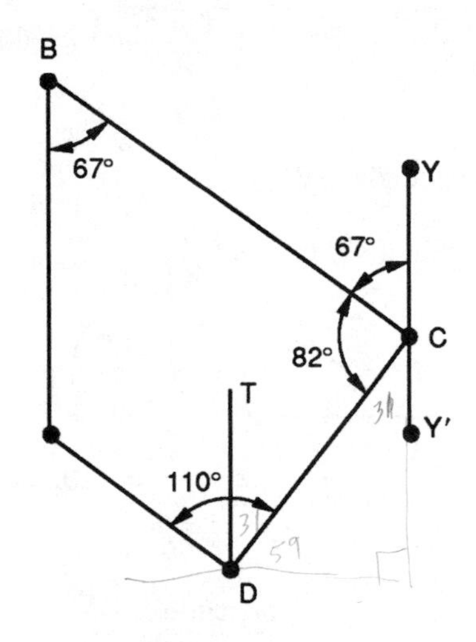

$$\angle BCY = 67°$$

$$\angle DCY' = 180° - 82° - 67° = 31°$$

$$\text{bearing}_{CD} = \boxed{\text{S}31°\text{W}}$$

(c)

$$\angle TDC = \angle DCY' = 31°$$

$$\text{bearing}_{DE} = 110° - 31° = 79°$$

$$= \boxed{\text{N}79°\text{W}}$$

(d)

$$\angle TDE = 110° - 31° = 79°$$

$$\angle NED = 180° - 79 = 101°$$

$$\angle AEN = 115° - \angle NED$$

$$= 115° - 101° = 14°$$

$$\text{bearing}_{EA} = \boxed{\text{N}14°\text{W}}$$

(e) The sum of the internal angles is

$$(n-2)(180°) = (5-2)(180°)$$

$$= 540°$$

$$\angle EAB = 540° - (124° + 82° + 110° + 115°)$$

$$= \boxed{109°}$$

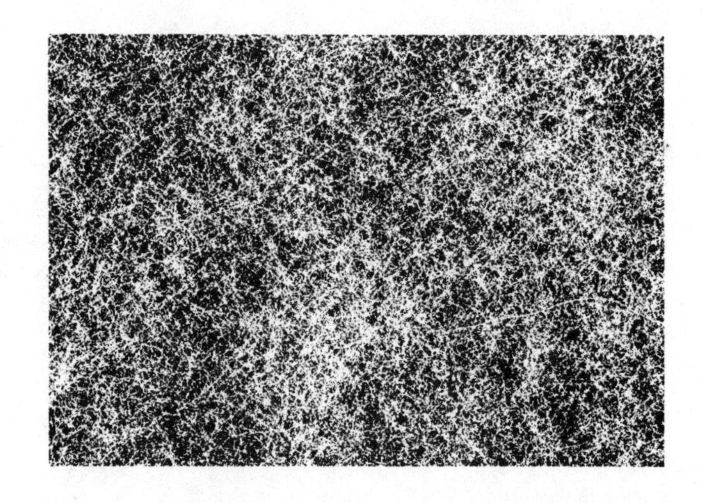

MASONRY

1. Satisfy UBC-91, Sec. 2406(e), Eq. 6-22 for compression on the masonry.

$$\frac{f_a}{F_a} + \frac{f_b}{F_b} \leq 1$$

In addition, tension stresses in the reinforcement must be less than 16,000 lbf/in^2. From UBC-91, Sec. 2406(c)2D, Eq. 6-4,

$$P_a = (0.2f'_m A_e + 0.65A_s F_{sc})\left[1 - \left(\frac{h'}{42t}\right)^3\right]$$

$$t = 15.5 \text{ in}$$

$$h = 120 \text{ in}$$

From UBC-91, Sec. 2407(b)4,

$$h' = (2)(120 \text{ in}) = 240 \text{ in}$$

$$f'_m = 1350 \text{ lbf/in}^2$$

$$n = 40$$

$$A_s = (4)(0.6 \text{ in}^2) = 2.4 \text{ in}^2$$

$$F_{sc} = 16{,}000 \text{ lbf/in}^2$$

$$A_e = (15.5 \text{ in})^2 = 240.25 \text{ in}^2 \quad \text{[fully grouted]}$$

Substituting numerical values,

$$P_a = \left[(0.2)\left(1350 \frac{\text{lbf}}{\text{in}^2}\right)(240.25 \text{ in}^2) + (0.65)(2.4 \text{ in}^2)\left(16{,}000 \frac{\text{lbf}}{\text{in}^2}\right)\right] \times \left(1 - \left[\frac{240 \text{ in}}{(42)(15.5 \text{ in})}\right]^3\right)$$

$$= 85{,}327 \text{ lbf (85.33 kips)}$$

From UBC-91, Sec. 2406(c)2D, Eq. 6-5,

$$F_a = \frac{P_a}{A_e} = \frac{85{,}327 \text{ lbf}}{240.25 \text{ in}^2} = 355 \text{ lbf/in}^2$$

Obtain the moment of inertia of the transformed section.

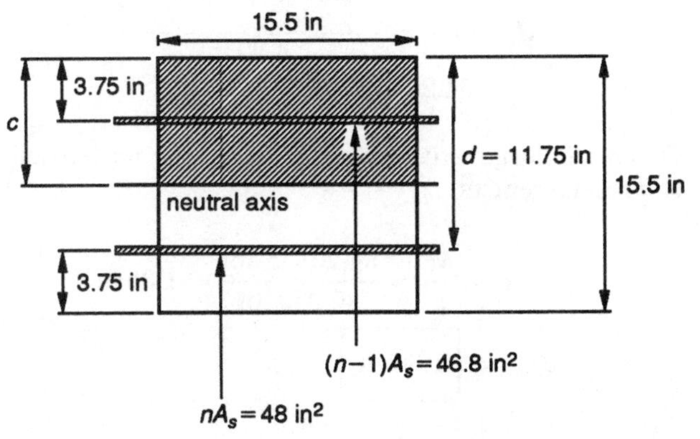

$$(15.5 \text{ in})\left(\frac{c^2}{2}\right) + (46.8)(c - 3.75 \text{ in}) = (48)(15.5 \text{ in} - 3.75 \text{ in} - c)$$

$$c = 5.41 \text{ in}$$

$$I = \left(\frac{1}{3}\right)(15.5 \text{ in})(5.41 \text{ in})^3 + (46.8 \text{ in}^2)(1.66 \text{ in})^2 + (48 \text{ in}^2)(6.34 \text{ in})^2 = 2876.4 \text{ in}^4$$

$$f_b = \frac{Mc}{I} = \frac{(100 \text{ in-kips})(5.41 \text{ in})}{2876.4 \text{ in}^4} = 0.188 \text{ kips/in}^2 \text{ (188 lbf/in}^2\text{)}$$

From UBC-91, Sec. 2406(c)3, Eq. 6-6,

$$F_b = 0.33f'_m = (0.33)\left(1350 \frac{\text{lbf}}{\text{in}^2}\right) = 450 \text{ lbf/in}^2$$

From UBC-91, Sec. 2406(e), Eq. 6-22,

$$\frac{f_a}{355 \frac{\text{lbf}}{\text{in}^2}} + \frac{188.1 \frac{\text{lbf}}{\text{in}^2}}{450 \frac{\text{lbf}}{\text{in}^2}} = 1$$

$$f_a = 207 \text{ lbf/in}^2$$

Compression on the masonry limits the load to

$$P_a = f_a A_e = \left(207 \frac{\text{lbf}}{\text{in}^2}\right)(240 \text{ in}^2) = 49{,}680 \text{ lbf (49.68 kips)}$$

Check the stress on the steel at this load.

$$f_s = \frac{-P_a n}{A_e + (n-1)A_s} + \frac{M(d-c)n}{I}$$

$$= -\frac{(49{,}680 \text{ lbf})(40)}{240 \text{ in}^2 + (40-1)(2.4 \text{ in}^2)} + \frac{(100 \text{ in-kips})(11.75 \text{ in} - 5.41 \text{ in})(40)\left(1000 \frac{\text{lbf}}{\text{kips}}\right)}{2876.4 \text{ in}^4}$$

$$= 2860 \text{ lbf/in}^2$$

Since $f_s < F_s(16{,}000 \text{ lbf/in}^2)$, tension in the steel does not control.

$$P_{max} = \boxed{49.68 \text{ kips}}$$

2.

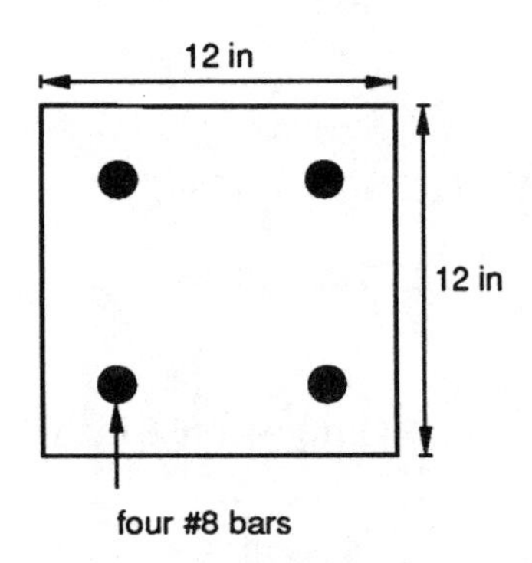

$$n = 42$$
$$F_s = 16{,}000 \text{ lbf/in}^2$$
$$f'_m = 1500 \text{ lbf/in}^2$$

(a) There is no explicit slenderness limit in UBC-91. The formula for axial compression (UBC-91, Sec. 2406(c)2D, Eq. 6-4) yields zero capacity at $h'/t = 42$, so

$$h' < \frac{(12 \text{ in})(42 \text{ ft})}{12 \text{ in}} = 42 \text{ ft}$$

A more reasonable limit is $h'/t = 30$, which is used by the UBC to require second-order analysis in walls (UBC-91, Sec. 2409(a)2).

Equating the allowable load to zero, $h \leq 42$ ft if supported at both ends; $h < 21$ ft if unrestrained at one end (sidesway permitted).

Reasonable practical limits are $h \leq 30$ ft if supported at both ends $h \leq 15$ ft; if unrestrained at one end (sidesway permitted).

(b) From UBC-91, Sec. 2406(c)2D, Eq. 6-4,

$$P_a = (0.2 f'_m A_e + 0.65 A_s F_{sc})\left[1 - \left(\frac{h'}{42t}\right)^3\right]$$

$$A_e = (12 \text{ in})(12 \text{ in}) = 144 \text{ in}^2$$
$$A_s = (4)(0.79 \text{ in}^2) = 3.16 \text{ in}^2$$
$$P_a = \left[(0.2)\left(1500 \frac{\text{lbf}}{\text{in}^2}\right)(144 \text{ in}^2) + (0.65)(3.16 \text{ in}^2) \times \left(16{,}000 \frac{\text{lbf}}{\text{in}^2}\right)\right]\left(1 - \left[\frac{(20 \text{ ft})\left(12 \frac{\text{in}}{\text{ft}}\right)}{(42)(12 \text{ in})}\right]^3\right)$$
$$= \boxed{67{,}851 \text{ lbf } (67.85 \text{ kips})}$$

(c) The condition to be satisfied for compression on the masonry is from UBC-91, Sec. 2406(e), Eq. 6-22.

$$\frac{f_a}{F_a} + \frac{f_b}{F_b} \leq 1$$

In addition, the tension stress in the steel must be less than 16,000 lbf/in². From (b),

$$P_a = 67.85 \text{ kips}$$
$$\frac{f_b}{F_b} \leq 1 - \frac{40 \text{ kips}}{67.85 \text{ kips}} = 0.410$$

Solving for f_b and noting that F_b is $0.33 f'_m$ from UBC-91, Sec. 2406(c)3, Eq. 6-6,

$$f_b \leq (0.410)(0.33)\left(1500 \frac{\text{lbf}}{\text{in}^2}\right) = 205 \text{ lbf/in}^2$$

Obtain the moment of inertia of the transformed section.

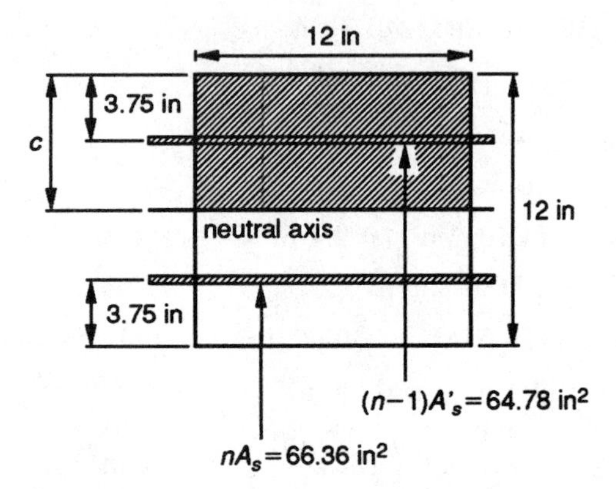

$$\frac{12c^2}{2} + (64.78 \text{ in}^2)(c - 3.75 \text{ in}) = (66.36 \text{ in}^2) \times (12 \text{ in} - 3.75 \text{ in} - c)$$
$$c = 4.92 \text{ in}$$

$$I = \left(\frac{1}{3}\right)(12 \text{ in})(4.92 \text{ in})^3 + (64.78 \text{ in}^2)(1.17 \text{ in})^2 + (66.36 \text{ in}^2)(3.33 \text{ in})^2 = 1301 \text{ in}^4$$

The allowable moment is

$$M = \frac{f_b I}{c} = \frac{\left(205 \frac{\text{lbf}}{\text{in}^2}\right)(1301 \text{ in}^4)}{4.92 \text{ in}} = 54{,}208 \text{ in-lbf}$$

Given the large axial force, there will be no tension in the reinforcement.

$$e = \frac{M}{p} = \frac{54{,}208 \text{ in-lbf}}{40{,}000 \text{ lbf}} = 1.36 \text{ in}$$
$$e_{\max} = \boxed{1.36 \text{ in}}$$

3.

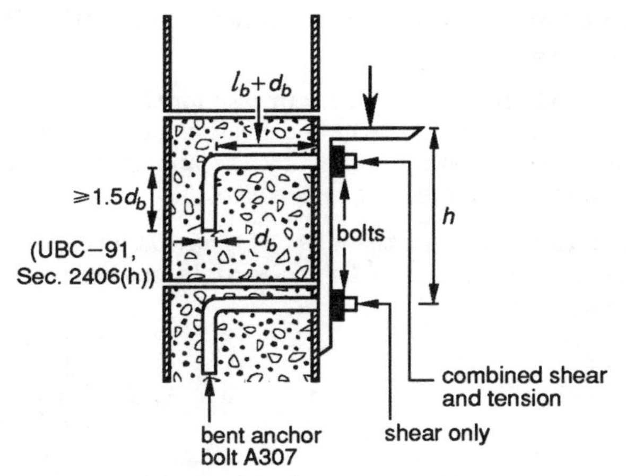

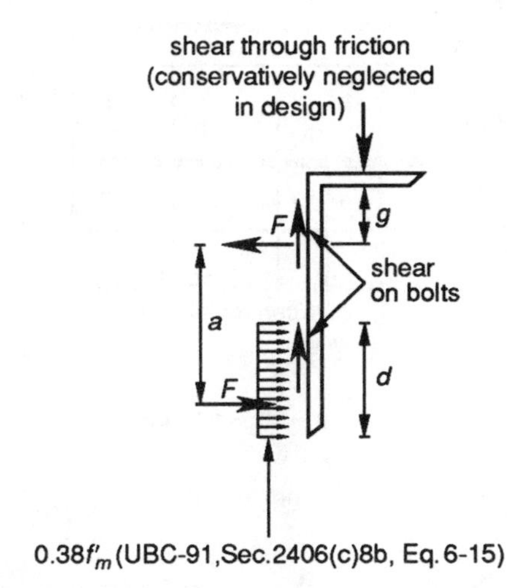

Assume l_b = 8 in. From UBC-91, Table 24-D-1, the allowable tension per bolt, B_t = 4500 lbf (without shear). From UBC-91, Table 24-D-2, choose 1-in bolts (yield capacity = 5650 lbf).

From UBC-91, Table 24-E, the allowable shear per bolt in the absence of tension is B_v = 2200 lbf.

The bolts on the upper layer are subjected to combined shear and tension forces. Assume the tension is kept at one-half of B_t; then from UBC-91, Sec. 2406(h)4, Eq. 6-33,

$$\frac{b_t}{B_t} + \frac{b_v}{B_v} \leq 1$$

From this,

$$\frac{b_v}{B_v} = 1 - \frac{b_t}{B_t}$$
$$= 1 - 0.5 = 0.5$$

The shear force that can be taken by a bolt of the top layer is

$$b_v = (0.5)(2200\text{ lbf}) = 1100\text{ lbf/bolt}$$

For bolts in the lower layer, $b_t \approx 0$, so $b_v = 2200$ lbf/bolt. The total shear capacity is

$$1100\text{ lbf} + 2200\text{ lbf} = 3300\text{ lbf/pair} \quad \begin{bmatrix}\text{one on top;}\\ \text{one on bottom}\end{bmatrix}$$

The number of pairs required is

$$\frac{9000\text{ lbf}}{3300\,\frac{\text{lbf}}{\text{pair}}} = 2.73\text{ pairs} \quad \begin{bmatrix}\text{say 3 pairs for a}\\ \text{total of 6 bolts}\end{bmatrix}$$

Distance a in the figure is needed to keep tension in the upper layer at $\frac{1}{2}B_t$.

$$\left[\frac{\left(4500\,\frac{\text{lbf}}{\text{bolt}}\right)(3\text{ bolts})}{2}\right] a = (9000\text{ lbf})(3\text{ in})$$
$$a = 4\text{ in}$$

From the figure, the approximate bearing depth, d, is

$$F = \left(2250\,\frac{\text{lbf}}{\text{bolt}}\right)(3\text{ bolts}) = 6750\text{ lbf}$$
$$F = (0.38 f'_m)(\text{width})d$$
$$6750\text{ lbf} = (0.38)\left(2000\,\frac{\text{lbf}}{\text{in}^2}\right)(8\text{ in})d$$
$$d = 1.11\text{ in}$$

The minimum vertical dimension for bracket h is

$$h \geq a + \frac{d}{2} + g$$

Assuming g is 2.5 in,

$$h \geq 4\text{ in} + \frac{1.11\text{ in}}{2} + 2.5\text{ in}$$
$$\geq 7.06\text{ in}$$

4. Assume that the wall is part of the facade and that the only vertical load acting is the wall's own weight. Assume also that the steel beams do not provide any rotational restraint to the wall and can be replaced by simple supports.

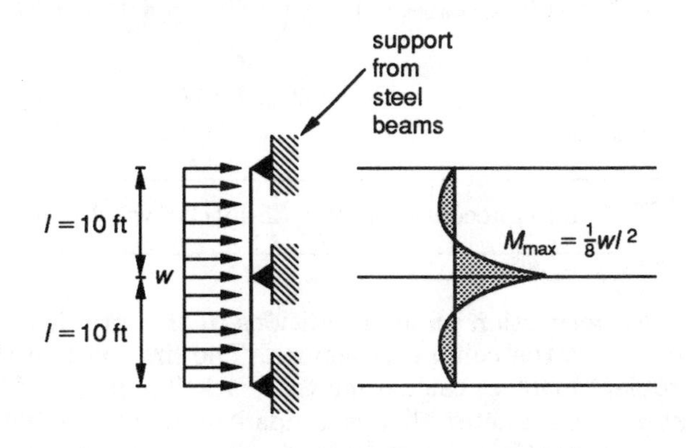

$$M_{\max} = \left(\frac{1}{8}\right)\left(35\,\frac{\text{lbf}}{\text{ft}}\right)(10\text{ ft})^2$$
$$= 437.5\text{ ft-lbf/foot of wall}$$

Calculate the cracked moment of inertia per foot of wall.

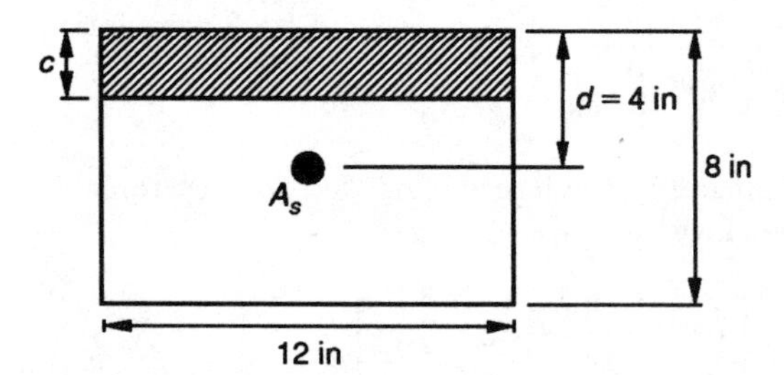

$$A_s = (0.31\text{ in}^2)\left(\frac{12\text{ in}}{48\text{ in}}\right) = 0.0775\text{ in}^2$$

$$nA_s = (44)(0.0775\text{ in}^2) = 3.41\text{ in}^2$$

$$\frac{12c^2}{2} = (3.41\text{ in}^2)(4 - c)$$

$$c = 1.25\text{ in}$$

$$I = \left(\frac{1}{3}\right)(12\text{ in})(1.25\text{ in})^3 + (3.41\text{ in}^2)(2.75\text{ in})^2 = 33.6\text{ in}^4$$

Compressive stresses in the steel due to self-weight are small and can be assumed to be zero.

Calculate tension on the steel, f_s.

$$f_s = \frac{M(d-c)n}{I} = \frac{(437.5\text{ ft-lbf})\left(12\,\frac{\text{in}}{\text{ft}}\right)(4\text{ in} - 1.25\text{ in})(44)}{33.6\text{ in}^4} = 18{,}906\text{ lbf/in}^2$$

Accounting for the one-third increase in allowable stress for checking against wind-induced stresses, the allowable stress is

$$(f_s)_{\text{allowable}} = \left(\frac{4}{3}\right)\left(16{,}000\,\frac{\text{lbf}}{\text{in}^2}\right) = 21{,}333\,\frac{\text{lbf}}{\text{in}^2} > f_s$$

The wall is acceptable for a 35-lbf/ft^2 wind load.

5. An inspection of the conditions in this problem reveals that the column is slender. The first part of the problem involves calculating the cracked moment of inertia, a parameter that is fundamental for a reliable assessment of the buckling load. The exact calculation of the cracked inertia requires accounting for the beneficial effect of the axial force. Although the computations to determine the exact extent of cracking are involved, they show the level of conservatism associated with an approximate solution based on neglecting the axial force.

The second part of the problem asks for the calculation of the maximum transverse deflection of the column. This deflection is influenced by the appearance of increased moments due to P-Δ effects.

(a) To account for the effect of the axial force on the extent of cracking, consider the partially cracked section shown.

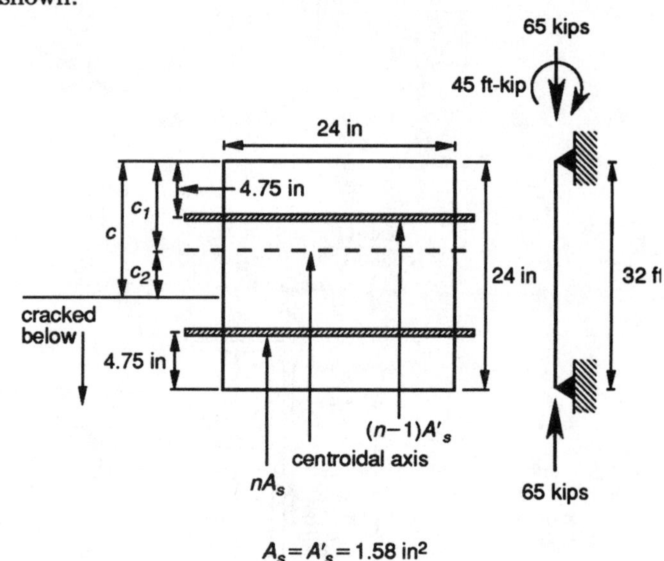

Obtaining the depth of the cracked region is best done iteratively. The solution is obtained when a depth c is established at which the stress calculated on the transformed section is zero at this depth. A fundamental complication results because the axial force, although in the centroid of the gross cross section, is eccentric with respect to the centroid of the partially cracked section, so the moment induced by its eccentricity must be included in the computation of stresses.

In the figure, the moment induced by the eccentricity of the axial load is $-(\text{load})(12\text{ in} - c)$; the negative sign indicates that this moment is in the opposite direction to the applied moment of 45.0 ft-kips. Assuming that the maximum moment occurs at the top end of the column,

$$M_{\max} = 45\text{ ft-kips} - (65\text{ kips})(12\text{ in} - c_1)$$

The iterative approach to find the exact neutral axis depth is

step 1: Select a value of c.

step 2: Compute the centroidal location of the transformed area c_1.

step 3: Compute the area, A, and the moment of inertia, I, of the transformed section.

step 4: Calculate M_{max} from Eq. 1.

step 5: Calculate the stress at the location of the neutral axis.

$$\sigma = -\frac{65 \text{ kips}}{A} + \frac{M_{max}c_2}{I} \quad \text{[Eq. 1]}$$

step 6: Repeat steps 1 through 5 until the value of σ is approximately zero.

The moment of inertia computed in the last iteration is the minimum cracked cross section.

From UBC-91, Sec. 2406(f)1B, Eq. 6-24,

$$E_m = 750f'_m = (750)\left(1800 \frac{\text{lbf}}{\text{in}^2}\right)$$
$$= 1{,}350{,}000 \text{ lbf/in}^2 \text{ (1350 kips/in}^2\text{)}$$

$$n = \frac{E_s}{E_c} = \frac{29{,}000 \frac{\text{kips}}{\text{in}^2}}{1350 \frac{\text{kips}}{\text{in}^2}}$$
$$= 21.5$$

With this value of n, the necessary expressions are

$$A = (24 \text{ in})c + 66.3 \text{ in}^2$$
$$I = (8 \text{ in})(c_1^3 + c_2^3) + (32.4 \text{ in}^2)(c_1 - 4.75 \text{ in})^2 + (33.9 \text{ in}^2)(19.25 \text{ in} - c_1)^2$$
$$c_1 = \left(\frac{1}{A}\right)[(12 \text{ in})c^2 + 807 \text{ in}^3]$$
$$c_2 = c - c_1$$

After sufficient iterations,

$$c = 14.1 \text{ in}$$
$$c_1 = 7.89 \text{ in}$$
$$c_2 = 6.21 \text{ in}$$
$$A = 404.7 \text{ in}^2$$
$$I = \boxed{10{,}548 \text{ in}^4}$$

For $P = 0$, the centroidal axis is located as follows:

$$\left(\frac{24 \text{ in}}{2}\right)c^2 + (32.4 \text{ in}^2)(c - 4.75 \text{ in})$$
$$= (33.9 \text{ in}^2)(19.25 \text{ in} - c)$$
$$c = 5.89 \text{ in}$$
$$I = \left(\frac{1}{3}\right)(24 \text{ in})(5.89 \text{ in})^3 + (32.4 \text{ in}^2)(1.14 \text{ in})^2 + (33.9 \text{ in}^2)(13.36 \text{ in})^2$$
$$= \boxed{7728 \text{ in}^4 \quad \text{[73\% of the exact value]}}$$

(b) Use UBC-91, Sec. 2411 to design the wall.

$$S = \left(\frac{1}{6}\right)(24 \text{ in})^3 = 2304 \text{ in}^3$$

$$f_r = (2.5)\sqrt{f'_m} = (2.5)\sqrt{1800}$$
$$= 106.1 \text{ lbf/in}^2 < 125 \text{ lbf/in}^2 \quad \text{[maximum]}$$

Use UBC-91, Sec. 2411(b)4, Eq. 11-13.

$$M_{cr} = Sf_r = (2304 \text{ in}^3)\left(106.1 \frac{\text{lbf}}{\text{in}^2}\right)$$
$$= 244.5 \text{ in-kips}$$

$$\Delta_s = \frac{5M_{cr}h^2}{48E_mI_g} + \frac{(5)(M_s - M_{cr})h^2}{48E_mI_{cr}}$$

$$I_g = \left(\frac{1}{12}\right)(24 \text{ in})^4 = 27{,}648 \text{ in}^4$$
$$E_m = 1350 \text{ kips/in}^2 \quad \text{[from (a)]}$$
$$I_{cr} = 7728 \text{ in}^4 \quad \text{[from (a)]}$$
$$h = (32 \text{ ft})\left(12 \frac{\text{in}}{\text{ft}}\right) = 384 \text{ in}$$
$$M_s = (45 \text{ ft-kips})\left(12 \frac{\text{in}}{\text{ft}}\right)$$
$$= 540 \text{ in-kips} \quad \text{[neglecting } P\text{-}\Delta \text{ to be adjusted]}$$

$$\Delta_s = \frac{(5)(244.5 \text{ in-kips})(384 \text{ in})^2}{(48)\left(1350 \frac{\text{kips}}{\text{in}^2}\right)(27{,}648 \text{ in}^4)} + \frac{(5)(540 \text{ in-kips} - 244.5 \text{ in-kips})(384)^2}{(48)\left(1350 \frac{\text{kips}}{\text{in}^2}\right)(7728 \text{ in}^4)}$$
$$= 0.536 \text{ in}$$

Adjust M_s.

$$M_s = 540 \text{ in-kips} + (65 \text{ kips})(0.536 \text{ in})$$
$$= 574.8 \text{ in-kips}$$

Δ_s with adjusted M_s:

$$\Delta_s = \frac{(5)(244.5 \text{ in-kips})(384 \text{ in})^2}{(48)\left(1350 \frac{\text{kips}}{\text{in}^2}\right)(27{,}648 \text{ in}^4)} + \frac{(5)(574.8 \text{ in-kips} - 244.5 \text{ in-kips})(384 \text{ in})^2}{(48)\left(1350 \frac{\text{kips}}{\text{in}^2}\right)(7728 \text{ in}^4)}$$
$$= 0.587 \text{ in}$$

One more iteration:

$$M_s = (540\text{ in-kips}) + (65\text{ kips})(0.587\text{ in})$$
$$= 578.2\text{ in-kips}$$

$$\Delta_s = \frac{(5)(244.5\text{ in-kips})(384\text{ in})^2}{(48)\left(1350\,\frac{\text{kips}}{\text{in}^2}\right)(27{,}648\text{ in}^4)} + \frac{(5)(578.2\text{ in-kips} - 244.5\text{ in-kips})(384\text{ in})^2}{(48)\left(1350\,\frac{\text{kips}}{\text{in}^2}\right)(7728\text{ in}^4)}$$

$$= \boxed{0.592\text{ in}}$$

NO POSTAGE NECESSARY IF MAILED IN THE UNITED STATES

BUSINESS REPLY MAIL

FIRST CLASS MAIL PERMIT NO. 33 BELMONT, CA

POSTAGE WILL BE PAID BY ADDRESSEE

PROFESSIONAL PUBLICATIONS INC
1250 FIFTH AVE
BELMONT CA 94002-9979

NO POSTAGE NECESSARY IF MAILED IN THE UNITED STATES

BUSINESS REPLY MAIL

FIRST CLASS MAIL PERMIT NO. 33 BELMONT, CA

POSTAGE WILL BE PAID BY ADDRESSEE

PROFESSIONAL PUBLICATIONS INC
1250 FIFTH AVE
BELMONT CA 94002-9979

NO POSTAGE NECESSARY IF MAILED IN THE UNITED STATES

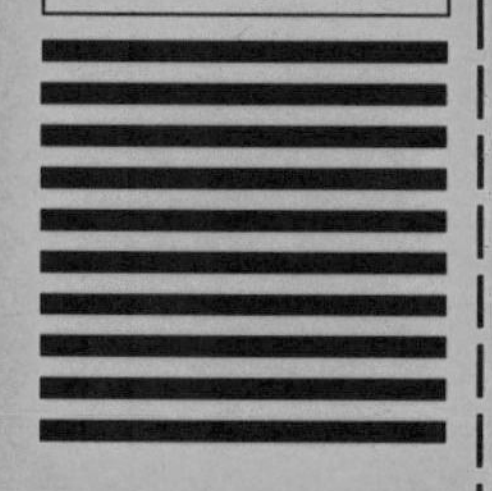

BUSINESS REPLY MAIL

FIRST CLASS MAIL PERMIT NO. 33 BELMONT, CA

POSTAGE WILL BE PAID BY ADDRESSEE

PROFESSIONAL PUBLICATIONS INC
1250 FIFTH AVE
BELMONT CA 94002-9979

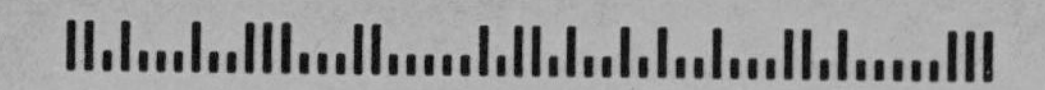